For EU product safety concerns:
IN TECH d.o.o., Prolaz Marije Krucifikse Kozulić 3, 51000 Rijeka, Croatia,
info@intechopen.com or visit our website at intechopen.com.

Meet the editors

Haikel Jelassi is a professor of physics and Head of the Research Laboratory on Energy and Matter for Nuclear Sciences Development, National Centre for Nuclear Sciences and Technologies, Tunisia.
His research focuses on nuclear and atomic physics. Prof. Jelassi is interested in many research fields in fundamental as well as applied physics such as: cold atoms, calculations of atomic processes for plasma diagnostic, as well as the study of irradiated dyes by gamma radiation. He has a special interest for the use of relativistic codes to calculate extended and accurate useful atomic data for plasma diagnostics and modeling (radiative and collisional properties of ions).

Djamel Benredjem is a professor in the Department of Physics at Université Paris-Sud (Orsay) and the director of the Laboratoire Aimé Cotton.
His research focuses on atomic physics in hot plasmas, with a special interest in radiative properties (radiative power losses, opacity, etc.). As a theoretician he models radiative properties and develops non-local thermodynamic equilibrium (NLTE) codes.

Contents

Preface IX

Section 1 Fundamentals of Plasma Physics 1

Chapter 1 **Noise-Free Rapid Approach to Solve Kinetic Equations for Hot Atoms in Fusion Plasmas 3**
Mikhail Tokar

Chapter 2 **The Behavior of Streaming Instabilities in Dissipative Plasma 21**
Eduard V. Rostomyan

Chapter 3 **Fundamentals of Plasma-Material Interactions in Magnetic Fusion Devices 47**
Jean Paul Allain and David N. Ruzic

Chapter 4 **Stopping Power of Ions in a Magnetized Plasma: Binary Collision Formulation 67**
Hrachya B. Nersisyan, Günter Zwicknagel and Claude Deutsch

Chapter 5 **Mode Transition and Hysteresis in Inductively Coupled Plasma Sources 89**
Shu-Xia Zhao

Chapter 6 **Numerical Modeling of Partial Discharge Development Process 109**
Cheng Pan, Ju Tang and Fuping Zeng

Chapter 7 **Numerical Approach to Dynamical Structure Factor of Dusty Plasmas 131**
Aamir Shahzad, Muhammad Asif Shakoori, Mao-Gang HE and Yan Feng

Section 2 Modern Applications of Plasma 145

Chapter 8 **A Test Facility to Investigate Sheath Effects during Ion Cyclotron Resonance Heating 147**
Kristel Crombe, Rodolphe D' Inca, Eric Faudot, Helmut Faugel, Ana Kostic, Mariia Usoltceva, Jean-Marie Noterdaeme, Anton Nikiforov, Helmut Fuenfgelder, Stephane Heuraux, Jonathan Jacquot, Fabrice Louche, Roman Ochoukov, Ilya Shesterikov and Dirk Van Eester

Chapter 9 **Plasma-Assisted Combustion 167**
Siyin Zhou, Haiqing Wang, Wansheng Nie and Xueke Che

Chapter 10 **Plasma Generation and Application in a Laser Ablation Pulsed Plasma Thruster 189**
Jianjun Wu, Yu Zhang, Yuqiang Cheng, Qiang Huang, Jian Li and Xiaobin Zhu

Chapter 11 **Physics of High-Density Radio Frequency Capacitively Coupled Plasma with Various Electrodes and Its Applications 209**
Yasunori Ohtsu

Chapter 12 **Space Plasma Interactions with Spacecraft Materials 225**
Daniel P. Engelhart, Elena A. Plis, Dale Ferguson, W. Robert Johnston, Russell Cooper and Ryan C. Hoffmann

Chapter 13 **Repetitive Nanosecond Volume Discharges under Airflows 247**
Jingfeng Tang, Liqiu Wei and Daren Yu

Chapter 14 **Modeling of Novel Plasma-Optical Systems 267**
Iryna Litovko and Alexey Goncharov

Chapter 15 **Plasma Damage on Low-k Dielectric Materials 291**
Yi-Lung Cheng, Chih-Yen Lee and Chiao-Wei Haung

Preface

Most of the matter in the universe is in a state called plasma. It exhibits complex and rich phenomena. Studying plasmas is essential for technology development and applications in tokamak reactors, medicine, and the environment. Plasma physics studies also allow an understanding of processes in the interstellar medium and in solar physics.

The aim of this book is to provide a basic knowledge of plasma science and technology to the common reader. The book will be useful for researchers and students. It provides a large overview from basic fundamentals to modern applications related to plasma science and technology. This was reached by presenting new experimental and theoretical results, and interdisciplinary and outstanding applications in plasma physics.

The editors hope that *Plasma Science & Technology: Basic Fundamentals and Modern Applications* will be useful to a large audience among researchers and students.

Prof. Dr. Haikel Jelassi
Laboratory "Energy and Matter for Nuclear Sciences Development" (LR16CNSTN02)
National Centre for Nuclear Sciences and Technologies
Sidi Thabet Technopark
Ariana, Tunisia

Prof. Djamel Benredjem
Laboratoire Aimé Cotton, CNRS and University Paris-Saclay
Orsay, France

Section 1

Fundamentals of Plasma Physics

Chapter 1

Noise-Free Rapid Approach to Solve Kinetic Equations for Hot Atoms in Fusion Plasmas

Mikhail Tokar

Additional information is available at the end of the chapter

http://dx.doi.org/10.5772/intechopen.76681

Abstract

At the first wall of a fusion reactor, charged plasma particles are recombined into neutral molecules and atoms recycling back into the plasma volume where charge exchange (*cx*) with ions. As a result hot atoms with chaotically directed velocities are generated which can strike and erode the wall. An approach to solve the kinetic equation in integral form for *cx* atoms, being alternative to statistical Monte Carlo methods, has been speeded up by a factor of 50, by applying an approximate pass method to evaluate integrals, involving the ion velocity distribution function. It is applied to two-dimensional transfer of *cx* atoms near the entrance of a duct, guiding to the first mirror for optical observations. The energy spectrum of hot *cx* atoms, escaping into the duct, is calculated and the mirror erosion rate is assessed. Computations are done for a molybdenum first mirror under plasma conditions expected in the fusion reactor DEMO. Kinetic modeling results are compared with those found with a diffusion approximation valid in very cold and dense plasmas. For ducts at the torus outboard a more rigorous kinetic consideration predicts an erosion rate by a factor up to 2 larger than the diffusion approximation.

Keywords: fusion, plasma, neutral, atoms, kinetic equation, numerical solution

1. Introduction

In devices for thermonuclear fusion research, for example, of the Tokamak type, particles of hydrogen isotopes, deuterium and tritium, are in the form of a hot fully ionized plasma [1, 2]. To avoid a destruction of the machine wall, a special region, the so-called scrape-off layer (SOL), is arranged at the plasma edge, where particles stream along the magnetic field to the special target plates [3]. Normally, it is done by using additional magnetic coils to form a divertor configuration (see **Figure 1a**).

By reaching the divertor target plates, plasma electrons and ions are recombined into neutral atoms and molecules which are finally exhausted from the device by pumps. However, in a future fusion reactor like DEMO [1, 2], only a minor fraction of 1% of neutrals generated at the targets will be pumped out. The rest of them is ionized again in the plasma near the targets. This "recycling" process significantly restrains the parallel plasma flow in the SOL [4]. Therefore, a considerable fraction of plasma particles lost from the plasma core will reach the vessel wall before they are exhausted into the divertor. Plasma fluxes to the wall saturate it with fuel particles in a time much shorter than the discharge duration, and a comparable amount of neutral species will recycle back from the wall into the plasma. Recycling neutrals are not confined by the magnetic field and penetrate at several centimeters into the SOL. Here, charge exchange (*cx*) collisions of them with ions generate atoms of energies much higher than that of primary recycling neutral particles. A noticeable fraction of such secondary *cx* atoms hit the vessel wall and erode it.

Statistical Monte Carlo methods [5] are normally used to model *cx* atoms at the edge of fusion devices. A crucial obstacle to apply these approaches for extensive parameter studies, for example, with the aim to optimize the duct geometry, has too long calculations needed to achieve reasonably small accident errors. This is, however, necessary, for example, to couple neutral parameters with the plasma calculations. In a one-dimensional geometry, an alternative approach, based on iteration procedure to solve the kinetic equation represented in an integral form, has been elaborated decades ago [6]. Being free from statistical noise and

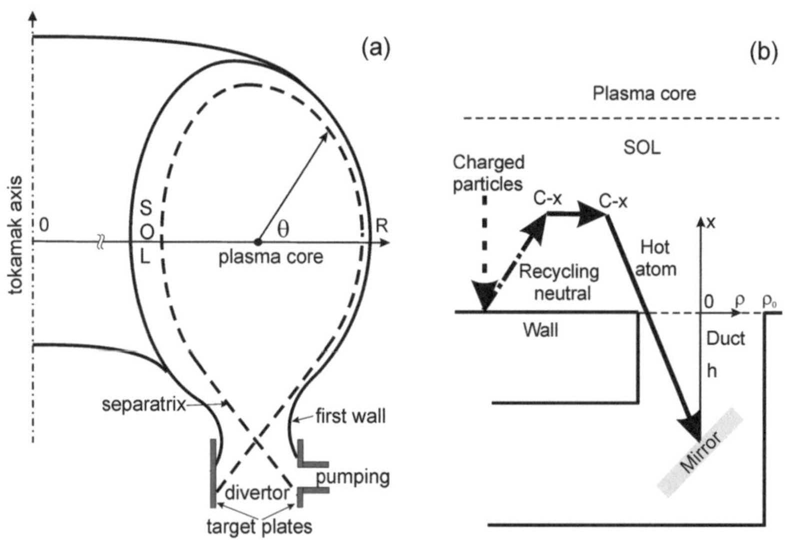

Figure 1. The cross section of toroidally symmetric fusion device of the Tokamak type with the SOL region formed by the presence of a divertor (a) and the processes near the opening in the wall for a duct guiding to a first mirror (b).

permitting the convergence of iterations to the error level defined by the machine accuracy, this method, nonetheless, is also time-consuming. The reason is the necessity to assess integrals in the velocity space from functions involving the ion velocity distribution function, additionally to integrations in the normal space. Recently [4], an approximate pass method has been applied to evaluate these integrals, and the acceleration of kinetic calculations by a factor of 50 has been achieved.

The amendments, outlined above, have allowed to perform calculations of the plasma parameters in the DEMO SOL with *cx* atoms described kinetically, by varying the input parameters, for example, the plasma transport characteristics, in a broad range [4]. In addition the results of these computations have been thoroughly compared with those obtained with *cx* species described in the so-called diffusion approximation. This approximation, often used in diverse edge modeling approaches to save CPU time (see, e.g., [7, 8]), is strictly valid under plasma conditions of low temperature and high density where the time between *cx* collisions of atoms with ions is much smaller than that till their ionization by electrons. Across the DEMO SOL, the plasma density and the temperatures of electrons and ions change, however, by orders of magnitude [4].

The usage of a diffusion approximation for *cx* atoms, generated from species recycling from the wall, becomes especially questionable by considering the situation near an opening in the vessel wall. Such openings will be made in a reactor for diverse purposes, for example, for ducts leading to the first mirrors, collecting light emitted by impurity species in the plasma (see **Figure 1b**). These installations are inaccessible for charged plasma particles, moving mostly along magnetic field lines and penetrating into the duct at a distance of 1 mm. However, hot *cx* atoms, unconfined by the magnetic field, can freely hit and erode these installations. The erosion rate is very sensitive to the energy spectrum of *cx* atoms which can be significantly dependent on the modeling approach. At the position of the duct opening, the inflow of recycling neutrals is actually absent, and the transfer of *cx* atoms has two- or even three-dimensional pattern. By calculating the density of *cx* atoms in the vicinity of a circular opening from a 2D diffusion equation, the erosion rate of a first mirror of Mo has been assessed in [9]. In the present paper, we extend the approaches, elaborated in [4] to model *cx* atoms kinetically in the 1D case, on a 2D geometry.

The results are compared with those of [9]. Although there *cx* atoms have been considered in the diffusion approximation to reduce CPU time, the new approach allows to perform more exact kinetic calculations even by orders of magnitude faster.

2. Basic equations

Although the concept of magnetic fusion is based on the idea that charged particles can be infinitely long confined within closed magnetic surfaces, there are diverse physical mechanisms leading to the losses across these surfaces (see, e.g., [10]). Therefore, electrons and ions escape through the separatrix into the SOL and may reach the first wall of the machine vessel.

Here, these species fast recombine into neutral atoms, with an energy comparable with that of ions. Partly, these atoms escape immediately back into the plasma, and these species are referred as backscattered (*bs*) atoms. The rest transmits the energy to the wall particles and may be bounded into molecules, desorbing back into the plasma as the wall, and is saturated with the working gas. Here, the densities of recycling backscattered atoms and desorbed molecules decay with the distance from the wall because diverse elementary processes such as ionization, charge exchange and dissociation in the latter case, caused by collisions with the plasma electrons and ions, lead to disintegration of neutrals [11]. By the dissociation and ionization of molecules, the so-called Franck-Condon *fc* atoms are generated of a characteristic energy E_{fc} of 3.5 eV [3, 11]. By charge exchange collisions of the primary neutral species, listed above, with the plasma ions, *cx* atoms of higher energies and with chaotically oriented velocities are generated. Although, namely, the latter are of our particular interest, to describe them firmly, all the species introduced have to be considered. In the vicinity of a circular opening in the wall (see **Figure 1b**), any neutral particle is characterized by the components V_x and V_ρ of its velocity, where x and ρ are distances from the wall and opening axis (see **Figure 1b**).

2.1. Recycling molecules and atoms

Recycling species, lunched from the wall, comprise backscattered atoms and desorbed molecules. Henceforth, we assume that the x components of their velocities have single values $V_{bs,m}$ and distinguish two groups of particles moving outward and toward the opening axis, with $V_\rho = V_{bs,m}$ and $V_\rho = -V_{bs,m}$, correspondingly. The magnitudes of $V_{bs,m}$ are predefined by the generation mechanism for the species in question. Within the plasma the point with certain coordinates x and ρ can be reached only by particles, coming from the wall points $(0, \rho + x)$ for $V_\rho < 0$ and $(0, |\rho - x|)$ for $V_\rho > 0$, respectively. By taking into account the attenuation processes with neutrals in the plasma, one gets

$$n_{bs,m}(x,\rho) = \frac{n_{bs,m}(0,\rho + x) + n_{bs,m}(0,|\rho - x|)}{2} \exp\left(-\int_0^x \frac{\nu_{a,m}}{V_{bs,m}} dy\right) \tag{1}$$

where $\nu_a = n\left(k_{ion}^a + k_{cx}^a\right)$ and $\nu_m = n\left(k_{dis}^m + k_{ion}^m + k_{cx}^m\right)$ are the decay frequencies for atoms and molecules, correspondingly, with n being the plasma density assumed the same for electrons and ions, $k_{ion}^{a,m}$ are the rate coefficients of the ionization by electrons, $k_{cx}^{a,m}$ those for the charge exchange with ions, and k_{dis}^m that for the dissociation of molecules.

2.2. Franck-Condon (*fc*) atoms

The *fc* atoms are born by the destruction of molecules and have positive and negative values both of V_ρ and of V_x. Henceforth, we distinguish four groups of these species with the velocity components $\left(V_x, V_\rho\right) = \left(\pm V_{fc}, \pm V_{fc}\right)$, where $V_{fc} = \sqrt{E_{fc}/m}$ and m is the atom mass. The source density of *fc* atoms of each group is as follows:

$$S_{fc} = n\left(2k_{dis}^{m} + k_{ion}^{m} + k_{cx}^{m}\right)n_m/4$$

and depends on ρ through n_m. The particle densities $n_{fc\pm\pm}$, where the first subscript $\pm$ corresponds to the sign of V_x and the second one – of V_ρ, change along the characteristics $\pm x \mp \rho = const$ according to the continuity equation:

$$\sqrt{2}V_x\partial_l n_{fc\pm\pm} = S_{fc}(l) - \nu_a n_{fc\pm\pm}$$

where l is the length of the corresponding characteristics. The boundary conditions take into account that *fc* atoms are reflected with the probability R_{fc} from the wall, $x = 0$, and are absent far from the wall, $x \to \infty$. For the total density of *fc* atoms, $n_{fc} = n_{fc++} + n_{fc+-} + n_{fc-+} + n_{fc--}$, one can obtain

$$\begin{aligned} n_{fc}(x,\rho) = &\int_0^\infty \frac{S_{fc}(y,|\rho - x + y|) + S_{fc}(y,|\rho + x - y|)}{V_{fc}}\mu_0 dy + \\ &+\frac{R_{fc}}{2}\Theta(\rho + x - \rho_0)\int_0^\infty \frac{S_{fc}(y,\rho + x + y) + S_{fc}(y,|\rho + x - y|)}{V_{fc}}\mu_1 dy + \\ &+\frac{R_{fc}}{2}\Theta(|\rho - x| - \rho_0)\int_0^\infty \frac{S_{fc}(y,|\rho - x| + y) + S_{fc}(y,||\rho - x| - y|)}{V_{fc}}\mu_1 dy \end{aligned} \quad (2)$$

with

$$\mu_0 = \exp\left(-\frac{|U_x - U_y|}{V_{fc}}\right), \mu_1 = \exp\left(-\frac{U_x + U_y}{V_{fc}}\right), U_{x,y} = \int_0^{x,y} \nu_a dz,$$

the Heaviside function $\Theta(\xi \geq 0) = 1, \Theta(\xi < 0) = 0$ and ρ_0 being the opening radius (see **Figure 1b**).

2.3. Charge exchange *cx* atoms

The velocity distribution function of *cx* atoms, $f_{cx}(x,\rho,V_x,V_\rho)$, is governed by the kinetic equation:

$$V_x\partial_x f_{cx} + \frac{V_\rho}{\rho}\partial_\rho(\rho f_{cx}) = \frac{S_{cx}}{\pi V_{th}^2}\exp\left(-\frac{V_x^2 + V_\rho^2}{V_{th}^2}\right) - \nu_a f_{cx}, \quad (3)$$

Here, $S_{cx} = S_{cx}^0 + S_{cx}^1$ is the total density of the source of *cx* atoms, with $S_{cx}^0 = n\left[k_{cx}^m n_m + k_{cx}^a (n_{bs} + n_{fc})\right]$ and $S_{cx}^1 = nk_{cx}^a n_{cx}$ being the contributions due to charge exchange collisions with ions of primary neutrals and *cx* atoms themselves, correspondingly, where

$$n_{cx} = \int_{-\infty}^{\infty}\int_{-\infty}^{\infty} f_{cx} dV_x dV_\rho;$$

is the density of the latter; the velocity distribution of the source is assumed as the Maxwellian one with the ion temperature T_i, and $V_{th} = \sqrt{2T_i/m}$ is the ion thermal velocity.

2.3.1. *Integral-differential equations for the density of* cx *atoms*

To solve Eq. (3), V_ρ is discretized as $\pm V_\rho^l$, with $V_\rho^l = \Delta V(l - 1/2)$, $\Delta V = V_{\max}/l_{\max}$, and $l = 1, \cdots, l_{\max}$. The V_x distribution functions $\varphi_l^\pm$ of particles with V_ρ in the range $\pm V_\rho^l - \Delta V/2 \leq V_\rho \leq \pm V_\rho^l + \Delta V/2$ are governed by the equations, following from the integration of Eq. (3) over these ranges:

$$V_x \partial_x \varphi_l^\pm \pm \frac{V_\rho^l}{\rho} \partial_\rho \left(\rho \varphi_l^\pm\right) = \frac{S_{cx}\delta_l}{2\sqrt{\pi} V_{th}} \exp\left(-\frac{V_x^2}{V_{th}^2}\right) - \nu_a \varphi_l^\pm, \tag{4}$$

with

$$\delta_l = \left\{\mathrm{erf}\left[-\left(\frac{V_\rho^l + \Delta V/2}{V_{th}}\right)\right] - \mathrm{erf}\left[-\left(\frac{V_\rho^l - \Delta V/2}{V_{th}}\right)\right]\right\} / \mathrm{erf}\left(\frac{V_{\max}}{V_{th}}\right).$$

From the equation above, one gets the following ones for the variables $\varphi_l = \varphi_l^+ + \varphi_l^-$ and $\gamma_l = V_\rho^l\left(\varphi_l^+ - \varphi_l^-\right)$:

$$V_x \partial_x \varphi_l + \frac{1}{\rho} \partial_\rho (\rho \gamma_l) = \frac{S_{cx}\delta_l}{\sqrt{\pi} V_{th}} \exp\left(-\frac{V_x^2}{V_{th}^2}\right) - \nu_a \varphi_l, \tag{5}$$

$$\partial_x \gamma_l + \frac{\left(V_\rho^l\right)^2}{V_x \rho} \partial_\rho (\rho \varphi_l) = -\frac{\nu_a}{V_x} \gamma_l, \tag{6}$$

The latter equation is straightforwardly integrated with respect to x, and for the ρ-component of the flux density, one gets

$$\gamma_l = -\frac{\left(V_\rho^l\right)^2}{V_x \rho} \int_{x_0}^{x} \partial_\rho (\rho \varphi_l) \exp\left(\frac{U_y - U_x}{V_x}\right) dy.$$

Here, x_0 is the position, where the boundary condition is imposed; since there is no influx of *cx* atoms from the wall, $\gamma_l(x = 0) = 0$ for $V_x > 0$, neutral species are absent deep enough in the plasma and $\gamma_l(x \to \infty) \to 0$ for $V_x < 0$. The found expression for γ_l can be made more convenient if (i) the x-variation of terms involved is approximated by linear Taylor series:

$$\partial_\rho(\rho\varphi_l)(y) \approx \partial_\rho(\rho\varphi_l)(x) + \partial_x\partial_\rho(\rho\varphi_l)(x)\cdot(y-x), U_y \approx U_x + \nu_a(x)\cdot(y-x),$$

and (ii) by taking into account that the width of the region, occupied by *cx* atoms, is of several mean free path lengths (MFPL), that is, $|x - x_0| > |V_x|/\nu_a$ typically. As a result we get

$$\gamma_l \approx -(D_l/\rho) \times \partial_\rho(\rho\varphi_l),$$

with $D_l = \left(V_\rho^l\right)^2/\nu_a$.

By substituting the last expression into Eq. (5), this is reduced to the following one:

$$V_x\partial_x\varphi_l - \frac{D_l}{\rho}\partial_\rho^2(\rho\varphi_l) = \frac{S_{cx}\delta_l}{\sqrt{\pi}V_{th}}\exp\left(-\frac{V_x^2}{V_{th}^2}\right) - \nu_a\varphi_l. \tag{7}$$

Equation (7) can be integrated as the first-order equation with respect to x, with the boundary conditions similar to those for γ_l, that is, $\varphi_l(x=0)=0$ for $V_x > 0$ and $\varphi_l(x \to \infty) \to 0$ for $V_x < 0$. Finally, for the density of *cx* atoms within the range $V_\rho^l - \Delta V/2 \le |V_\rho| \le V_\rho^l + \Delta V/2$, $\eta_l = \int_{-\infty}^{\infty} \varphi_l dV_x$, one obtains the following integral-differential equation:

$$-\frac{D_l}{\nu_a\rho}\partial_\rho^2(\rho\eta_l) = \int_0^\infty \frac{S_{cx}\delta_l}{\sqrt{\pi}V_{th}} I_\alpha dy - \eta_l, \tag{8}$$

where $I_\alpha = \int_0^\infty F_\alpha du$, with $F_\alpha(u) = u^{-1}\exp(-u^2 - \alpha/u)$ and $\alpha = |U_y - U_x|/V_{th}(y)$. For the total density of *cx* atoms, one has $n_{cx} = \sum_{l=1}^{l_{max}} \eta_l$.

2.3.2. *Diffusion approximation*

By neglecting the ρ derivative and by considering a single V_ρ value, Eq. (8) is reduced to an integral one obtained in the 1D case (Eq. (16) in Ref. [6] or Eq. (11) in Ref. [4]). In this case $l \equiv 1, \delta_l = 1$ and $n_{cx} = \eta_1$. As it was demonstrated in Ref. [4], this integral equation can be reduced to an ordinary differential diffusion equation, if (i) the electron temperature is low enough and the ionization rate is much smaller than that of the charge exchange, $k_{ion}^a \ll k_{cx}^a$, and (ii) the characteristic dimension for the plasma parameter change is much larger than the typical MFPL for *cx* atoms V_{th}/ν_a. The same procedure can be performed in the case under the consideration, providing the following 2D diffusion equation:

$$-\frac{D_{th}}{\rho}\partial_\rho\left(\rho\partial_\rho n_{cx}\right) - \partial_x\left[\frac{\partial_x(n_{cx}T_i)}{\nu_a m}\right] = S_{cx}^0 - k_{ion}^a n n_{cx}, \tag{9}$$

with $D_{th} = V_{th}^2/(2\nu_a)$. The boundary conditions to Eq. (9) imply as follows: (i) *cx* atoms leave the plasma with the velocity close to the ion thermal one, and (ii) neutral species are absent far from the wall, that is, $n_{cx}(x \to \infty) = 0$. Additionally, $\partial_\rho \eta_l = 0$ and $\partial_\rho n_{cx} = 0$ for $\rho = 0$ and $\rho \to \infty$ for Eqs. (8) and (9), respectively.

2.3.3. *Assessment of the velocity space integral I_α*

By inspecting Eq. (8), one can see the cause for large calculation time for kinetic computations for *cx* atoms. Even in the 1D case, for each grid point, one has to calculate enclosed double integrals over the ion velocity space and over the whole plasma volume, $0 \le y \le r_w$, and repeat this, in an iterative procedure, with respect to the whole profiles of $n_{cx}(x)$. In a 2D situation, the runs through the grid points in the (ρ, V_ρ) phase subspace are involved additionally into calculations. These can be, however, efficiently parallelized. The integral I_α is appeared due to the generation of *cx* atoms with the ion Maxwellian velocity distribution and consequent attenuation by ionization and *cx* collisions. With a high accuracy, it can be assessed by an approximate pass method [12]. The function $F_\alpha(u)$ is shown in **Figure 2a** for $\alpha = 0.3$, 1, and 3. One can distinguish four u intervals, where $F_\alpha(u)$ behaves principally differently: $u \le u_1$, $u_1 < u \le u_m$, $u_m < u \le u_2$, and $u_2 < u$. The characteristic points $u_{1,2}$ and u_m are defined by the conditions $F'_\alpha(u_m) = 0$ and $F''_\alpha(u_{1,2}) = 0$ for the derivatives $F'_\alpha = -F_\alpha g_1/u^2$ and $F''_\alpha = F_\alpha g_2/u^4$ with

$$g_1(u) = 2u^3 + u - \alpha, \;\; g_2(u) = (g_1 + u)^2 - (2 + 6u^2)u^2.$$

For u_m one can use the Cardano formula for the real root of a cubic equation $u_m = \sqrt[3]{b+a} - \sqrt[3]{b-a}$ with $a = \alpha/4$ and $b = \sqrt{a^2 + 1/216}$. By searching for $u_{1,2}$, we notice that for the roots of interest the equation $g_2(u_{1,2}) = 0$ reduces to the following ones:

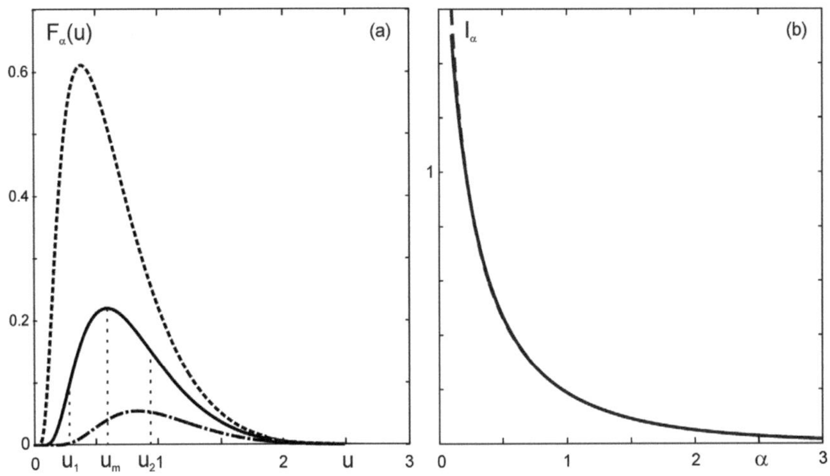

Figure 2. The functions $F_\alpha(u) = u^{-1}\exp(-u^2 - \alpha/u)$ for $\alpha = 1$ (solid curve), $\alpha = 0.3$ (dashed curve), and $\alpha = 3$ (dashed-dotted curve) (a); the integral $I_\alpha = \int_0^\infty F_\alpha(u)du$ estimated by the pass method (solid curve) and computed numerically (dashed curve) (b).

$$g_1(u_1) + u_1 + u_1\sqrt{2 + 6u_1^2} = 0, g_1(u_2) + u_2 - u_2\sqrt{2 + 6u_2^2} = 0.$$

With properly selected initial approximations accurate enough, solutions of these equations are found after three to five iterations with the Newton approach.

To assess the integrals I_α, we approximate function $F_\alpha(u)$ as a linear one at $u \le u_1$ and by polynomials of the fifth order at the intervals $u_1 < u \le u_m$ and $u_m < u \le u_2$. The coefficients are selected to reproduce $F_{\alpha m,1,2} = F_\alpha(u_{m,1,2})$, $F'_{\alpha 1,2} = F'_\alpha(u_{1,2})$, $F''_{\alpha m} = F''_\alpha(u_m)$, and $F'_\alpha(u_m) = F''_\alpha(u_{1,2}) = 0$. In addition, for $u > u_2$ the factor $\exp(-u^2)$ leads to a very fast decay of F_α with increasing u. Therefore, for the aim to assess the corresponding contribution to I_α, we approximate F_α by $\exp(-u^2 - \alpha/u_2)/u_2$. This results in the following expression:

$$\begin{aligned} I_\alpha \approx & \frac{F_{\alpha 1}^2}{2F'_{\alpha 1}} + \frac{F_{\alpha 1}\delta_1 + F_{\alpha m}(u_2 - u_1) + F_{\alpha 2}\delta_2}{2} + \\ & + \frac{F'_{\alpha 1}\delta_1^2 - F'_{\alpha 2}\delta_2^2}{10} + F''_{\alpha m}\frac{\delta_1^3 + \delta_2^3}{120} + \frac{\sqrt{\pi}}{2}\frac{1 - erfu_2}{u_2}\exp\left(-\frac{\alpha}{u_2}\right), \end{aligned} \tag{10}$$

where $\delta_{1,2} = |u_m - u_{1,2}|$. **Figure 2b** shows I_α versus α found with the pass method, outlined above and evaluated numerically. By approximating the numerical results very accurately, the approximate pass method procedure allows to reduce the CPU time by a factor of 50 compared to direct estimates.

Alternatively to the usage of formula (10), one can calculate the integral I_α in advance and save the results in a table. Then, for any particular α, the integral I_α is interpolated from the values in this table. However, some time is needed to find the proper α interval, and this time grows up with $\alpha \to 0$ since $I_\alpha \to \infty$ as $-\ln\alpha$. Thus, the density of the data in the table has to be increased as $\alpha \to 0$, and it is not obvious that such a way is more time-saving than formula (10).

2.3.4. Numerical procedure

The procedure to solve Eq. (9) numerically is organized as follows: For the problem in question, with the plasma profiles assumed unaffected by the processes in the vicinity of the duct opening, the integral $I_\alpha(x, y)$ can be assessed once in advance. Then,

i. the source density $S_{cx}(x, \rho)$ is calculated, in the first approximation with $n_{cx} = 0$;

ii. for all x the first term on the right-hand side of the equation is computed as a function of ρ, and the second-order ordinary Eq. (8) is solved by the method outlined in [10] for all V_l;

iii. the next approximation for $n_{cx}(x, \rho)$ and $S_{cx}(x, \rho)$ is computed, and the procedure is repeated till the relative error in, for example, $n_{cx}(0, 0)$, becomes smaller than a desirable one.

A typical behavior of the error with the iteration number for calculations presented in the next section is shown in **Figure 3**. One can see that with consequent iterations the calculation error reduces steadily without any noise and the machine accuracy can be actually achieved.

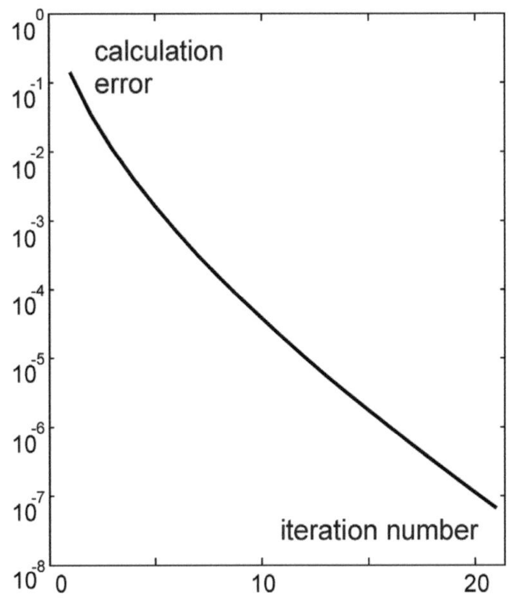

Figure 3. The reduction of the relative error in calculation with the iteration number.

Normally, a diffusion approximation is used to describe *cx* atoms to reduce CPU. The present approach to solve kinetic equation makes this unnecessary. To get the density of *cx* atoms, $n_{cx}(x, \rho)$, with a relative error below 10^{-6}, kinetic calculations require much less CPU time than it is needed to solve the diffusion equation. Thus, for 500 and 30 grid points in the x and ρ directions, respectively, $l_{\max}$ =10 and $V_{\max} = 3V_{th}$, the CPU times for a 1 GHz processor are 30 s and 112 min, correspondingly. Of course, the time for kinetic calculations increases significantly if neutrals affect the plasma background, and $I_\alpha(x, y)$ has to be evaluated by each iteration. Even in this case kinetic computations for *cx* atoms are done faster by factor of 4.

3. Results of calculations

Figure 4 shows the profiles of the plasma parameters, assumed homogeneous on the flux surfaces, across the surfaces, computed in [13] with the input parameters from the European DEMO project [1, 2]. The distance x from the first wall to a certain flux surface depends on the poloidal angle θ (see **Figure 1a**) and in **Figure 4** x corresponds to the torus outboard, $\theta = 0$, where this distance is minimal. The profiles of the *cx* atom density found with these parameters and for the duct radius $\rho_0 = 0.05\,\text{m}$, at the opening axis, $\rho = 0$, and far from it, $\rho = 0.25\,\text{m}$ are shown in **Figure 5**. There is a noticeable difference between the profiles of n_{cx} calculated in the diffusion approximation (5a) and kinetically (5b). In particular, the difference between the density profiles at these two positions is significantly larger in the diffusion approximation than for those computed kinetically.

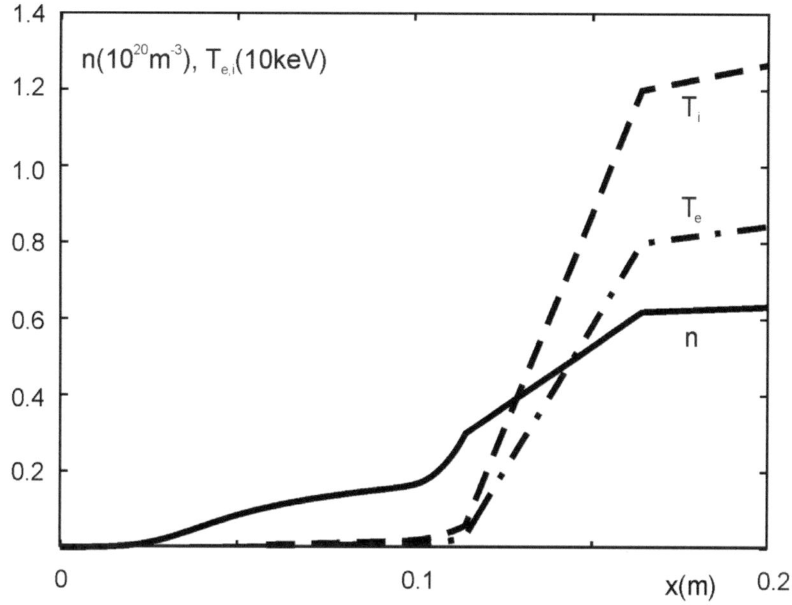

Figure 4. The profiles of the plasma density (solid curve), the ion (dashed curve), and electron (dashed-dotted curve) temperatures versus the distance from the wall at the DEMO torus outboard.

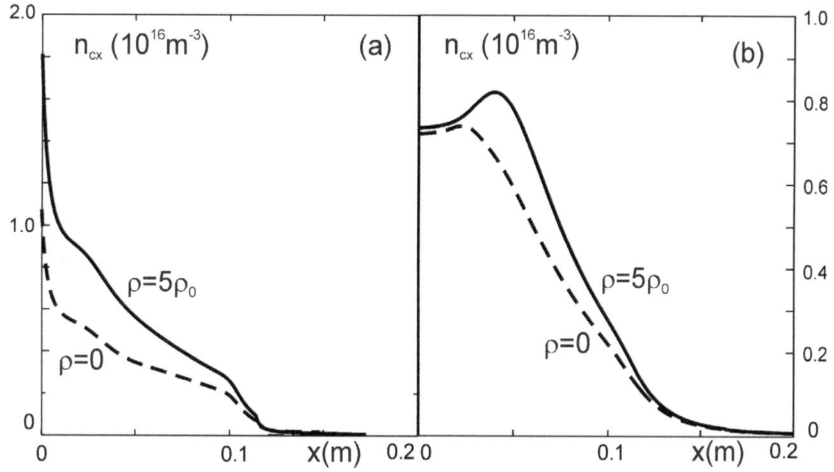

Figure 5. The profiles of the *cx* atom densities far from the opening (solid curve) and at its axis (dashed curve) computed in the diffusion approximation (a) and kinetically (b).

In the latter case, n_{cx} is still noticeable in much deeper and hotter plasma regions. This circumstance is of importance for the erosion of installations in ducts due to physical sputtering because this is very sensitive to the energy of impinging particles. To demonstrate this the erosion rate of a Mo mirror, imposed into the duct of three different radii ρ_0, at the distance h from its opening (see **Figure 1b**) is assessed. To do such assessment, consider an infinitesimally thin toroid within the plasma, with a square cross section of the width $d\rho$, thickness dx, and the radius ρ, situated at the distance x from the wall. Cx atoms with the energies within the range $E, E + dE$ are generated in the toroid with the rate:

$$dR_{cx}(x, \rho, E) = S_{cx}(x, \rho)\varphi_i(x, E)2\pi\rho d\rho dx dE. \tag{11}$$

where

$$\varphi_i(x, E) = \frac{2\sqrt{E}}{\sqrt{\pi}T_i^{1.5}(x)}\exp\left[-\frac{E}{T_i(x)}\right]. \tag{12}$$

is the Maxwellian energy distribution function of ions.

Since cx atoms move in all directions, some of them hit the mirror inside the duct (see **Figure 1b**). Henceforth, we analyze the surface position at the duct axis. One can see that cx atoms generated only in toroids with $\rho \le \rho_{\max} = \rho_0(1 + x/h)$ can hit this mirror point. The contribution of the plasma volume in question to the density of the cx atom flux perpendicular to the mirror surface is

$$dj_{cx}(x, \rho, E) = dR_{cx}\frac{s \cdot \exp(-\lambda/s)}{4\pi\left[(x+h)^2 + \rho^2\right]}, \tag{13}$$

where $s = \left[1 + \rho^2/(x+h)^2\right]^{-1/2}$ is the cosine of the atom incidence angle with respect to the duct axis. The exponential factor in Eq. (13), with $\lambda(x, E) = U_x/\sqrt{2E/m}$, takes into account the destruction of cx atoms in ionization and charge exchange collisions on their way through the plasma. By ionization the atom disappears. By charge exchange a new cx atom is generated. The latter process is taken into account by the contribution S_{cx}^1 in the source density S_{cx}.

The energy spectrum of cx atoms, hitting the mirror, is characterized by their flux density $\gamma = \int dj_{cx}/dE$ in the energy range dE, where the integration is performed over ρ and x. By proceeding from ρ to s, according to the relation $\rho = (x+h)\sqrt{1/s^2 - 1}$, one gets

$$\gamma(h, \rho_0, E) = \int_0^{r_w} \varphi_i(x, E)dx \int_{h/\sqrt{h^2+\rho_0^2}}^{1} \frac{S_{cx}}{2}\exp\left(-\frac{\lambda}{s}\right)ds. \tag{14}$$

The density of the outflow of mirror particles eroded by physical sputtering with cx atoms is as follows:

$$\Gamma_{sp}(h, \rho_0) = \int_0^{\infty} dE \int_0^{r_w} \varphi_i(x, E)dx \int_{h/\sqrt{h^2+\rho_0^2}}^{1} \frac{S_{cx}}{2}\exp\left(-\frac{\lambda}{s}\right)Y_{sp}(E, s)ds \tag{15}$$

Here, Y_{sp} is the sputtering yield, whose dependence on E and s is calculated by applying semiempirical formulas from Ref. [14]. The erosion rate, measured henceforth in nm per full-power year (nm/pfy), is $h_{sp} = \Gamma_{sp}/n_{sp}$, with the particle density of molybdenum n_{sp}. It is shown in **Figure 6** versus the distance h between the wall position and the mirror point in question. One can see that in the former case the erosion rate is by a factor of 2 larger than in the latter one. It is of importance to notice that h_{sp}, found with n_{cx} computed kinetically, does not unavoidably excides that obtained with n_{cx} calculated in the diffusion approximation. The results above have been gotten for a mirror positioned in a duct at the plasma outboard, that is, at the largest major radius R (see **Figure 1a**). Here, the local gradients of the plasma parameters are the largest because the distance between two particular flux surfaces has the minimum. Therefore, cx atoms penetrate, before they are ionized, into plasma regions with higher ion temperatures. The situation is different at the torus top where the corresponding distance has maximum and neutral species are attenuated already at significantly low n and T_i. This leads to noticeably smaller h_{sp} (see **Figure 7**). In this case, oppositely to the situation at the outboard, h_{sp} is larger if n_{cx} is computed in the diffusion approximation.

Due to technical requirements, an acceptable mirror erosion rate in a future fusion reactor should not exceed $1nm/pfy$. As one can see in the case of a duct at the torus outboard, this target level is exceeded significantly for all h and ρ_0 under consideration. Seeding of the working gas into the duct is considered as a possible way to diminish the erosion rate below the maximum level allowed. The energy of cx atoms, coming into the duct, is reduced through elastic collisions with gas molecules, before cx atoms hit the mirror. Elastic collisions between neutral species lead to scattering on large angles. Consider a cx atom which, without gas in the duct, can hit the mirror directly. If the gas is seeded, in a narrow duct in question, with $\rho_0 \ll h$, any collision of the cx atom with gas molecules leads to such a change of the atom velocity that with an overwhelming probability the atom will many times strike the duct wall before it gets the mirror. Through the collisions with the wall, the cx atom loses its energy so dramatically

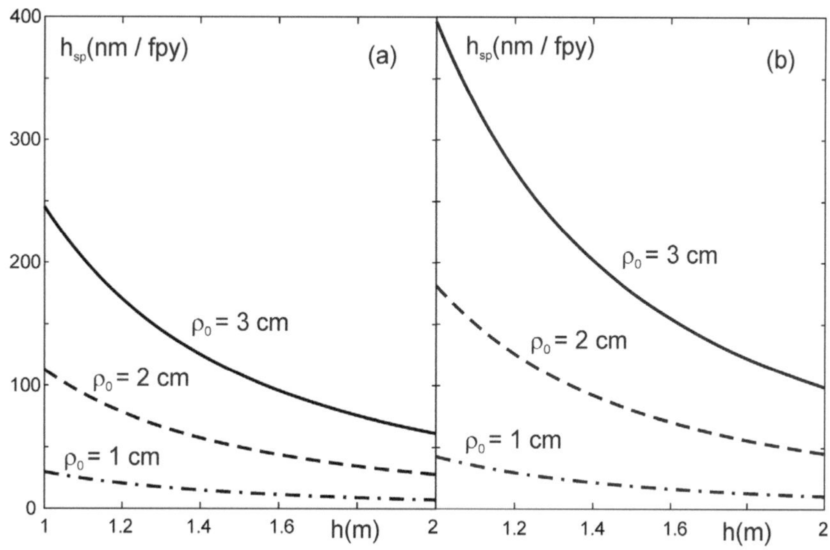

Figure 6. The erosion rate of a Mo mirror versus the distance from the wall position to the mirror surface for ducts positioned at the torus **outboard** computed with the density of cx atoms found in the diffusion approximation (a) and kinetically (b).

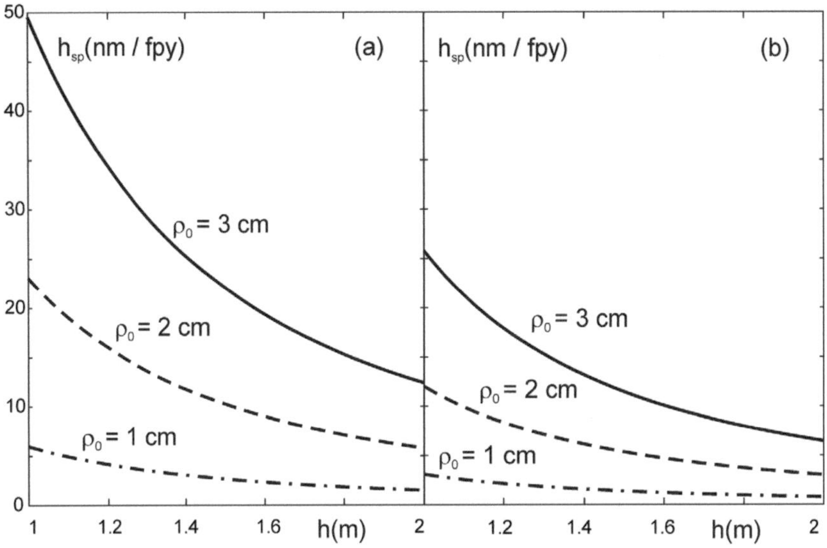

Figure 7. The erosion rate of a Mo mirror versus the distance from the wall position to the mirror surface for ducts positioned at the torus **top** computed with the density of *cx* atoms found in the diffusion approximation (a) and kinetically (b). (Note that the h_{sp} scale is different than in **Figure 6**).

that it cannot contribute to mirror erosion. The reduction of the flux j_E of incoming *cx* atoms with the energy E by collisions with the gas in the duct is governed by the equation:

$$dj_E/dl = -\sigma_{el}(E)n_g j_E,$$

where $\sigma_{el} \approx 3.8 \cdot 10^{-19} E^{-0.14} \left[m^2, eV\right]$ is the cross section for elastic collisions between *cx* atoms and molecules of hydrogen isotopes [15]. Consequently, the density of the outflow of mirror particles eroded by physical sputtering with *cx* atoms is modified, compared with Eq. (15), as follows:

$$\Gamma_{sp}(h, \rho_0) = \int_0^\infty dE \int_0^{r_w} \varphi_i dx \int_{h/\sqrt{h^2+\rho_0^2}}^{1} \frac{S_{cx}}{2} \exp\left(-\frac{\lambda + \sigma_{el} n_g h}{s}\right) Y_{sp}(E, s) ds \quad (16)$$

In **Figure 8**, the calculated dependences of the erosion rate h_{sp} on h and ρ_0 are shown for several magnitudes of the density n_g of the working deuterium gas in the mirror duct. One can see that the enhancement of n_g above a level of $2 \cdot 10^{19} \mathrm{m}^{-3}$ should lead to the reduction of h_{sp} below the target level of $1nm/pfy$. The question, to what extent the local plasma parameters may be changed by the outflow of the gas from the duct, has to be investigated in the future on the basis of approaches developed in [16]. There, it has been demonstrated that the ionization of the gas outflowing into the SOL can lead to dramatic growth of the local density and cooling of the plasma to a temperature of 1 eV. Such cold dense plasma cloud can affect the transfer of *cx* atoms in the plasma near the opening in the wall.

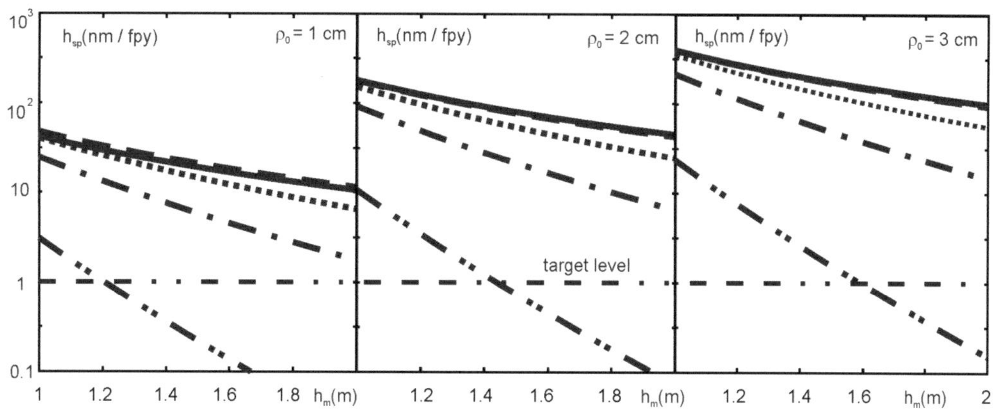

Figure 8. The erosion rate of a Mo mirror versus the distance from the wall position to the mirror surface for ducts positioned at the torus outboard computed with *cx* atoms treated kinetically for different gas densities in the duct: $n_g = 0$ (solid line), $3 \cdot 10^{17} \mathrm{m}^{-3}$ (dashed line), $3 \cdot 10^{18} \mathrm{m}^{-3}$ (dotted line), $10^{19} \mathrm{m}^{-3}$ (dashed-dotted line), and $3 \cdot 10^{19} \mathrm{m}^{-3}$ (dashed-double dotted line).

4. Conclusions

The iteration approach to solve 1D kinetic equation for *cx* atoms, proposed decades ago [6], has been elaborated further to describe the transport of these species in a 2D geometry, in the vicinity of a circular opening in the wall of a fusion reactor. Unlike the Monte Carlo methods, this approach does not generate statistical noise so that calculation errors can be reduced to the level restricted by the machine accuracy. In order to perform calculations for a broad range of input parameters and do a thorough comparison with the results, obtained in the diffusion approximation for *cx* atoms [13], the solving procedure has been accelerated by a factor of 50, by applying an approximate pass method to assess integrals in the velocity space from functions, involving the Maxwellian velocity distribution of plasma ions.

The found possibility to speed up kinetic calculations is of importance, in particular, to perform firm assessments of the erosion rate of the first mirrors in future fusion reactors like DEMO. For a mirror located at the torus outboard, more accurate kinetic calculations predict by a factor of 2 higher erosion rate than the approximate diffusion approach. The erosion rate can be reduced very strongly either by putting the mirror duct at the torus top or by seeding the working gas into the duct. In the latter case, the elastic collisions with molecules in the gas reduce significantly the fraction of *cx* atoms which can hit and erode the mirror.

Nomenclature

$D_{l,\,th}$	Diffusivity of *cx* atoms in kinetic and diffusion descriptions, correspondingly

E	Energies of ions and atoms
$F_\alpha(u) \equiv \frac{1}{u}\exp\left(-u^2 - \frac{\alpha}{u}\right), I_\alpha \equiv \int_0^\infty F_\alpha[u]du$	
h	Distance from duct opening to the mirror
h_{sp}	Mirror erosion rate in nm per full-power year
$\gamma(h, \rho_0, E)$	Flux density of cx atoms onto the mirror
$\Gamma(h, \rho_0)$	Outflow density of sputtered mirror particles
$k_{cx}^{a,m}$	Charge exchange rate coefficients
k_{dis}^{m}	Molecule dissociation rate coefficient
$k_{ion}^{a,m}$	Ionization rate coefficients
$\lambda(x, E) = U_x/\sqrt{2E/m}$	
m	Mass of atoms
n	Plasma density
$n_{m,bs,fc,cx}$	Densities of different neutral species—desorbed molecules, backscattered atoms, Franck-Condon, and charge exchanged atoms, respectively
$\nu_{m,a}$	Decay frequencies for molecules and atoms
$R_{cx}(x, \rho, E)$	The generation rate of cx atoms in the energy range $E, E + dE$ within an infinitesimally thin plasma toroid of the width $d\rho$, thickness dx, the radius ρ, and situated at the distance x from the wall
ρ	Distance from the opening axis
ρ_0	The opening radius
s	Cosine of the cx atom incidence angle with respect to the duct axis
$S_{fc,cx}$	Source densities of fc and cx atom species
$S_{cx}^{0,1}$	cx atom source density contributions due to charge exchange of primary neutral species and cx atoms
σ_{el}	Cross section for elastic collisions between cx atoms and molecules of hydrogen isotopes
$T_{e,i}$	Electron and ion temperatures
$U_{x,y} \equiv \int_0^{x,y} \nu_a dz$	

$V_{m,bs,fc}$	Absolute velocities of neutral species
$V_{x,\rho}$	Velocity components perpendicular to the wall and to the opening axis, respectively
$V_{th} = \sqrt{2T_i/m}$	Thermal velocity of ions
$\varphi_i(x, E)$	Energy distribution function of ions
x	Distance from the wall

Author details

Mikhail Tokar

Address all correspondence to: m.tokar@fz-juelich.de

Institute for Energy and Climate Research – Plasma Physics (IEK-4), FZJ, Jülich, Germany

References

[1] Federici G, Kemp R, Ward D, Bachmann C, Franke T, Gonzalez S, Lowry C, Gadomska M, Harman J, Meszaros B, Morlock C, Romanelli F, Wenninger R. Mint: Overview of EU DEMO design and R&D activities. Fusion Engineering and Design. 2014;**89**:882-889. DOI: 10.1016/j.fusengdes.2014.01.070

[2] Wenninger R, Arbeiter F, Aubert J, Aho-Mantila L, Albanese R, Ambrosino R, Angioni C, Artaud JF, Bernert M, Fable E, Fasoli A, Federici G, Garcia J, Giruzzi G, Jenko F, Maget P, Mattei M, Maviglia F, Poli E, Ramogida G, Reux C, Schneider M, Sieglin B, Villone F, Wischmeier M, Zohm H. Mint: Advances in the physics basis for the European DEMO design. Nuclear Fusion. 2015;**55**:063003. DOI: 10.1088/0029-5515/55/6/063003

[3] Stangeby PC. The Plasma Boundary of Magnetic Fusion Devices. Bristol and Philadelphia: Institute of Physics Publishing; 2000. p. 717. ISBN: 0 7503 0559 2

[4] Tokar MZ. Mnt: Scrape-off layer modeling with kinetic or diffusion description of charge-exchange atoms. Physics of Plasmas. 2016;**23**:122512. DOI: 10.1063/1.4972538

[5] Landau DP, Binder K. A Guide to Monte Carlo Simulations in Statistical Physics. Cambridge: University Printing House; 2015. p. 519. ISBN: 9781107074026

[6] Rehker S, Wobig H. Mint: A kinetic model for the neutral gas between plasma and wall. Plasma Physics. 1973;**15**:1083-1097

[7] Rognlien TD, Brown PN, Campbell RB, Kaiser TB, Knoll DA, McHugh PR, Porter GD, Rensink ME, Smith GR. Mint: 2-D Fluid transport simulations of gaseous/radiative divertors. Contributions to Plasma Physics. 1994;**34**:362-367

[8] Weber S. Mint: Aiming at 3-dimensional plasma fluid modelling of the W7-X divertor. Contributions to Plasma Physics. 1998;**38**:43-48

[9] Tokar MZ, Beckers M, Biel W. Mint: Erosion of installations in ports of a fusion reactor by hot fuel atoms. Nuclear Materials and Energy. 2017;**12**:1298-1302. DOI: 10.1016/j.nme.2017.01.006

[10] Tokar MZ. Mint: Modelling of profile evolution by transport transitions in fusion plasmas. In: Ahsan A, editor. Two Phase Flow, Phase Change and Numerical Modeling. Rijeka: InTech; 2011. pp. 149-172. ISBN: 978-953-307-584-6

[11] Janev RK, Langer WD, Evans KJ, Post DEJ. Mint: Elementary Processes in Hydrogen-Helium Plasmas. Hamburg: Springer; 1987. p. 328. ISBN: 978-3-642-71935-6

[12] Mathews J, Walker RL. Mint: Mathematical Methods of Physics. 2nd ed. Menlo Park: Benjamin; 1970. p. 515. ISBN: 0805370021

[13] Tokar MZ. Mint: An assessment for the erosion rate of DEMO first wall. Nuclear Fusion. 2018;**58**:016016. DOI: 10.1088/1741-4326/aa92dd

[14] Eckstein W, García-Rosales C, Roth J, Ottenberger W. Mint: Sputtering data. Garching: Institute of Plasma Physics; IPP 9/82 [Internet]. 1993. Available from: http://pubman.mpdl.mpg.de/pubman/item/escidoc:2131245:1/component/escidoc:2131244/IPP 9 82.pdf

[15] McGuire P, Krüger H. Mint: Elastic and rotationally inelastic H-H2 and H-D2 collisions. The Journal of Chemical Physics. 1975;**63**:1090-1094

[16] Tokar MZ. Mint: Modeling of localized impulsive injection of neutrals and plasma response. Plasma Physics and Controlled Fusion. 2017;**59**:055005-055012. DOI: 10.1088/1361-6587/aa6219

Chapter 2

The Behavior of Streaming Instabilities in Dissipative Plasma

Eduard V. Rostomyan

Additional information is available at the end of the chapter

http://dx.doi.org/10.5772/intechopen.79247

Abstract

An approach is presented that allows getting detailed information on the behavior of streaming instabilities (SI) from the dispersion relation (DR). The approach is based on general assumptions and does not refer to any particular model and/or type of the stream interaction with background system (Cherenkov, cyclotron, etc.). The basis of the approach is transformation of the DR to an equation for slowly varying amplitude of the developing waveform. The solution of the equation actually presents results of the important problem of time evolution of initial perturbation and gives detailed information on the instability behavior. Most of the information is unavailable by other methods. For particular SI, only two parameters should be specified. The expression for the fields' structure shows that with increase in level of dissipation, SI gradually turns to dissipative streaming instability (DSI). Two new, previously unknown types of DSI are presented: DSI of overlimiting electron beam and DSI under weak beam-plasma coupling. Growth rates of these DSI depend on dissipation more critically than usual. Presented approach is valid for a large class of SI: beam-plasma instabilities of various types (Cherenkov, cyclotron, etc.) including over-limiting e-beam instabilities, the instability in spatially separated beam-plasma systems, Buneman instability, etc.

Keywords: streaming instability, dissipative instability, space–time evolution, slowly varying amplitude, transformation to dissipative instability

1. Introduction

Plasma is rich in instabilities. Many of them are a result of relative motion of plasma components. These, streaming instabilities (SI) are the most common in space and laboratory plasmas.

A well-known example is the beam-plasma instability [1], in which the directed motion of a small group of fast electrons passing through the background plasma excites potential oscillations with high growth rate near the plasma frequency. Close attention to this instability is due mainly to design of high power sources of electromagnetic radiation based on this instability. The sources have many advantages as compared to well-known vacuum devices [2, 3]. Another example (we mention these two only) is the Buneman instability [4], in which plasma electrons move with respect to ions. The instability plays an important role in many scenarios in space physics and geophysics. A striking example of plasma with relative electron-ion motion is current-carrying plasma. This object is often considered in plasma physics. The instabilities which are due to relative electron-ion motion play an important role in physics of controlled fusion also.

A clear understanding of physical nature of the SI, their role and influence on various processes in plasma requires substantial efforts. Physics of interaction of plasma components moving relatively to each other is essentially based on the concept of negative energy wave (NEW) [5]. This requires account of all factors which lead to NEW growth. Among them, dissipation plays an important role. Dissipation leads to energy losses for the growth of NEW. Influence of dissipation on the instabilities of streaming type is unique. Dissipation never suppresses the instabilities completely regardless on its level. Dissipation of high-level transforms the SI to dissipative streaming instability (DSI) [1]. These instabilities have a number of features: comparatively low growth rate, comparatively low level of excited oscillations, etc. For a few decades, DSI have been widely discussed, and it is supposed that they can be applied to explain various phenomena in space and laboratory plasma. Up to recently only one type of DSI was known, and it was believed that all types of electron stream instabilities (e.g., Cherenkov type, cyclotron type etc.) transform to the single known type of DSI. However, it turned out that other types of DSI also exist [6–8]. Changes in some basic physical parameters and/or system geometry lead to significant changes in physical nature of e-stream interaction with plasma. This changes result in two new, previously unknown types of DSI: DSI of over-limiting electron beam and DSI under weak coupling of the stream with the plasma. In both cases, the growth rate depends on dissipation more critically: $1/\nu$ instead of conventional $1/\sqrt{\nu}$ (here ν is the frequency of the collisions).

The transformation of the SI to dissipative type makes their behavior in the presence of dissipation of particular interest. In order to understand how instability turns to another type, it is necessary to investigate the evolution of its fields in space and time [9, 10]. Simultaneously, the expressions for fields' evolution give all available information on the SI: growth rates (spatial and temporal) under arbitrary level of dissipation, character of the instability (absolute/convective), range of unstable perturbations' velocities, influence of dissipation on the instability, etc. These details help to understand how the instability turns to DSI, how it transforms given equilibrium of background plasma, predict the level and/or scale of the changes, how nonlinear phenomena arise as well as predict possible saturation mechanisms, etc. In general, the character of the fields' development in space and time is one of the most important aspects of every instability.

The character of space–time evolution of given instability is an important issue in many branches of physics. In plasma physics, we firstly note theory of amplifiers and oscillators in the microwave range based on interaction of e-beam with wave, where obvious progress is achieved [2, 3]. These studies are also important for research on plasma instabilities associated with research on nuclear fusion, astrophysics, etc.

The mathematical solution of the problem of initial perturbation evolution reduces to calculation of the integral with a complete dispersion relation (DR) in the denominator of the integrand. An overall view on the character of the instability may be obtained by investigation of the asymptotic behavior of the Green's function. In order to derive analytical expression for the fields' space–time distribution, the DR should be specified and solved before integration. In this way, essential difficulties appear which usually cannot be overcome. One must apply approximate methods to obtain results. Presented here (see also [11]) approach is similar to traditional approach in many respects, but, in the same time, advantageously differs from it. Representation of the fields in form of wave train with slowly varying amplitude (SVA) allowed to overcome the difficulties and to obtain the space–time structure of the fields without reference on any particular model. Thereby, the approach singles out intrinsic peculiarities of various types of SI. The results show that all types of the beam-plasma instabilities (Cherenkov, cyclotron, etc.) have similar dynamics of development. By specifying only two parameters in the unified expression one can investigate given particular case of beam instability. With increase in level of dissipation all SI gradually turn to DSI.

This review considers all these aspects: getting detailed information on SI, their space–time evolution and transformation to DSI. Presented approach shows that the DR which usually describes given SI can serve not only for solution of the well-known (and very simplified) initial and boundary problems. Its application is much wider. It can give much more information on the instability. Namely, it actually gives the solution of the well-known (and very important [9]) problem of time evolution of initial perturbation. The DR can give space–time structure of the fields at the instability development. In its turn, the fields' structure contains complete information on the instability. Most of this information is unavailable by other methods. The expressions for fields' evolution also show in detail the transformation of SI to dissipative type. Two new, previously unknown types of DSI are presented.

Large variety of SI characterize by various types of the interaction with background systems (plasma-filled or not), various values of streaming currents, etc. From this follows various types of their DR and ensuing equation for SVA. They are considered separately. In Section 2, the evolution of various types of beam instabilities (Cherenkov, cyclotron, and the instability in periodical structure) are considered. All they characterize by small contribution of the beam in DR and this fact allowed generalizing the consideration. Section 3 gives the evolution of over-limiting e-beam instability. Due to influence of the beam space charge, the instability of such beams has other physical nature as compared to instability of conventional e-beams. In Sections 4 and 5, the instability in spatially separated beam-plasma system and the Buneman instability are considered. The peculiarity of last case is in the role of plasma ions.

2. The behavior beam-plasma instabilities in dissipative plasma

2.1. Equation for slowly varying amplitude

Consider an electrodynamical system of arbitrary geometry (plasma filling is not obligatory) and let a monoenergetic relativistic electron beam penetrate it. The general form of the dispersion relation (DR) of such system is

$$D_0(\omega, \mathbf{k}) + D_b(\omega, \mathbf{k}) = 0 \tag{1}$$

where ω is the frequency of perturbations and $\mathbf{k}$ is the wave vector. $D_0(\omega, \mathbf{k}) = 0$ is the "cold" DR describing proper frequencies of the systems in the absence of the beam (its main part), and $D_b(\omega, \mathbf{k})$ is the beam contribution. We also assume that the beam density is small enough to satisfy the condition $|D_b(\omega, \mathbf{k})| << |D_0(\omega, \mathbf{k})|$. In following consideration, we will not specify the form of $D_0(\omega, \mathbf{k})$. Beam electrons interact with proper oscillations of background system and this interaction leads to instability. The interaction may be of various types: Cherenkov, cyclotron, interaction with periodical structure, etc. The general form of $D_b(\omega, \mathbf{k})$ may be written as

$$D_b(\omega, \mathbf{k}) = -\frac{\omega_b^2 A(\omega, \mathbf{k})}{\gamma^3 (\omega - \mathbf{k}\mathbf{u} - f)^2} \tag{2}$$

where $\mathbf{u}$ is the velocity of the beam electrons, ω_b is the Langmuir frequency of streaming electrons, γ is the relativistic factor of the beam electrons, and $A(\omega, \mathbf{k})$ is a polynomial with respect to ω and $\mathbf{k}$ of degree no higher than two. The expression for f depends on the type of the beam interaction with plasma:

$$f = \begin{cases} 0, & \text{if the interaction is of Cherenkov type} \\ n\Omega/\gamma, & \text{if the interaction is of cyclotron type} \\ k_{\text{cor}} u, & \text{if the beam interacts with periodical structure} \end{cases}, \tag{3}$$

where Ω is the cyclotron frequency, $n = 1, 2, 3\ldots$, $k_{cor} = 2\pi/l_0$ l_0 is the spatial period of the structure. Below, we will show that properties of the instabilities follow from the general form (2) and do not depend on the expression for f.

Let an initial perturbation arises in point $z = 0$ (electron stream propagates in the direction $z > 0$) at instant $t = 0$ and the instability begins developing. Our aim is to obtain shape of the perturbation (i.e., space–time structure of the fields) at arbitrary instant t and based on the expression, investigate the behavior of the instability. In following consideration, we interest in longitudinal structure of the field (their dependence on z and t only). We single out two arguments: the frequency ω and longitudinal wavelength k. Other arguments play no part in following. To avoid overburdening of the formulas below, they are omitted. The transversal structure of the fields may be obtained in regular way by expansion on series of eigenfunctions of given system.

The development of wave pulse in its linear stage obeys the DR (1). The beam instability reveals itself most effectively on frequencies, closely approximating to roots of the main part of Eq. (1) and simultaneously to the beam proper oscillations (e.g., space charge wave). This means that following two conditions must be satisfied:

$$D_0(\omega, k) = 0 \quad ; \quad \omega - ku - f = 0. \tag{4}$$

Therefore, it would appear reasonable to assume that developing fields form a wave train of following type

$$E(z, t) = E_0(z, t) \exp\{-i\omega_0 t + ik_0 z\}, \tag{5}$$

where the carrier frequency ω_0 and k_0 satisfy the conditions (4). We also assume that the amplitude $E_0(z, t)$ is slowly varying as compared to ω_0 and k_0 that is,

$$\left|\frac{\partial E_0}{\partial t}\right| \ll \omega_0 E_0 \quad ; \quad \left|\frac{\partial E_0}{\partial z}\right| \ll k_0 E_0 \tag{6}$$

In such formulation, the problem of the instability evolution reduces to determination of the slowly varying amplitude (SVA) $E_0(z, t)$. As the fields vary near ω_0 and k_0, one can use following formal substitutions to derive an equation for SVA

$$\omega \to \omega_0 + i\frac{\partial}{\partial t} \quad ; \quad k \to k_0 - i\frac{\partial}{\partial z}, \tag{7}$$

Expanding the DR (1) in power series near ω_0 and k_0, one can obtain the equation for SVA

$$\left(\frac{\partial}{\partial t} + u\frac{\partial}{\partial z}\right)^2 \left(\frac{\partial}{\partial t} + \mathrm{v}_0\frac{\partial}{\partial z} + \nu\right) E_0(z, t) = i|\delta_0|^3 E_0(z, t) \tag{8}$$

where

$$\delta_0 = \left\{\frac{\omega_b^2 A(\omega, k)}{\partial D_0/\partial\omega}\right\}_{\substack{\omega = \omega_0 \\ k = k_0}} \quad ; \quad \nu = \left\{\frac{\operatorname{Im} D_0}{\partial D_0(\omega,k)/\partial\omega}\right\}_{\substack{\omega = \omega_0 \\ k = k_0}} \quad ; \quad \mathrm{v}_0 = -\left\{\frac{\partial D_0(\omega,k)/\partial k}{\partial D_0(\omega,k)/\partial\omega}\right\}_{\substack{\omega = \omega_0 \\ k = k_0}}$$

$\operatorname{Im}\delta_0$ is the maximal growth rate of the beam instability, ν describes dissipation in the system (its coincidence to collision frequency is not obligatory), and v_0 is the group velocity of resonant wave in the "cold" system.

The Eq. (8) describes the evolution of SVA E_0 (z, t) in space and time for all systems those may be described by the DR in form (1). Eq. (8) may be solved by using Fourier transformation with respect to spatial coordinate z and Laplace transformation with respect to time t. The corresponding equation for the transform $E_0(\omega, k)$ is

$$\left\{(\omega - ku)^2(\omega - k\mathrm{v}_0 + i\nu) - |\delta_0|^3\right\} E_0(\omega, k) = J(\omega, k) \tag{9}$$

$$E_0(\omega, k) = \int_0^{\infty} dt \int_{-\infty}^{\infty} dz \; E_0(z, t) \exp(i\omega t - ikz) : \tag{10}$$

where the function $J(\omega, k)$ is determined by initial conditions. Its power with respect to ω and k is no higher than the power of the origin equation. The specific form of this function is not essential for following. It is only necessary that $J(\omega, k)$ be smooth and not equal to zero identically. The amplitude of the wave train can be obtained by inverse transformation

$$E_0(z, t) = \frac{1}{(2\pi)^2} \int_{C(\omega)} d\omega \int_{-\infty}^{\infty} \frac{dk J(\omega, k) \exp(-i\omega t + ikz)}{(\omega - ku)^2(\omega - k\mathrm{v}_0 + i\nu) - |\delta_{Bn}|^3} \tag{11}$$

Here, $C(\omega)$ is the contour of integration over ω. For given case, it is a straight line that lies in the upper half plane of the complex plane $\omega = \mathrm{Re}\omega + i\,\mathrm{Im}\,\omega$ and passes above all singularities of the integrand. Thus, the problem has been reduced to the integration in Eq. (11). It is convenient to transform the variables ω and k to another pair ω and $\omega' = \omega - ku$. The first integration (over ω) may be carried out by residue method and the integration contour must be closed in the lower half plane. The pole is

$$\omega_1\left(\omega'\right) = \left(1 - \frac{\mathrm{v}_0}{u}\right)^{-1} \left(\frac{|\delta_0|^3}{\omega'^2} - \omega' \frac{\mathrm{v}_0}{u} - i\nu\right) \tag{12}$$

The second integration (over ω') cannot be carried out exactly, and we are forced to restrict ourselves by approximate, steepest descend method. That is, Eq. (11) will be worked out in asymptotic limit of comparatively large t. In this case, the integration contour should be deformed in order to pass through the saddle point in needed direction. The saddle point is

$$\omega'_s = \delta_0 \left\{ \frac{2(ut - z)}{(z - \mathrm{v}_0 t)} \right\}^{1/3} \exp(2\pi i/3) \tag{13}$$

As a result of the integration, we obtain following expression for the SVA [11].

$$E_0(z, t) = \frac{J_0}{2\sqrt{\pi}} \frac{\exp\left\{\chi_\nu^{(\mathrm{und})}(z, t)\right\}}{\sqrt{(u - \mathrm{v}_0) f(z, t)}} \exp\{i\varphi(z, t)\} \tag{14}$$

$$\chi_\nu^{(\mathrm{und})}(z, t) = \chi_0^{(\mathrm{und})}(z, t) - \nu \frac{z - \mathrm{v}_0 t}{u - \mathrm{v}_0} \quad ; \quad \chi_0^{(\mathrm{und})}(z, t) = \frac{3\sqrt{3}}{4} \frac{\delta_0}{u - \mathrm{v}_0} \left\{ 2(ut - z)(z - \mathrm{v}_0 t)^2 \right\}^{1/3}$$

$$f(z, t) = 3\delta_0^3 (ut - z) \quad ; \quad \varphi(z, t) = \frac{\chi(z, t)}{\sqrt{3}} + \frac{\pi}{4}$$

and J_0 is the value of $J(\omega, \omega')$ at the points. $\omega = \omega_1(\omega'_s)$, $\omega' = \omega'_s$.

2.2. Analysis of the fields' dynamics

We have arrived to very complex expressions (14). However, the field's structure (i.e., the instability behavior) may be determined by analyzing the factor

$$\exp \chi_{\nu}^{(\text{und})}(z,t) \tag{15}$$

The information, which are available from the analysis are much more detailed and complete as compared to results of well-known initial and boundary problems. The analysis gives: growth rate(s), the velocities of unstable perturbations, the character of the instability and influence of the dissipation on it, etc. The expression (15) shows that along with exponential increasing the field covers more and more space. In the absence of dissipation, the velocities of unstable perturbations range from v_0 to u. The length of the wave train increases depending on time $l \sim (u - \mathrm{v}_0)t$. One can easily see convective character of streaming instabilities in laboratory frame, as well as in other frames moving at velocities $v < \mathrm{v}_0$ and $v > u$. If the observer's velocity is within the range $\mathrm{v}_0 < v < u$, the instability is absolute (see **Figure 1**, where the dependence of the SVA on z at various instants t_1, t_2 and t_3 is presented; the leading edge moves at velocity u, but the back edge moves at velocity $\mathrm{v}_0 < u$).

The peak (and the field's properties in it) may be determined from the equation

$$\frac{\partial}{\partial z}\chi_{\nu}^{(\text{und})} = 0 \tag{16}$$

Its solution in the absence of dissipation gives $z = w_g t$, where

$$w_g = (1/3)(2u + \mathrm{v}_0) \tag{17}$$

That is, the peak places on 1/3 of the train's length from the front and moves at the velocity w_g. Actually, w_g represents group velocity of the generated wave, with account of the beam contribution in the DR. The field's value in the peak exponentially increases and the growth

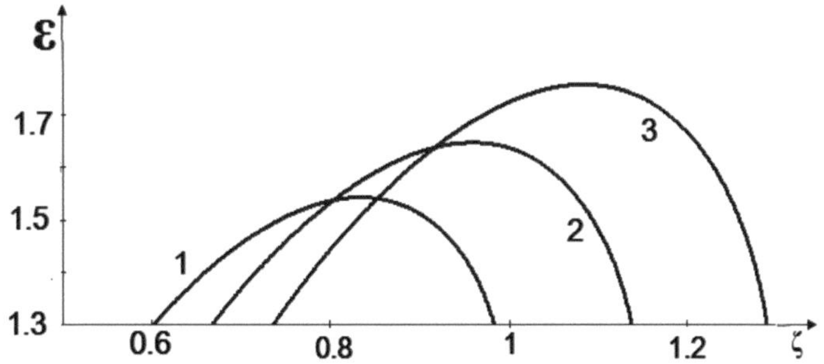

Figure 1. Asymptotic shapes of beam instability $\varepsilon = \exp \chi_0^{(\text{und})}(z,t)$ depending on longitudinal coordinate $\zeta = z\delta_0/u$ at various instants $\tau_1 = \delta_0 t_1 < \tau_2 = \delta_0 t_2 < \tau_3 = \delta_0 t_3$.

rate is equal to $\delta_m = (\sqrt{3}/2)|\delta_0|$ that is, coincides to solution of the initial problem. However, the initial problem can not specify the point, where the maximal growth occurs. The advantage of this approach is evident.

In a fixed point z, the field first increases and attains maximum at instant $t = z/w_a$ where

$$w_a = \frac{3u\mathrm{v}_0}{u + 2\mathrm{v}_0} \tag{18}$$

Then, the field falls off and at the time $t \geq z/\mathrm{v}_0$ the train passes the considered point. The velocity w_a is the group velocity of the resonant wave upon amplification with account of the beam contribution in the DR. For given z, the field's maximum is

$$E_0 \sim \exp \delta_m z/\left(u^2 \mathrm{v}_0\right)^{1/3} \tag{19}$$

The exponent $\delta_m/\left(u^2\mathrm{v}_0\right)^{1/3}$ coincides to solution of the boundary problem as it is the maximal spatial growth rate. The coincidence to the results of well-known initial and boundary problems testifies presented approach. It may appear that this way of instability analysis is a bit more complicate. However, it must be admitted that along with growth rates we have obtained much other information. The information obviously clarifies the picture of the instability and makes it realistic. One can easily see the merits of presented approach.

The relations between characteristic velocities are

$$\mathrm{v}_0 < w_a < w_g < u \tag{20}$$

At fixed instant t, perturbations exist only at distances $\mathrm{v}_0 t \leq z \leq ut$. The wave train passes given point z during the time $z/u \leq t \leq z/\mathrm{v}_0$. In a fixed point, the amplitude attains maximum at the instant, when the peak has already passed it (see **Figure 1**). The reason is that the perturbations with smaller velocities reach considered point in longer time, and they grow more efficiently. Perturbations with velocity w_a are the most efficiently enhanced perturbations.

Generally, the dependence of the perturbations' amplitudes on their velocity v has a form $E \sim \exp \Gamma(v)t$, where

$$\Gamma(v) = \frac{3}{2^{2/3}} \frac{\delta_m}{u - \mathrm{v}_0} \left\{(v - \mathrm{v}_0)^2 (u - v)\right\}^{1/3} \tag{21}$$

The character of spatial growth depending on v is

$$E \sim \exp \Gamma(v) z/v \tag{22}$$

Presented above analysis is true if we neglect dissipation. Dissipation essentially changes the instability behavior. It suppresses slow perturbations. The threshold velocity is

$$V_{thr} = \frac{\lambda u + \mathrm{v}_0}{(1 + \lambda)} \quad ; \quad \lambda = \frac{2^{5/2}}{3^{9/4}} \left(\frac{\nu}{|\delta_0|}\right)^{\frac{2}{3}} \tag{23}$$

Only perturbations moving at higher velocities $v > V_{thr}$ develop. The wave train shortens. Dissipation decreases the field growth

$$\Gamma(v) \to \Gamma_\nu(v) = \Gamma(v) - \nu\frac{u - v}{u - \mathrm{v}_0} \tag{24}$$

The dynamics of the field in the peak may be obtained by analyzing the Eq. (16). It takes following form

$$(z - \mathrm{v}_0 t)(ut - z)^2 = \left(1/\lambda^2\right)\left(z - w_g t\right)^3 \tag{25}$$

If $v << \delta_0$, this equation leads to small corrections to the expressions (17) and (18) for characteristic velocities and for the maximal growth rate in the peak. In the opposite case of high-level dissipation, only the perturbations are unstable, whose velocity is close to the beam velocity u. In this approximation, the solution of Eq. (25) is $z = u - \Delta u$ where

$$\Delta u = 3^{-3/2}\lambda^{-1}(u - \mathrm{v}_0), \tag{26}$$

and the expression for maximal growth rate takes the form $\Gamma_{\nu\to\infty} = \left(\delta_m^3/\nu\right)^{1/2}$. Obviously, this case corresponds to dissipative streaming instability (DSI). The same expression for $\Gamma_{\nu\to\infty}$ can be obtained from Eq. (1) by direct usage of the initial problem [1]. If one specifies δ_m, he can obtain the growth rate of DSI in unbound beam-plasma system, in magnetized beam-plasma waveguide, etc.

In general, by substitution of two parameters only: growth rate and the group velocity of resonant wave in "cold" system one can obtain the behavior of specific e-beam instability.

It is not superfluous to repeat once again that the expression (14) and resulting analysis is valid for all types of e-beam instabilities: Cherenkov, cyclotron, beam instability in periodical structures, etc. Also, the analysis does not depend on specific geometry, external fields, etc.

3. The behavior of overlimiting electron beam instability

The picture described above is valid for e-beams, instability of which is due to induced radiation of the system proper waves by the beam electrons. However, it is known that with increase in beam current the physical nature of e-beam instabilities changes [6, 7, 12–14]. This is a result of influence of the beam space charge. It sets a limit for the beam current in vacuum systems. The limit may be overcome, for example, in plasma filled waveguide. The instability of over-limiting e-beams (OB) is due either to aperiodical modulation of the beam density in media with negative dielectric constant or to excitation of the NEW. In this section, we consider behavior of the first type of OB instability. It develops, for example, in uniform cross-section magnetized beam-plasma waveguide. It is clear that the change of the physical nature of the instability affects on its behavior. This instability sharply differs from the instability of conventional (underlimiting) e-beams: (1) its growth rate attains maximum at the point of exact

Cherenkov resonance, (2) it is of nonradiative type, and (3) with increase in dissipation, it turns to a new type of DSI [6, 14].

3.1. Statement of the problem: analysis of the DR

Mathematical description of OB is not so well-known as for underlimiting beams, and in order to catch the differences, we consider both cases simultaneously. Consider a cylindrical waveguide, fully filled by cold plasma. A monoenergetic relativistic electron beam penetrates it. The external longitudinal magnetic field is assumed to be strong enough to freeze transversal motion of the beam and the plasma electrons. For simplicity, we assume that the beam and plasma radii coincide to the waveguide's radius and consider only the symmetrical E-modes with nonzero components E_r, E_z, and B_φ. It is known [1] that the system under consideration is described by the following DR

$$k_\perp^2 + \left(k^2 - \frac{\omega^2}{c^2}\right)\left(1 - \frac{\omega_p^2}{\omega(\omega + i\nu)} - \frac{\omega_b^2}{\gamma^3(\omega - ku)^2}\right) = 0 \tag{27}$$

ω and k are the frequency and the longitudinal (along z axis) wave vector, $k_\perp = \mu_{0s}/R$. R is the waveguide's radius, μ_{0s} are the roots of Bessel function J_0:$J_0(\mu_{0s}) = 0$, s = 1,2,3…, $\omega_{p,b}$ are the respective Langmuir frequencies for the beam and the plasma, u is the velocity of the beam, $\gamma = (1 - u^2/c^2)^{-1/2}$, c is speed of light. The DR (27) determines the growth rates of the beam-plasma instability. As we have mentioned earlier, the character of the beam-plasma interaction changes depending on the beam current value. This change must reveal itself in the solutions of the DR (27). In order to consider the solutions, we look them in the form $\omega = ku + \delta$, $\delta << ku$. The DR (27) reduces to [1, 6].

$$x^3 + i\frac{\nu}{\omega_0}\frac{\omega_p^2 \mathrm{v}_0}{u\gamma^2\omega_\perp^2}x^2 + \frac{\alpha \mathrm{v}_0 u}{\gamma^2 c^2}x = \frac{\alpha}{2\gamma^4}\frac{\mathrm{v}_0}{u} \tag{28}$$

where $x = \delta/ku$, $\alpha = \omega_b^2/k_\perp^2 u^2\gamma^3$, $\beta = u/c$, $\omega_\perp^2 = k_\perp^2 u^2\gamma^2$, and $\mathrm{v}_0 = u\mu/(1+\mu)$ is the group velocity of the resonant wave in "cold" system, $\mu = \gamma^2\omega_\perp^2/\omega_0^2$; $\omega_0 = \left(\omega_p^2 - \omega_\perp^2\right)^{1/2}$ is the resonant frequency of the plasma waveguide that is, ω_0 satisfies following conditions

$$D_0(\omega, k) = 0 \quad ; \quad \omega = ku \tag{29}$$

The solutions of Eq. (28) depend on the value of parameter α. This parameter actually serves as a parameter that determines the beam current value and the character of beam-plasma interaction. It corresponds (correct to the factor γ^{-2}) to the ratio of the beam current to the limiting current in vacuum waveguide [14] $I_0 = mu^3\gamma/4e$, that is, $\alpha = (I_b/I_0)\gamma^{-2}$ (I_b is the beam current). The values $\alpha \ll \gamma^{-2}$, correspond to underlimiting beam current $I \ll I_0$ and the instability in this case is caused by induced radiation of system proper waves by the beam electrons. Neglecting the second and third terms one can obtain the well-known growth rate of resonant beam instability in plasma waveguide

$$\delta_{\text{und}} = \frac{\sqrt{3}}{2}\frac{\omega_0}{\gamma}\left(\frac{\omega_b^2}{2\omega_0^2(1+\mu)}\right)^{1/3} \tag{30}$$

However, if dissipation exceeds growth rate, the instability turns to DSI with the growth rate

$$\delta_{\text{und}}^{(\nu)} = \frac{\omega_b\omega_0}{2\gamma^{3/2}\omega_p}\left(\frac{\omega_0}{\nu}\right)^{1/2} \tag{31}$$

If the beam current increases and became higher than the limiting vacuum current that is,

$$\gamma^{-2} \ll \alpha \ll 1, \tag{32}$$

the instability has the same nature as the instability in medium with negative dielectric constant. If the beam is underlimiting, this effect is slight and is not observed. But now, this effect is dominant. Its distinctive peculiarity is that this effect attains its maximum in the point of exact Cherenkov resonance. The growth rate differs from Eq. (30) and is equal [13].

$$\delta_{\text{ovl}} = \frac{\omega_b\beta}{\gamma^{1/2}(1+\mu)^{1/2}} \tag{33}$$

The different dependence of the growth rates of Eqs. (30) and (33) on beam density should be noted.

If, along with the beam current, dissipation also increases the instability turns to DSI of overlimiting e-beam with the growth rate [6].

$$\delta_{\text{ovl}}^{(\nu)} = \frac{\beta^2}{\gamma}\frac{\omega_b^2}{\omega_p^2}\frac{\omega_0^2}{\nu} \tag{34}$$

We emphasize new dependence on ν, that is, actually we have new type of DSI. More critical dependence on ν is due to superposition of two factors those lead to NEW excitation.

Higher values of parameter α (that is, $\alpha >> 1$) correspond to very high currents. For example, in the case of a cylindrical waveguide this condition leads to $I_b \geq 1,4\left(mc^3/e\right)\beta^3\gamma^3$ and means that the beam current is more than the limiting Pierce current. Until now such high currents beams have not been used in beam-plasma interaction experiments.

3.2. Equation for SVA and its solution: transition to the new type of DSI

In order to consider the evolution of an initial perturbation in a magnetized plasma waveguide penetrated by an OB, we proceed from the DR (27). Our steps coincide to those for the case of underlimiting e-beams: expand the DR (27) in series near ω_0 and k_0 (see (29) and derive an equation for SVA. Making use the condition of OB $2\beta^2\gamma^2\delta/k_0u \geq 1$ [13], one can obtain [6, 12].

$$\left(\frac{\partial}{\partial t} + u\frac{\partial}{\partial z}\right)\left(\frac{\partial}{\partial t} + \mathrm{v}_0\frac{\partial}{\partial z} + \nu\right)E_0(z,t) = \delta_{\text{ovl}}^2 E_0(z,t), \tag{35}$$

(the denotations coincide to those in (8)). The Eq. (35) for SVA may be solved by analogy to solution of Eq. (8). Without delving into details, we present here the results [6, 12].

$$E_0(z,t) = -\frac{J_0}{2\sqrt{\pi}} \frac{\exp \chi^{(ovl)}(z,t)}{(u-v_0)^{1/2}\delta_{ovl}^{1/2}(ut-z)^{1/2}} \tag{36}$$

$$\chi^{(ovl)}(z,t) = \chi_0^{(ovl)}(z,t) - \nu\frac{ut-z}{u-v_0} \quad ; \quad \chi_0^{(ovl)}(z,t) = \frac{2\delta_{ovl}}{u-v_0}\{(z-v_0t)(ut-z)\}^{1/2}$$

The analysis of the expression (36) is similar to previous case. It again reduces to the analysis of the exponent $\chi^{(ovl)}(z,t)$. The analysis shows that unstable perturbations vary through the same range from v_0 to u. The analysis of the instability character (absolute/convective) fully coincides to that for underlimiting e-beams. However, in this case, the waveform is symmetric with respect to its peak. The peak places in the middle at all instants and moves at average velocity

$$w_{go} = 1/2(u+v_0) \tag{37}$$

The field's value in the peak exponentially increases and the growth rate is equal to maximal growth rate for OB δ_{ovl} (33) (or, the same, to solution of the initial problem).

At fixed point z the SVA attains its maximum $\sim \exp \delta_{ovl} z/(uv_0)^{1/2}$ at the instant $t = z/w_a$, where

$$w_{ao} = \frac{2uv_0}{u+v_0} \tag{38}$$

The expression $\delta_{ovl}/(uv_0)^{1/2}$ is the maximal spatial growth rate at wave amplification by OB, and coincides to result of the boundary problem. The SVA depends on the perturbations' velocity v as

$$E_0(z=vt,t) \sim \exp\{\Gamma_0(v)t\} \quad ; \quad \Gamma_0(v) = 2\delta_{ovl}\frac{\sqrt{(u-v)(v-v_0)}}{u-v_0} \tag{39}$$

The character of the space growth depending on perturbations' velocity is $\sim \exp \Gamma_0(v) z/(uv_0)^{1/2}$.

Dissipation fundamentally changes this picture of the instability. For given velocity v the dependence of the SVA on the dissipation level becomes

$$\Gamma_0(v) \to \Gamma_\nu(v) = \Gamma_0(v) - \nu\,\frac{u-v}{u-v_0} \tag{40}$$

Dissipation suppresses slow perturbations. Only high-velocity perturbations can develop. The threshold velocity is

$$V_{th}^{(ovl)} = \frac{\lambda u + v_0}{1+\lambda} \quad ; \quad \lambda = \nu^2/4\delta_{ovl}^2 \tag{41}$$

The dynamics of the peak in the presence of dissipation may be obtained by analyzing the equation

$$(z - w_{go}t)^2 - \lambda(ut - z)(z - v_0 t) = 0 \tag{42}$$

The solution of Eq. (42) presents the peak's coordinate z_m

$$z_m = w_{go}t\left\{1 + \sqrt{\frac{\lambda}{1+\lambda}\left(1 - \frac{uv_0}{w_{go}^2}\right)}\right\} \tag{43}$$

Substitution of z_m into $\chi^{(\mathrm{ovl})}$ gives the maximal growth rate under arbitrary $\nu/\delta_{\mathrm{ovl}}$

$$E_0(z = z_m, t) \sim \exp\left(\delta_{\mathrm{ovl}}t \cdot f(\lambda)\right) \quad ; \quad f(x) = \sqrt{1 + x^2} - x \tag{44}$$

In limit of high-level dissipation, we have

$$E \sim \exp \delta_{\mathrm{ovl}}^{(\nu)} t \tag{45}$$

where $\delta_{\mathrm{ovl}}^{(\nu)}$ is given by Eq. (34). That is, with increase in level of dissipation the instability of OB transforms to the new type of DSI. The shapes of the waveform for OB instability for various level of dissipation are plotted in **Figure 2**. **Figure 3** presents the curve $f(x)$.

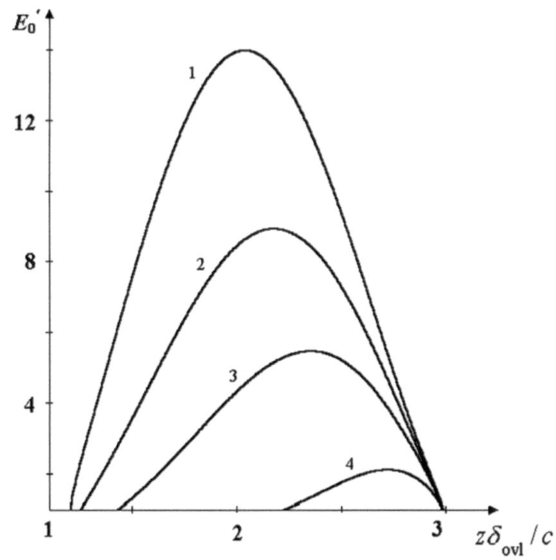

Figure 2. Shapes of the waveform versus longitudinal coordinate at fixed instant $t = 3/\delta_{ovl}$ for various values of parameter $k = \nu/\delta_{ovl}$ $k_1 = 0$, $k_2 = 1$, $k_3 = 2$, $k_4 = 4$.

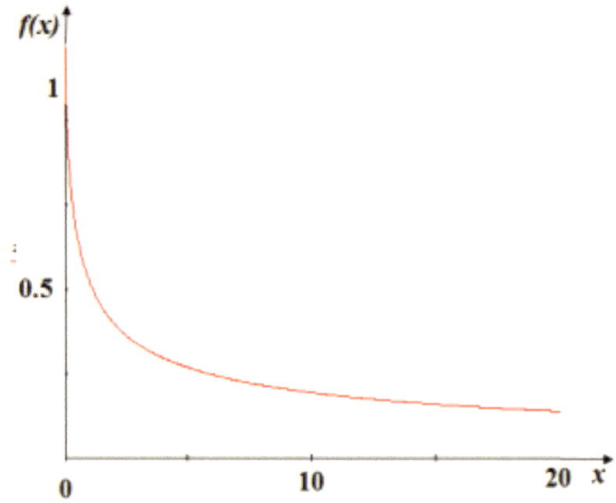

Figure 3. The function $f(x)$gives the dependence of maximal growth rate on dissipation level.

4. The behavior of the instability in spatially separated beam-plasma system

4.1. Statement of the problem: the dispersion relation

There is a factor which significantly influences on the physics of beam-plasma interaction. The factor is the level of overlap of the beam and the plasma fields. The well-known beam-plasma instability corresponds to full overlap of the beam and the plasma fields (strong beam-plasma coupling). In this case, physical nature of developing instability is due to induced radiation of the system's normal mode oscillations by the beam electrons. The oscillations are determined by plasma alone, as its density is assumed much higher than the beam density. The beam oscillations are actually suppressed and do not reveal themselves. Excited fields are actually detached from the beam in that they exist in beam absence.

The opposite case when the beam and plasma fields are overlapped slightly is the case of weak beam-plasma coupling. It may be realized, for instance, if the beam and the plasma are spatially separated in transverse direction. This transverse geometry provides conditions for increasing the role of the beam's normal mode oscillations. In this case, the beam-plasma interaction has other physical nature. Electron beam is actually left to its own. Its oscillations come into play. Account of the beam's normal mode oscillation leads to substantially new effects. Moreover, there is NEW among beam proper waves. Its growth causes instability due to the sign of energy. The growth rate of this instability attains maximum in resonance of plasma wave with NEW. Resonance of this (wave–wave) type comes instead of wave-particle resonance (conventional Cherenkov Effect) and was named "Collective Cherenkov Effect" [14, 15].

Consider weak interaction of monoenergetic electron beam and plasma in waveguide in general form [8, 14]. The only assumption is following. The beam and plasma are separated spatially, which implies weak coupling of the beam and the plasma fields. For a start, we do not particularize the cross sections. The beam current is assumed to be less than the limiting vacuum current. Dissipation in the system is taken into account by introducing collisions in plasma. We restrict ourselves by the case of strong external longitudinal magnetic field that prevents transversal motion of beam and plasma particles.

In strong external magnetic field, perturbations in plasma and beam have longitudinal components only. In such system, it is expedient to describe perturbations by using polarization potential ψ [14]. This actually is a single nonzero component of well-known Hertz vector.

We proceed from equations for ψ and for the beam and the plasma currents $j_{p,b}$.

$$\frac{\partial}{\partial t}\left(\Delta_\perp + \frac{\partial^2}{\partial z^2} - \frac{1}{c^2}\frac{\partial^2}{\partial t^2}\right)\psi = -4\pi\left(j_{bz} + j_{pz}\right) \quad ; \quad E_z = \frac{\partial^2\psi}{\partial z^2} - \frac{1}{c^2}\frac{\partial^2\psi}{\partial t^2} \tag{46}$$

$$\left(\frac{\partial}{\partial t} + u\frac{\partial}{\partial t}\right)^2 j_b = \frac{\omega_b^2\gamma^{-3}}{4\pi}\frac{\partial}{\partial t}E_z \quad ; \quad \left(\frac{\partial}{\partial t} + \nu\right)j_p = \frac{\omega_p^2}{4\pi}E_z.$$

Here $j_{bz}(\mathbf{r}_\perp, z, t) = p_b(\mathbf{r}_\perp)j_b(z,t)$ and $j_{pz}(\mathbf{r}_\perp, z, t) = p_p(\mathbf{r}_\perp)j_p(z,t)$ are perturbations of the longitudinal current densities of the beam and the plasma. Functions $p_{b,p}(\mathbf{r}_\perp)$describe transverse density profiles for beam and plasma. For homogeneous beam/plasma $p_{b,p} \equiv 1$, for infinitesimal thin beam/plasma $p_{b,p} \sim \delta(r - r_{b,p})$ (δ is Dirac function). $\Delta_\perp$ is the Laplace operator over transverse coordinates, z is longitudinal coordinate, t is the time, c is speed of light, $\omega_{p,b}$ are the Langmuir frequencies for plasma and beam respectively, ν – is the collision frequency in plasma, γ is the relativistic factors of the beam electrons, u is the beam velocity.

In general, the analytical treatment of the problem may be developed in different ways. The traditional way is to consider a multilayer structure of given geometry. With increase in number of layers this way leads to a very cumbersome DR. However, in the case of weak coupling (namely when the integral describing the overlap of the beam and the plasma fields (see below) is small), the interaction may be considered by another approach. The approach is perturbation theory over wave coupling [14]. Parameter of weak beam-plasma coupling serves as a small parameter that underlies this approach. This way leads to a DR of much simpler form, which, in addition, clearly shows the interaction of the beam and the plasma waves. Also, the procedure is not associated with a specific shape/geometry; that is, obtained results may be easily adapted to systems of any cross-section.

The set of Eq. (46) reduces to following eigenvalue problem

$$\Delta_\perp\psi - \kappa^2\left[1 - p_p(\mathbf{r}_\perp)\delta\varepsilon_p - p_b(\mathbf{r}_\perp)\delta\varepsilon_b\right]\psi = 0 \quad ; \quad \psi|_\Sigma = 0 \tag{47}$$

where ψ is the proper function of the problem, Σ means the surface of the waveguide (it is not specified yet).

$$\kappa^2 = k^2 - \frac{\omega^2}{c^2} \quad ; \quad \delta\varepsilon_p = \frac{\omega_p^2}{\omega(\omega + i\nu)} \quad ; \quad \delta\varepsilon_b = \frac{\omega_b^2}{\gamma^3(\omega - ku)}. \tag{48}$$

ω and $k$are the frequency and longitudinal wave vector, ν is the frequency of plasma collisions. As we have mentioned earlier, direct solution of the problem (47) presents considerable difficulties. However, in case of spatially separated beam and plasma that is, when $p_b(\mathbf{r}_\perp)p_p(\mathbf{r}_\perp) = 0$ and the integral describing the overlap of the fields (see below) is small, it is possible to apply perturbation theory. It assumes that in zero order approximation the beam and the plasma are independent and they may be described by two independent eigenvalue problems for plasma and beam respectively [14].

$$\Delta_\perp \psi_\alpha - \kappa^2 \left[1 - p_\alpha(\mathbf{r}_\perp)\delta\varepsilon_\alpha\right] \psi_\alpha = 0 \quad ; \quad \psi_\alpha\big|_\Sigma = 0 \quad ; \quad \alpha = p, b \tag{49}$$

Proper functions ψ_p and $\psi_b$of these zero-order problems as well as the zero-order DR for the beam and the plasma are assumed to be known. If one applies perturbation theory to the zero-order problems those are described by the DR

$$\left\{D_p(\omega, k)\right\}_{\substack{\omega = \omega_0 \\ k = k_0}} = 0 \quad ; \quad \left\{D_b(\omega, k)\right\}_{\substack{\omega = \omega_0 \\ k = k_0}} = 0 \tag{50}$$

(the point $\{\omega_0, k_0\}$ is the intersection point of the plasma and the beam curves) and search the solution of Eq. (47) in the form $\psi = A\psi_p + B\psi_b$, $A, B = const$, he can obtain in first order approximation the following DR

$$D_p(\omega, k)D_b(\omega, k) = G\left(\kappa^4 \delta\varepsilon_p \delta\varepsilon_b\right)_{\substack{\omega = \omega_0, \\ k = k_0}} \tag{51}$$

where

$$D_{p,b}(\omega, k) = k_{\perp p,b}^2 - \kappa^2 \delta\varepsilon_{p,b} = 0. \tag{52}$$

G is the coupling coefficient. It shows the efficiency of beam-plasma interaction, $k_{\perp p,b}$ are the actual transverse wavenumbers for the beam and the plasma respectively (see also [8])

$$G = \frac{\left(\iint\limits_{S_w} p_p \psi_p \psi_b d\mathbf{r}_\perp\right)\left(\iint\limits_{S_w} p_b \psi_p \psi_b d\mathbf{r}_\perp\right)}{\left(\iint\limits_{S_w} p_p \psi_p^2 d\mathbf{r}_\perp\right)\left(\iint\limits_{S_w} p_b \psi_b^2 d\mathbf{r}_\perp\right)} > 0 \tag{53}$$

$$k_{\perp p,b}^2 = \left(\iint\limits_{S_w} \left(\left(\nabla_\perp \psi_{p,b}\right)^2 + \kappa^2 \psi_{p,b}^2\right) d\mathbf{r}_\perp\right)\left(\iint\limits_{S_w} p_{p,b}(\mathbf{r}_\perp)\, \psi_{p,b}^2 d\mathbf{r}_\perp\right)^{-1}$$

Mathematically, G is expressed in terms of integrals those represent the overlap of the beam and the plasma fields. Physically, it determines as far the field of plasma wave penetrates into beam and vice versa. According to our consideration, G is small $G \ll 1$. One more condition of validity of presented consideration is homogeneity of the beam and the plasma inside the cross sections.

4.2. The growth rates

The spectra of the beam waves are given by D_b Eq. (52) and have following form

$$\omega_{\pm} = ku(1 + x_{\pm}) \quad ; \quad x_{\pm} = \frac{\sqrt{\alpha}}{\gamma}\left(\pm\sqrt{\beta^4\gamma^2\alpha + 1} - \beta^2\gamma\sqrt{\alpha}\right) \tag{54}$$

where $\alpha = \omega_b^2/k_{\perp b}^2 u^2\gamma^3$ is the parameter that determines the beam current value (see previous section) $\beta = u/c$. The beam-plasma interaction in the absence of dissipation leads to conventional beam instability that is caused by excitation of the system normal mode waves by the beam electrons. Its maximal growth rate depends on beam density as $n_b^{1/3}$. With increase in level of dissipation the conventional beam instability is gradually converted to that of dissipative type. Its maximal growth rate depends on dissipation as $\sim 1/\sqrt{\nu}$. For these instabilities the normal mode oscillations of the beam are neglected. The concept of the NEW is invoked only to explain the physical meaning of DSI. These results are valid only for the case of strong beam-plasma coupling. The decrease in beam-plasma coupling leads to exhibition of the beam's normal mode oscillation. In this case, the instability is caused by the excitation of the NEW. Specific features of weak beam-plasma interaction should appear themselves in solutions of Eq. (51). If one looks them in the form$\omega = ku(1 + x)$, then Eq. (51) becomes

$$(x + q + i\nu/ku)(x - x_+)\ (x - x_-) = G\alpha/2\gamma^4 \tag{55}$$

where $q = (1/2\gamma^2)\left(k_{\perp p}^2 u^2\gamma^2/\omega_p^2 - 1\right)$. The usual Cherenkov resonance of the beam electrons with plasma wave corresponds to the condition $q = 0$; however, the resonance between the beam slow wave and plasma wave (collective Cherenkov effect) corresponds to $q = -x_-$.The interaction of the beam and plasma waves leads to instability. Mathematically, it is due to corrections to the expression for NEW. Using the condition of collective Cherenkov resonance one can obtain

$$\left(x' + i\frac{\nu}{2\gamma^2 ku}\right)x' = -\frac{G\sqrt{\alpha}}{4\gamma^3}. \tag{56}$$

where $x' = x - x_-$. In the absence of dissipation the growth rate of instability caused by NEW growth is

$$\delta_{\text{new}}^{(\nu=0)} = \frac{ku}{2\gamma}\sqrt{\frac{G\sqrt{\alpha}}{\gamma}}. \tag{57}$$

It depends on beam density as $n_b^{1/4}$. Under conventional Cherenkov resonance the system is stable. Dissipation exhibits itself as additional factor that intensifies growth of the NEW. Eq. (56) gives following expression for the growth rate upon arbitrary level of the dissipation [8].

$$\delta(\lambda) = \delta_{\text{new}}^{(\nu=0)}\left\{\sqrt{1 + \lambda^2/4} - \lambda/2\right\} \tag{58}$$

where $\lambda = \nu/\left(2\delta_{\text{new}}^{(\nu=0)}\gamma^2\right)$. The expression (58) shows gradual transition of no dissipative instability to that of dissipative type with increase in level of dissipation. This dependence on dissipation coincides to that depicted in **Figure 3**. In the limit of strong dissipation$\lambda >> 1$, Eq. (58) becomes

$$\delta\left(\nu >> \delta_{\text{NEW}}^{(\nu=0)}\right) = \delta_{\text{NEW}}^{(\nu\to\infty)}\left[1 - \frac{G\gamma\sqrt{\alpha}(ku)^2}{\nu^2}\right], \tag{59}$$

where

$$\delta_{\text{NEW}}^{(\nu\to\infty)} = \frac{2\gamma^2\left(\delta_{\text{NEW}}^{(\nu=0)}\right)^2}{\nu} = \frac{G\sqrt{\alpha}}{2\gamma}\frac{(ku)^2}{\nu} \tag{60}$$

$\delta_{\text{NEW}}^{(\nu\to\infty)}$ presents the maximal growth rate of the new type of dissipative instability, shown up in [8]. It also follows from Eq. (56) by neglecting first term in parentheses. The new type of dissipative beam-plasma instability is now substantiated for beam and plasma layers in waveguide. The cross-sections of the layers and the waveguide are arbitrary. The instability of new type results from the superposition of dissipation on the instability that is already caused by the growth of the NEW. The instability comes instead of the conventional DSI (with growth rate $\sim 1/\sqrt{\nu}$) when beam-plasma coupling becomes small. The dependence on dissipation becomes more critical. The same instability can be substantiated in finite external magnetic field also [18].

4.3. The space–time dynamics of the instability in spatially separated beam and plasma

We have already obtained some properties of the instability in system with spatially separated beam and plasma. Consider now the behavior of this instability in detail. In so doing, we consider the evolution of an initial perturbation in system with spatially separated e-beam and plasma. We proceed from the DR (51). The successive steps are known: to derive the equation for SVA, solve it and analyze the solution. As a result, we have following equation for SVA:

$$\left(\frac{\partial}{\partial t} + \mathrm{v}_b\frac{\partial}{\partial z}\right)\left(\frac{\partial}{\partial t} + \mathrm{v}_\mathrm{p}\frac{\partial}{\partial z} + \nu^*\right)E_0(z,t) = \delta_0^2 E_0(z,t). \tag{61}$$

where $\delta_0 \equiv \delta_{\text{NEW}}^{(\nu=0)}$ (57), $\mathrm{v}_{p,b}$ are group velocities of the plasma wave and the NEW of the beam, respectively, and $\nu^* = \operatorname{Im} D_p\left(\partial D_p/\partial\omega\right)^{-1}$ is proportional to collision frequency $\nu^* = const\cdot\nu$.

The Eq. (61) is actually the same Eq. (35). This implies that the fields' space–time evolution at the instability development in spatially separated beam-plasma system qualitatively coincides to that of over-limiting e-beam instability. It remains to repeat briefly the milestones of the analysis above for behavior of OB instability in new terms (assuming $\mathrm{v}_b > \mathrm{v}_p$) and, where it is needed, to interpret results according new denotations. For this, we first rewrite the analyzing expression in new denotations

$$\chi_\nu^{(\text{ovl})} \to \chi_\nu^{(\text{ss})} = \frac{2\delta_0}{\mathrm{v}_b - \mathrm{v}_p}\sqrt{(z - \mathrm{v}_p t)(\mathrm{v}_b t - z)} - \nu^*\frac{\mathrm{v}_b t - z}{\mathrm{v}_b - \mathrm{v}_p}. \tag{62}$$

For the instability under weak beam-plasma coupling the velocities of unstable perturbation vary through the range $v_p \le v \le v_b$, The character of the instability is determined by group velocities of plasma wave and the NEW. The statements on the character of the instability (convective or absolute) remain valid with account of replacements $v_0 \to v_p$ and $u \to v_b$. The place and the velocity of the peak of the wave train can be obtained, as earlier, by solving the equation

$$\frac{\partial}{\partial z} \exp \chi_\nu^{(ss)} = 0. \tag{63}$$

In the absence of dissipation, the peak places in the middle of the train at all instants that is, it moves at the average velocity $w_{gs} = (1/2)(v_b + v_p)$. The field value in the peak exponentially increases and the growth rate is equal to $\delta_{\mathrm{NEW}}^{(\nu=0)}$ (57). In the absence of dissipation, the waveform is symmetric with respect to its peak at all instants.

Dissipation suppresses slow perturbations. The threshold velocity is (compare to previous subsection)

$$V_{th}^{(ss)} = \frac{\lambda' v_b + v_p}{1 + \lambda'} \quad ; \quad \lambda' = \left(\frac{\nu^*}{2\delta_0}\right)^2 \tag{64}$$

The wave train shortens. Only high velocity perturbations (at velocities in the range $V_{th} < v < v_b$) develop. Herewith the behavior of the fields in the peak (and the place/velocity of the peak) may be obtained by analyzing Eq. (63). If one takes into account the dissipation, the solution of (63) yields $z = w_0 t$, where

$$w_0 = \frac{1}{2}\left\{ v_b + v_p + \sqrt{\frac{\lambda'}{1+\lambda'}}(v_b - v_p) \right\} > w_{gs} \tag{65}$$

The peak shifts to the front of wave train. For high-level dissipation, we have $w_0, V_{th}^{(ss)} \to v_b$ that is, one can conclude: the group velocity of perturbation of the new DSI is equal to the group velocity of the NEW. This distinguishes the DSI under weak coupling from the DSI of OB (where the velocity of perturbations was equal to the beam velocity).

Substitution of Eq. (65) into $\chi_\nu^{(ss)}$ gives us the dependence of the growth rate on dissipation of arbitrary level. The field value in the peak depends on dissipation as

$$E_0 \sim \exp \delta_0 t \left(\sqrt{1 + \lambda'^2/4} - \lambda'/2 \right) \tag{66}$$

This result agrees to Eq. (58). This coincidence actually serves as an additional proof of the correctness of the approach based on analysis of developing wave train (i.e., correctness of the initial assumptions, derived equation for SVA, its solution etc.). Analogous coincidence exists in case of underlimiting e-beams (see Section 2), but very cumbersome expressions (solutions of third-order algebraic equation) prevent showing it obviously.

In conclusion to present section, we can state that two various types of e-beam instabilities: (1) the OB instability and (2) the instability under weak beam-plasma coupling have similar behavior. Both these instabilities transform to dissipative instabilities with the maximal growth rate $\sim 1/\nu$. In spite of their different physical nature, these instabilities have similar mathematical description. The contribution of the OB in the DR is given by expression having first order pole. The DR of the systems with spatially separated beam and plasma also may be reduced to analogous form. For comparison: the contribution of underlimiting e-beam is given by an expression with second-order pole for all types beam instabilities (Cherenkov, cyclotron etc.). This leads to their similar behavior. However, a difference between these two DSI also exists. In system with OB dissipation shifts the velocities of unstable modes to the beam velocity u. In the second case, the velocities are approximately equal to group velocity of NEW.

5. The behavior of the Buneman instability in dissipative plasma

5.1. Statement of the problem: the equation for SVA

The physical essence of the Buneman instability (BI) [4] is in the fact that the proper space charge oscillations of moving electrons due to the Doppler Effect experience red shift, and this greatly reduced frequency becomes close to the proper frequency of ions. Actually, the BI is due to resonance of the negative energy wave with the ion oscillations. For future interpretations and comparisons, we present the well-known [1, 4] DR and the maximal growth rate for the simplest case of the BI (cold e-stream, heavy ions, and accounting for collisions)

$$1-\frac{\omega_{Le}^2}{(\omega-\mathbf{ku})(\omega+i\nu_{Bn}-\mathbf{ku})}-\frac{\omega_{Li}^2}{\omega^2}=0 \quad ; \quad \delta_{Bn}^{(m)}=\frac{\sqrt{3}}{2}\omega_{Le}\left(\frac{m}{2M}\right)^{1/3} \tag{67}$$

($\mathbf{u}$ is the velocity of streaming electrons, ω_{Le} and ω_{Li} are Langmuir frequencies for electrons and. ions respectively, $\nu_{Bn}$is the frequency of collisions). The BI develops if $\omega_{Le} \geq \mathbf{ku}$, and the growth rate attains its maximum under $\omega_{Le} \approx \mathbf{ku}$.

Now consider a plasma system, the DR of which may be written as

$$D_0(\omega,\mathbf{k})+\Delta D=0 \tag{68}$$

where $\Delta D=-\omega_{Li}^2/\omega^2$ describes the contribution of ions in the DR, while $D_0(\omega,\mathbf{k})$ describes contribution of moving electrons as well as collisions/dissipation in the system. In following consideration, we do not specify the form of $D_0(\omega,\mathbf{k})$. As $\omega_{Li} << \omega_{Le}$ we have $|\Delta D| << |D_0|$ and the ions in Eq. (68) play a role under small ω that is, $\omega_{Li} >> \omega \to 0$. One can at once see imaginary roots of the Eq. (68). The system becomes unstable (low frequency instability) and the growth rate may be obtained from

$$|\omega(k)|^3=\omega_{Li}^2\left[\left(\frac{\partial D_0(\omega,k)}{\partial\omega}\right)_{\omega\to 0}\right]^{-1} \tag{69}$$

An initial perturbation arises and the instability begins to develop in point $z = 0$ (electron stream propagates in the direction $z > 0$) at instant $t = 0$. Our aim is to obtain the shape of the perturbation and investigate in detail the behavior of the BI. The procedure for obtaining the equation for SVA is known. Applying this procedure, we arrive to following Eq. [16]

$$\frac{\partial^2}{\partial t^2}\left(\frac{\partial}{\partial t} + \mathrm{v}_0\frac{\partial}{\partial z} + \nu\right)E_0(z,t) = i|\delta_{Bn}|^3 E_0(z,t) \tag{70}$$

$$|\delta_{Bn}|^3 = \left(\frac{\omega_{Li}^2}{\partial D_0/\partial\omega}\right)_{\substack{\omega\to 0\\ k=k_0}} \quad ; \quad \mathrm{v}_0 = -\left(\frac{\partial D_0/\partial k}{\partial D_0/\partial\omega}\right)_{\substack{\omega\to 0\\ k=k_0}} \quad ; \quad \nu = \left(\frac{\mathrm{Im}\, D_0}{\partial D_0/\partial\omega}\right)_{\substack{\omega\to 0\\ k=k_0}} + ik_0\mathrm{v}_0 \equiv \nu' + ik_0\mathrm{v}_0$$

$\mathrm{Im}\,\delta_{Bn}$ is the general form of the resonant growth rate of the low-frequency BI [1, 4] (compare to Eq. (67)); v_0 is the group velocity of the resonant wave in the system. Here, it is equal to velocity of streaming electrons; ν' actually presents dissipation. In unbound plasma, the main cause of dissipation is collisions of plasma particles. Equality of the ν' in this form to collision frequency is not obligatory.

Eq. (70) may be solved in known manner: that is, by using the Fourier and Laplace transformations. The problem reduces to integration in the inverse transformation. All these steps are known. So as not to repeat, we at once present resulting expression for the SVA [16]

$$E_0(z,t) = \frac{J_0}{\sqrt{2\pi}}\frac{\exp\left\{\chi_{Bn}(z,t) - \nu'\frac{z}{\mathrm{v}_0}\right\}\, e^{i\left(\frac{\chi_{Bn}+\sqrt{3}k_0 z}{\sqrt{3}} - \frac{\pi}{6}\right)}}{\left(6\mathrm{v}_0 z|\delta_{Bn}|^3\right)^{1/2}} \tag{71}$$

$$\chi_{Bn}(z,t) = \frac{3\sqrt{3}}{4}|\delta_{Bn}|\left\{\frac{2z\tau^2}{\mathrm{v}_0}\right\}^{\frac{1}{3}};\ \tau = t - z/\mathrm{v}_0,$$

5.2. Analysis of the Buneman instability behavior

As earlier, the structure of the fields is basically determined by the factor [16].

$$\exp\left\{\frac{3\sqrt{3}}{4}|\delta_{Bn}|\left\{\frac{2z\tau^2}{\mathrm{v}_0}\right\}^{1/3} - \nu'\frac{z}{\mathrm{v}_0}\right\}. \tag{72}$$

In the absence of dissipation the velocities of unstable perturbations range from 0 to the group velocity v_0. The length of the induced wave train increases as $l \approx \mathrm{v}_0 t$. The condition

$$\frac{\partial}{\partial z}\left(\chi_{Bn} - \nu'\frac{z}{\mathrm{v}_0}\right) = 0 \tag{73}$$

(compare to Eq. (16)) determines the peak's movement. In the absence of dissipation the peak disposes on 2/3 of the train's length from its front and moves at velocity $\mathrm{v}_0/3$. Substitution of $z = \mathrm{v}_0 t/3$ into Eq. (72) gives the field's behavior in the peak. It grows exponentially

$E_0 \sim \exp\left[(\sqrt{3}/2)|\delta_{Bn}|t\right]$ and the growth rate is equal to the maximal growth rate of the BI obtained earlier as a result of initial problem (e.g., see [1, 4] and Eq. (67)). However, in contrary to this approach, the initial problem does not give the point of the maximal growth. This approach gives the point. In addition, it gives the rates of the field growth in every point of the wave train (in the presence of dissipation also).

Dissipation changes the fields' dynamics and mode structure. It is easily seen from Eq. (72) that dissipation suppresses fast perturbations. The threshold velocity v_{th} can be obtained from the equation $\chi_{Bn}(z,t) = vz/v_0$ and is equal

$$v_{th} = \frac{v_0}{1+\lambda_0^{3/2}} \quad ; \quad \lambda_0 = \frac{2^{5/3}}{3^{-3/2}} \frac{\nu'}{|\delta_{Bn}|} \tag{74}$$

The wave train shortens. Actually the pulse slows down. Dissipation influences on the peak location/movement. Its place $z = z_{max}$ can be obtained from the equation

$$(v_0 t - 3z)^3 = (3\lambda_0)^3 z^2 (v_0 t - z) \tag{75}$$

The solution of this third-order algebraic equation gives location and velocity of the peak under arbitrary ratio ν'/δ_{Bn}. To avoid cumbersome expressions, we present here the solution only in the most interesting limit of high dissipation $\lambda_0 \to \infty$.

$$z = z_{max} = \left(\frac{3^{3/4}}{2^{5/2}}\right)\left(\frac{|\delta_{Bn}|}{\nu'}\right)^3 v_0 t \tag{76}$$

Substitution of this expression into $\chi_{Bn}(z,t)$ gives the field's behavior in the peak under high-level dissipation. The field's value increases exponentially

$$E_0 \sim \exp\{\delta_\nu \ t\} \tag{77}$$

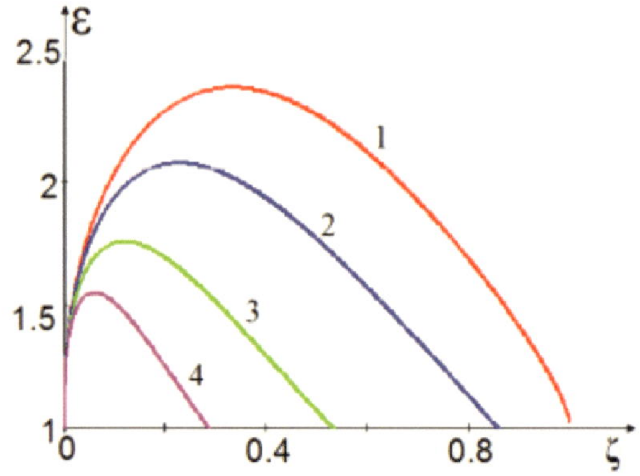

Figure 4. The shapes of initial perturbation for various level of dissipation. The dimensionless distance $\zeta = z\delta_{Bn}/v_0$, and the dimensionless field $\varepsilon = E_0/(J_0/(v_0\delta_{Bn}))$ are marked along the axes. Curve 1 corresponds to $\lambda' = \nu/|\delta_{Bn}| = 0$; curve 2 – To $\lambda' = 0.5$; curve 3 – To $\lambda' = 1.5$; curve 4 – To $\lambda' = 3$.

where the growth rate $\delta_\nu = \sqrt{|\delta_{Bn}|^3/2\nu'}$ is nothing else, as the growth rate of DSI of conventional type [1, 16, 17]. This once again justifies that high-level dissipation transforms the BI to DSI.

In addition, the expression for $\chi_{Bn}(z,t)$ gives much other information on the character of BI development. For example, by substituting $z = vt$ one can investigate the behavior of the perturbation, moving at given velocity v and determine the rate of their growth

$$E_0(z = vt, t) \sim \exp G(v)t \quad ; \quad G(v) = \frac{3\sqrt{3}}{\mathrm{v}_0}|\delta_{Bn}|\left\{2v(\mathrm{v}_0 - v)^2\right\}^{1/3} - \nu'\frac{v}{\mathrm{v}_0} \tag{78}$$

Figure 4 presents shapes of induced wave train for various levels of dissipation.

6. Conclusion

Now, we can generalize the properties of the SI. Originated perturbations form a wave train, carrier frequency and wave vector of which are determined by resonant conditions. The expression for space–time distribution of the fields gives much information on the behavior of the instability in limit of comparatively large times. The solutions of conventional initial and boundary problems follow from the expression by itself. The growth rate in the peak is equal to maximal growth rate of resonant instability δ, which usually describes given instability. The initial value problem gives the same growth rate without specifying where the growth takes place. That is, the approach gives realistic picture of the SI development. Dissipation leads to shortening of the wave train. With increase in level of dissipation the SI gradually turns to dissipative type. In the limit $\nu \gg \delta$ (ν is the collision frequency) the growth of the fields takes place according to dissipative instability. The approach gives also information on the growth rate for arbitrary δ/ν. Obvious expression may be obtained by solving algebraic equation of second/third order.

The approach justifies existence of two new, previously unknown types of DSI. For these DSI, the role of the beam's space charge and/or proper oscillation becomes decisive. For both DSI, the growth rates have more critical dependence on dissipation as compared to conventional. Presented approach obviously shows the transition to the new types of DSI.

Actually the approach presents solution of the well-known problem of time evolution of initial perturbation in systems those undergo the instabilities of streaming type. The importance of the problem is doubtless. Its traditional solution is restricted by mathematical difficulties. Presented methods allows without any difficulties obtain result for various SI in spite of their different mathematical description (e.g., the description of Buneman instability differs from the instability in spatially separated beam-plasma system and from beam-plasma instabilities; herewith, the description various types of beam-plasma instabilities (Cherenkov, cyclotron, and other) also differs from each other). The approach by itself unified the differences. For beam-plasma instabilities results of the approach are unified even more and their usage is not more difficult than usage of the result of the initial and boundary problems (in spite of presented approach gives incomparably more data). In this sense, the approach can be used instead of the problems. It could seem that the procedure is a bit more difficult. However, this difficulty only seems.

The general character of presented approach should be emphasized once more. It is based on very general assumptions and does not refer on any particular model. The approach transforms the general form of the DR to an equation for SVA of the developing wave train. For a large class of beam-plasma instabilities (Cherenkov, cyclotron, etc.), the equation for SVA is actually the same. Its solution gives analytical expression describing evolution of initial perturbation. Various SI evolve in similar manner. This emphasizes identity of their physical nature (induced radiation of the system proper waves by the beam electrons). For given instability, one should specify two parameters only: the resonant growth rate and the group velocity of the resonant wave. Obtained expression gives detailed information on the instability. The information is: the shape of developing wave train (envelope), velocities of unstable perturbations, the type of given instability (absolute or convective), location of the peak and the character of its movement, the rate of field's growth in the peak, temporal and spatial growth rates, the rate of growth for perturbation moving at given velocity. Most of these data are unavailable by other methods.

Validity limitations also should be mentioned. Obtained results may not be applied to the systems where beam instability is caused by finite longitudinal dimension, for example, Pierce instability.

Presented approach has neither inner contradictions, no contradictions to previous results of the beam-plasma interaction theory. Its results fully coincide to those obtained by direct analysis of the DR. In some cases, (e.g., for overlimiting e-beam instability and the instability in spatially separated beam-plasma system) obvious analysis is possible due to comparatively simple contribution of the beam in the DR (namely when the contribution has first (but not second) order pole).

The results of presented approach actually are continuation and further development of the results of the initial and boundary problems. In its turn, the results of the problems have been repeatedly tested and rechecked experimentally. This actually can serve as confirmation of validity of the approach.

In [19, 20] the nonlinear dynamics of the beam-plasma instability was investigated numerically at no stationary beam injection into plasma-filled systems. The results show that at the initial stage of instability development the field has a shape matching reasonably to presented results.

Obtained results on SI evolution help to understand how the instability transforms given equilibrium of background plasma, estimate the level and/or scale of originated irregularities clear up how the nonlinear stage arises and predict saturation mechanisms. The systems, to which this may be applied are numerous, as the SI are the most common instabilities: from the Earth ionosphere to current carrying plasma (where the Buneman instability plays important role). Not to mention relativistic microwave electronics etc.

Author details

Eduard V. Rostomyan

Address all correspondence to: eduard_rostomyan@mail.ru

Institute of Radiophysics and Electronics Armenian National Academy of Sciences, Astarack, Armenia

References

[1] Alexandrov AF, Bogdankevich LS, Rukhadze AA. Principles of Plasma Electrodynamics. In: Rukhadze AA, editor. Berlin: Springer-Verlag; 1984. 490pp

[2] Kuzelev MV, Rukhadze AA. Plasma Physics Report. 2000;**26**:231

[3] Kuzelev MV, Loza OT, Rukhadze AA, Strelkov PS, Shkvarunets AG. Plasma Physics Report. 2001;**27**:669

[4] Buneman O. Physical Review Letters. 1958;**1**:8

[5] Briggs RJ. Electron-Stream interaction with Plasma. Cambridge, Mass: M.I.T Press; 1964. 187pp

[6] Rostomyan EV. IEEE Transactions on Plasma Sciences. 2003;**31**(6):1278

[7] Rostomyan EV. EPL. 2007;**77**:45001

[8] Rostomyan EV. Journal of Plasma Physics. 2012;**78**(5):531

[9] Bers A. Space-Time Evolution of Absolute and Convective Plasma instabilities. In Basics of Plasma Physics. Edited by Rosenbluth MN, Sagdeev RZ. North Holland, Amsterdam. 1983;**1**:451

[10] Kuzelev MV. Plasma Physics Report. 2006;**32**:572

[11] Rostomyan EV. Physics of Plasmas. 2000;**7**:1595

[12] Rostomyan EV. European Journal of Applied Physics. 2001;**14**(3):177

[13] Aizatski NI. Soviet Journal of Plasma Physics. 1980;**6**:597

[14] Kuzelev MV, Rukhadze AA. Plasma Free Electron Lasers. Paris: Edition Frontier; 1995

[15] Kuzelev MV, Rukhadze AA. Behavior of Streaming Instabilities in Dissipative Plasma. Soviet Physics Uspekhi. 1987;**30**:507

[16] Rostomyan EV. Physics of Plasmas. 2017;**24**:102102

[17] Rostomyan EV. Physics of Plasmas. 2016;**23**:102115

[18] Rostomyan EV. Physics Letters A. 2009;**373**:2581

[19] Rostomian EV, Filipichev DS. Soviet Journal of Plasma Physics. 1993;**19**:66

[20] Vedenin PS, Roukhlin VG, Tarakanov VP. Soviet Journal of Plasma Physics. 1989;**15**:1246

Chapter 3

Fundamentals of Plasma-Material Interactions in Magnetic Fusion Devices

Jean Paul Allain and David N. Ruzic

Additional information is available at the end of the chapter

http://dx.doi.org/10.5772/intechopen.77157

Abstract

The interaction of plasmas and materials has a long history in the modification of condensed matter. Plasma-material interaction (PMI) can govern how low-temperature and high-temperature plasmas interact and modify materials surfaces. In magnetic fusion devices, PMI can also influence the operation of the fusion device. For example, incident energetic charged particle on fusion wall material surfaces can release target atoms via sputtering and can implant fuel particles in the lattice. Implanted energetic particles can mix fuel and influence recycling of fuel back to the plasma. Sputtered target atoms can become ionized in the magnetic sheath and re-deposit at the wall surface. The magnetic sheath will influence the energy and angular distribution of incident energetic particles and influence the implantation and release of fusion fuel.

Keywords: plasma-material interactions, sputtering, reflection, retention, magnetic sheath

1. Introduction

The interaction of plasmas and materials is one of the most interesting and critical subjects in the field of plasma technology. In fact, plasma processing has been the hallmark of plasma technology and its impact on the semiconductor industry. Plasmas are ubiquitous in nature and are responsible for some of the most fundamental interactions known to man. For example, in astrobiology, the early formation of the stars from interstellar media (ISM) dust and the presence of hydrogen molecules in interstellar space are conjectured to be governed by carbonaceous dust grain surface interactions with plasma energetic particles (hydrogen, helium) forming more complex organic molecules [1, 2]. Ion-induced etching in modern high-density plasma-processing tools is driven by the complex energetic multi-particle interaction with material surfaces, and fluorescent lamps operate by a balance of the plasma-material interaction (PMI) under high gas pressure containers.

Fusion reactors also witness significant modification of wall materials when exposed to high-density high-temperature magnetized plasmas. The nuclear fusion reaction in the core plasma generates a plethora of energetic particles including He and neutrons for hydrogen-based fuels. Both highly energetic He and neutrons can penetrate deep inside the material structure. However, the charged particles that remain inside the reactor are controlled by the boundary plasma with the reactor wall, and through what is known as a magnetic sheath, the charged particles are driven to the surface by several mechanisms that dictate both their incidence angle and energy to the material surface. Fusion devices are only experimental and operate under a pulsating configuration that enables only pulsed fusion plasmas that range from a few seconds up to about 50–60 s for the most modern, super-conducting fusion devices. During the pulse, the plasma can drive many complex interactions at the plasma edge and at the wall surface can implant energetic particles that range from a few eV up to several kilo-electron volts (keV). These energetic particles are mostly fuel particles such as hydrogen or deuterium but also in some cases helium. Other particles are mostly impurities from the ambient (e.g., water) such as oxygen and material atoms sputtered and re-deposited in the fusion device. In this chapter, we examine in detail the interactions of these energetic particles and material surfaces to understand the basic mechanisms that drive plasma-material interactions.

Beyond energetic particle interaction on the wall material surface, there are additional complex interactions that take place and evolve during plasma-material interaction. The surface kinetics of impurity-driven modification can change the surface chemistry and morphology significantly. Over time, fusion devices also can drive transient events that consist of instabilities from the plasma, releasing large amounts of energy in the form of heat to the wall material surface. In a fusion reactor, deuterium-tritium plasma is confined by strong magnetic fields at a temperature of hundreds of millions of degrees Celsius. Therefore, in the exhaust of such a reactor, the so-called divertor, the plasma-facing surfaces are subjected to extremely high and intermittent heat loads (10 MW/m^2 time-averaged, with periodic excursions in the GW/m^2 level on sub-millisecond timescales), while simultaneously being bombarded by extreme fluxes of energetic particles (hydrogen isotopes, helium, neutrons). The radiation interaction with matter will be dynamic, imposing time-dependent changes on the structure, composition, and chemistry of both bulk and surface region of material components. Performance and lifetime limits of nuclear fusion materials will ultimately need to survive >100-dpa and >1000-appm He production over the high-duty cycle operation of the reactor. Currently, no material can meet such requirements namely limited by critical material properties including creep resistance, fracture toughness, surface erosion/re-deposition, corrosion, chemistry, thermal conductivity, and many others.

Although progress has been made in the last decade in establishing an understanding of plasma-material interactions, there remain critical knowledge gaps as it relates to predicting and designing for the behavior at the plasma-material interface under the so-called "reactor-relevant" plasma conditions anticipated in a future plasma-burning neutron-dominated environment. Ultimately, a magnetically confined fusion plasma must be able to not only operate under "burn" conditions (e.g., more power out than power in) but when used to generate electricity operate at high-duty cycles (e.g., weeks or months of continuous power).

The plasma-material interface is one important factor to the realization of nuclear fusion power. At this interface, high particle and heat flux from the fusion plasma can limit the material's lifetime and reliability and therefore hinder operation of the fusion device. This region is critical to the operation of a nuclear fusion reactor since material can be emitted both atomistically (e.g., evaporation, sputtering, etc.) and/or macroscopically (i.e., during transient events, such as disruptions or edge-localized modes). The environmental conditions of a future nuclear fusion reactor interacting with the plasma-material interface are extreme. The incident plasma will carry heat fluxes of the order of hundreds of MWm^{-2} and particle fluxes that can average 10^{24} m^{-2} s^{-1}. The fusion reactor wall would need to operate at high temperatures near 800°C, and the incident energy of particles will vary from a few eV ions to MeV neutrons. To exacerbate this, another challenge is the management of damage over the course of time. Operating at reactor-relevant conditions means the wall material would need to perform over the course of not just seconds or minutes (i.e., as in most advanced fusion devices today and the near future) but months to years. Therefore, plasma-material interface is a dynamic, evolving, reconstituted region of material that is constantly eroded and re-deposited a million times over rendering our current understanding of material damage quite limited.

Another important factor is the limited attention given not only to the structural properties of refractory metals such as tungsten but especially surface-dominating properties (e.g., erosion, ion mixing, hydrogen- and helium-induced bubbles and swelling at the surface, surface diffusion, surface chemistry, morphology, and nanoscale patterning) that ultimately dictate particle recycling emitted back to the edge plasma consequently cooling the fusion plasma. In addition, understanding the relevant plasma-facing component issues, which vary with respect to fusion device design, is also lacking in fusion material's R&D efforts. For example, the differences of plasma edge conditions for first wall versus divertor fusion materials are quite different. Incident He fluxes to the first wall of future plasma-burning devices may vary between 10^{18} and 10^{19} m^{-2} s^{-1} and energies 100 and 1000 eV and at the divertor, fluxes between 10^{22} and 10^{23} m^{-2} s^{-1} and energies 5 and 100 eV. These flux and energy regimes induce distinct damage mechanisms that must be understood in the development of advanced fusion materials. Of the various material options at the plasma-material interface (i.e., graphite, liquid metals, etc.), refractory metals (molybdenum, tungsten, etc.) are attractive for use during steady-state, high-temperature (700–1000°C) operation with heat flux ranging between 10 and 20 MW/m^2. However, one major challenge for the application of solid refractory metals in future burning plasmas is the large production of helium ash in a fusion reactor. He implantation and generation of cavities, bubbles, surface morphology (e.g., fuzz), and blisters is of major concern for the application of commercial-grade tungsten materials. The surface nano- and mesoscale morphology could be detrimental to the performance of nuclear fusion reactor operation given the possible micro- and macroscale emission of W particulate (dust) into the plasma.

Given the broad area of plasma-material interactions and its importance to many different disciplines, we limit our treatment in this chapter to the interactions and plasmas in magnetic fusion devices. We also limit our coverage of PMI to one specific set of mechanisms mainly focused on physical sputtering and incident-particle reflection.

2. PMI basics

The basic interaction of a plasma bounded by a material surface is the balance of charged particles that arrive at a given time. In a magnetized plasma, the incident ions gyro-orbit around magnetic field lines that intersect material surfaces at ultra-shallow angles between 1 and 3° with respect to the surface, resulting in the so-called "Chodura sheath" where incident ions arrive with a distribution of incident angles and energies ranging from 20 to 60° with respect to surface normal and energies between 10 and 100 eV, respectively. The incident ions (mostly hydrogen fuel particles) implant at depths between a few nm to hundreds of nm. Sputtering of the wall material will depend on these conditions for both light and heavy mass target materials. The incident hydrogen particles will also reflect or backscatter from the surface and carry a finite amount of energy also resulting in a balance of implanted versus recycled fuel particles in a fusion device. In this section, we briefly discuss some of the most salient sputtering and reflection mechanisms with realistic (e.g., rough) surfaces found in a fusion device.

2.1. Ion-surface interactions

Ion-surface interactions are one of the most important effects in fusion research devices. Open field lines terminate at divertor plates or strike walls at very grazing angles. The ion trajectories, which spiral around these field lines, direct energetic ions onto the wall material. Therefore, ion-solid and more recently ion-liquid interactions are the critical reaction at the boundary and therefore the most important to understand. The incident ion could reflect back into the plasma or could become embedded in the surface. Perhaps more importantly, the ion could knock some of the wall material into the plasma, thus leading to sputtering. Since sputtered species are usually electrically neutral, they ignore magnetic field lines and can penetrate a significant distance into the plasma before becoming ionized. Therefore, the energy and angular distribution of sputtered material becomes crucial to predicting edge plasma behavior, and the behavior of the edge plasma is often a controlling factor on the behavior of the core plasma. The interaction of energetic ions with wall materials can also result in not only erosion and re-deposition of post-ionized material wall particles but also could drive composition and morphology changes that over time significantly affect materials' surface properties. Both composition and morphology changes on the surface can result in significant changes both to the plasma-material interactions and consequently to the plasma edge, which can have effects on the core plasma performance.

2.2. Simulating ion-surface interactions

The number of variables that could go into a single ion-surface interaction is numerous. Consider the incident ion. What is its mass, its atomic number, its energy? What angle does it strike the surface with respect to the surface normal? Now, consider the target material. What is its composition, and how does that composition vary with depth? What does the surface roughness look like and at what scale lengths? What is the chemical binding energy of the variety of constituents that may be present, and with what energy is each constituent bound to each other? The information we wish to know has many facets as well. What species is liberated? What is its energy? What angle does it leave the surface with respect to the surface normal and with respect to the incoming trajectory? Does it come off as a neutral, an ion, a dimer, or a molecule? With so many variables, exhaustive experimental determination of these quantities is impossible. What is possible is a computer model based on the physics of the interactions and then tested against experimental data. If a model can be shown to agree with experiments over a wide variety of ion-target pairs, there is some confidence that it will accurately predict PMI variables even for situations that may be impossible to directly measure. Such a computer scheme exists—Monte-Carlo simulations based on a binary collision approximation. Monte-Carlo simulations are ideal for ion-surface interactions. The physics of any one interaction is straightforward. Stringing many together while randomizing the impact parameter according to the physical parameters of the situation can be done with relative ease. Both the incident particle and all particles, which receive more than some pre-set amount of kinetic energy, are then followed after the collision. After every particle in this cascade is tracked until they come to rest or leave the surface, the final location and velocity of each atom is recorded. The transport of ions in matter (TRIM) simulation code has been one of the most successful PMI codes to simulate the interaction of energetic particles with surfaces and in the context of PMI-simulating effects such as ion implantation and sputtering [3].

2.3. Effects of roughness on PMI

Expanding from the successful TRIM simulation platform, many variances have emerged over many decades and one of them is the incorporation of fractal geometry to mimic realistic surfaces [4].

Figure 1a shows the reflection of 50 eV H from an Ni surface as a function of fractal dimension [5]. R_n is the fraction of particles that reflect and R_e is the fraction of energy that is reflected. Note the precipitous drop in both R_n and R_e when some roughness is added, especially when the incident particle strikes the surface at a grazing incidence (75° from the normal). Initially, roughness reduces reflection as expected. The gradual rise in reflection for very rough surfaces is attributed to there being less of a chance for an upward-moving atom to be recaptured due to the lower average density of the material near the surface. Planar TRIM is akin to the D = 2.00 case for normal incidence. Some interesting comparisons [4] are shown in **Figure 1b**. Here, D is fixed at 2.30 and planar TRIM is compared to fractal TRIM (FTRIM) for three different incident energies as a function of incident angle. Specular reflection tendencies are clearly seen with TRIM but not in FTRIM. At 50 eV, the calculation was repeated using both generator A and generator B (see **Figure 1**) to show that the results did not depend on the generator, just on the dimension. Finally, a comparison is made with a molecular dynamic simulation at 10 eV. The similarity of those results shows that FTRIM can be used with some confidence even at low energies. Comparisons of predicted reflection to experiment are not possible because reflection measurements have not been done in this energy range. The reflected particles come off neutral and are very difficult to detect. A better comparison to experiment can be made when sputtering is considered.

Figure 2a shows the FTRIM prediction [6] for physical sputtering of 300 eV H on C and normal incidence and at 60° incidence. Two experimental points [7] are also shown where the fractal

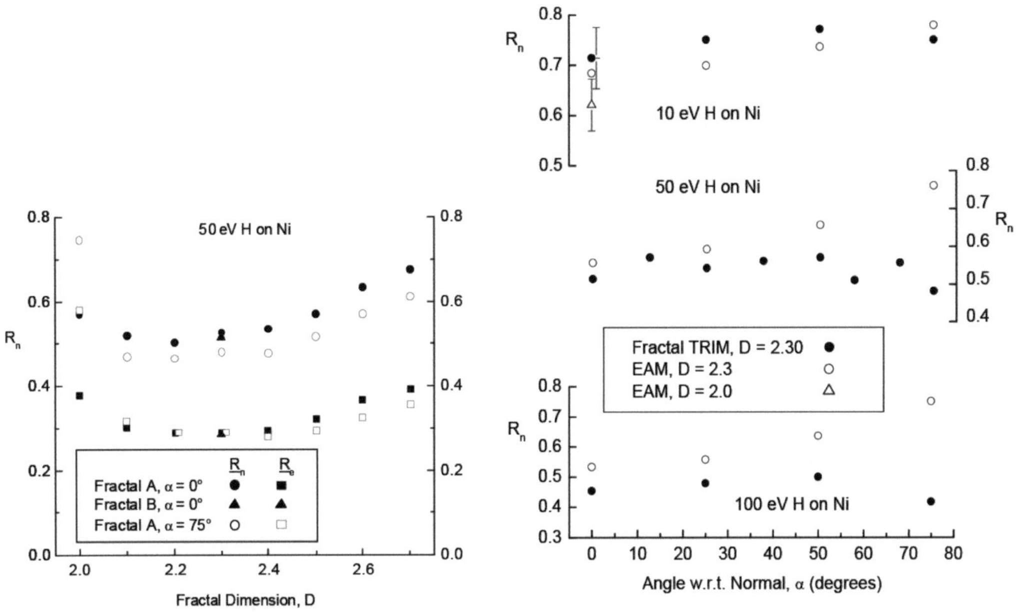

(a) (b)

Figure 1. (a) R_n and R_e, particle and energy reflection coefficients, versus fractal dimension D for normal incidence (α = 0°) and grazing incidence (α = 75°) for three different surfaces. H is incident on Ni at 50 eV. The error bars are the size of the data points. Note the large drop-off of R at the grazing incidence when D is greater than 2.00. (b) Reflection coefficients for 10, 50, and 100 eV H on Ni as a function of the incident angle (with respect to normal, w.R.T.) for a fractal Ni surface with dimension 2.30. Planar TRIM results and a molecular dynamics calculation using the embedded atom method (EAM) are also shown. Note that planar TRIM predicts reflection at the grazing incidence to be two to three times more likely than the fractal TRIM results. Statistical errors in the fractal and planar TRIM reflection coefficients are generally less than 5%.

dimension of the surface is known [8] and the agreement is very good. **Figure 2b** shows the sputtering yield of 100 eV C on C as a function of the fractal dimension for a variety of incident angles. Initially, adding roughness increases the yield. This is due to the ability of the incident ions to knock off target atoms, which may protrude from the surface. Therefore, as expected, higher angles of incidence show an even greater rise in sputtering. The sputtering goes down for a very high roughness because the sputtered atoms are recaptured by overhanging features. Planar TRIM results are also shown. Note that the effective surface roughness for planar TRIM is only 2.00 at the normal incidence. The algorithm described earlier picks the location of the initial collision partner. If that partner would be above the surface, it is not used. The set of initial collision partner locations has a nonuniform depth distribution if the incident angle is not perpendicular to the surface. An equivalent roughness to the fractal surfaces can then be assigned. FTRIM predicts less sputtering than TRIM at higher angles of incidence and more sputtering at normal incidence. This result is significant, in that the realistic surfaces that evolve due to plasma-induced erosion and re-deposition in a fusion device with enhanced roughness will likely impact the amount of net erosion and reflected energy of fuel particles that can influence the operational regimes in these devices.

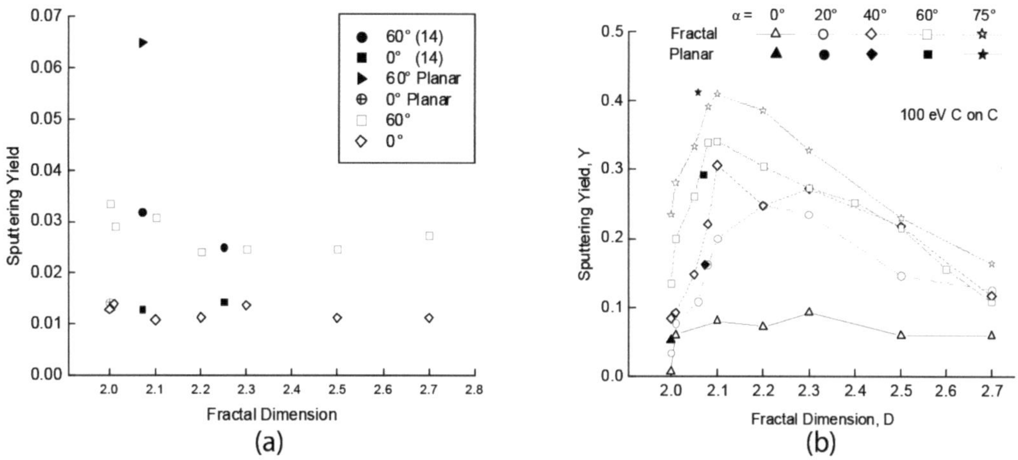

Figure 2. (a) Sputtering yield of 300 eV H on C as a function of fractal dimension, D. The D of the experimental points by Haasz et al. [7] is based on the work by Avnir et al. [8]. (b) Sputtering yield of target material as a function of the fractal dimension and the angle of incidence for 100 eV C incident on a C target. The normal incidence is 0°. Statistical errors in the yield are generally less than 5%.

3. Sputtering from plasma-material interactions

3.1. Solid-phase sputtering of fusion-relevant materials

One of the fundamental interactions between plasmas and material surfaces is physical sputtering or ion-induced desorption. The plasma edge magnetic field lines in a magnetic fusion device such as a tokamak directs energetic charged particles to gyro-orbit and bombards surfaces at both incident angle distributions and energy distributions that are linked with the operational regime of the fusion device. Therefore, the average incident angle will be close to 45° with respect to the normal and low-incident-particle energies. Fusion device PMI can be divided between two overall types of solid-state materials: low-Z materials and high-Z materials. There is a trade-off in the selection of materials for the first wall in fusion devices. The cooling of the plasma due to eroded particles from the device wall material goes as ~Z^2. Therefore, low-Z materials are attractive; however, these materials tend to have low surface binding energies, which result in high-sputter yields. The sputtering yield is defined as the ratio of flux of sputtered particles over the flux of incident energetic particles. High-Z materials can sputter orders of magnitude lower than low-Z materials; however, this must be balanced against plasma cooling losses. As stated earlier in fusion devices, the material surface will evolve where roughness can become significant [9–11].

3.1.1. Low-Z material sputtering by light incident particles

Low-Z material sputtering is relevant to plasma-surface interaction physics in fusion devices from the standpoint of minimizing fractional impurity levels in fusion plasmas [12]. **Figure 3a** shows experimental measurements and VFTRIM-3D simulations of sputtering yields for Li, He, and D bombardment at 45° incidence on deuterium-treated solid-phase lithium [13].

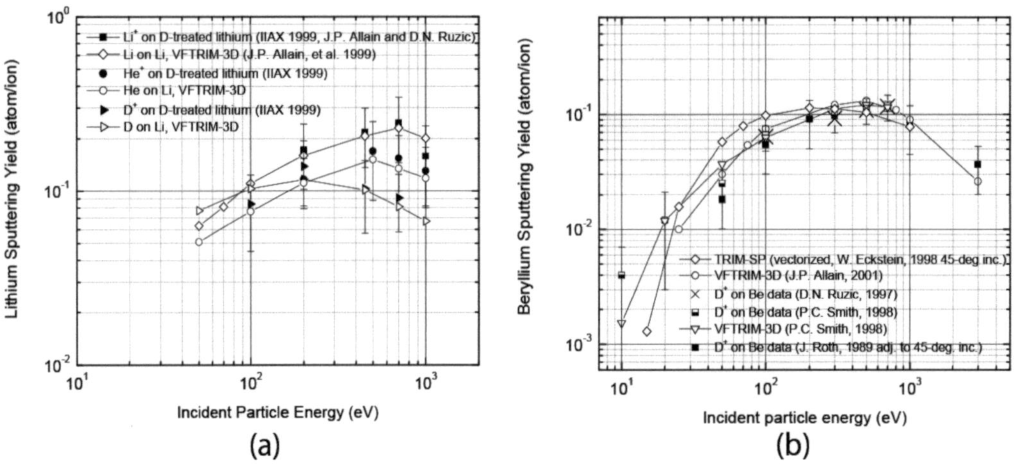

Figure 3. (a) Experimental and VFTRIM-3D simulation data for li, D, and he bombardment of solid-phase D-treated lithium at 45° incidence. (b) Experimental and simulation results for D^+ bombardment of D-treated beryllium at 45° incidence. Normal incidence data are adjusted to 45° using Yamamura's formula for oblique incidence.

Experimental data were taken at the Ion-surface Interaction Experiment (IIAX) facility, which is an ion beam experimental device able to measure, among other things, physical sputtering yields for low-energy, light-particle interaction. The data is done at 45° incidence, based roughly on the average angle of incidence a sheath-accelerated, gyrating particle makes where the magnetic field lines cross the divertor plates at oblique incident angles [14].

Computational runs were modeled using a surface, which consisted of 50 a/o Li and 50 a/o D, consistent with deuterium concentration measurements [15]. The model used a surface binding energy of 1.68 eV based on the heat of sublimation for solid lithium. The value of 1.68 eV for the surface binding energy of Li has been measured in plasma-surface interaction experiments in PISCES-B [16]. The bond energy (*BE*)—the energy to break a bond in the bulk—was taken as *BE = 0.1 SBE*. Deuteration of the solid lithium surface is done with a deuterium plasma from a hollow cathode source with a flux of 10^{16} ions/cm^2/s for 20 min. This flux is sufficient to saturate the surface and have enough atomic percentage of deuterium to assume a 50/50 composition at the surface over a range at least the depth of origin of sputtered species.

The sputtering yield behavior shown in **Figure 3a** is as expected. Due to the ineffective transfer of energy between hydrogen isotopes and lithium compared to helium or lithium itself, lithium sputtering due to hydrogen isotopes is relatively low. For the same incident energy, hydrogen atoms will penetrate farther into the lithium bcc lattice. Therefore, hydrogen isotope bombardment of lithium will reach a maximum sputtering yield at a lower incident energy than for helium or lithium bombardment. At greater energies, the penetration depth is deep enough that the net backward momentum distributed to surface atoms is not sufficient to overcome the surface binding energy, and thus the lithium-sputtering yield begins to decrease. Self-sputtering of lithium will be discussed in Section 4.2, following which we discuss lithium sputtering from deuterium and helium bombardment.

VFTRIM-3D simulation results for deuterium bombardment are shown in **Figure 3a** with open triangles (pointing toward right), while closed triangles represent IIAX data. The solid line serves to guide the eye. The experimental and simulated yields versus incident-particle energy diverge with a decreasing energy primarily in the low 10–100 eV range, although the error bars are relatively large. At these lower energies, the range of incoming deuterium ions extends only to a few monolayers. Over the period of the dose, the surface may be enriched with more deuterium, leading to a lower amount of lithium sputtered than predicted. In addition, at these lower energies, the influence of surface roughness on the sputtering yield is enhanced. This occurs due to D atoms segregating to protruding regions of the surface where the net attractive force to the bulk/surface goes as r^{-3} (where r is the distance from the surface) [16] and thus the effective binding energy to the surface for these atoms drops. Studies have shown that hydrogen atoms will tend to segregate to interstitial sites in a metal lattice [17–19]. In addition, the diffusion of hydrogen atoms has been measured in lithium experiments investigated by Sugai [15]. Such diffusion is not modeled by TRIM-SP, only that a continuous distribution of D atoms exists in the lithium bcc lattice. Thus, the ability for diffusion and segregation of deuterium atoms around the protruding regions of the lithium surface adds to the probability that less amount of lithium is sputtered since a larger amount of deuterium is preferentially sputtered. The yield reaches a maximum around 200–300 eV. At an incident-particle energy of 200 eV, where the yield is a maximum, the mean sputtered energy of lithium atoms is 9.0 eV as predicted by TRIM-SP.

Figure 3a also shows the experimental and computational results for He^+ bombardment of D-treated lithium at 45° incidence. The line with open circles represents the TRIM-SP simulation data. The solid circles represent IIAX data. The prediction made by the computational model falls within the experimental error. The functional behavior shows a maximum of the sputtering yield of lithium at 500 eV. The decrease of lithium sputtering due to deuteration of lithium is stronger for helium bombardment than for deuterium. This is due to the effective transfer of energy from the incident hydrogen isotopes to the implanted deuterium atoms, leading to a relatively larger net momentum imparted to lithium surface atoms.

Beryllium sputtering has been studied quite extensively ranging from ion-beam experiments to experiments from magnetized linear plasma devices, such as PISCES-B [20–29]. **Figure 3b** shows both experimental data and simulation data for deuterium bombardment of beryllium. Experimental data are shown in x's and half-filled squares for IIAX data by Ruzic et al. [24], filled squares for Roth et al. [20], which are adjusted to 45° incidence by an empirical formula given by Yamamura et al. [30]. The empirical relation is shown as Eq. (1) and the fitting parameters used are obtained from the quoted reference for the factors f and α_{opt} where α_{opt} is the nominal incidence angle at maximum yield [31]. Simulated data for TRIM-SP are shown in open circles and triangles [23]. TRIM-SP simulation is shown with open diamonds for 45° incidence [32]

$$\frac{Y(E_o,\alpha)}{Y(E_o,\alpha=0)} = \frac{\exp\left(f\left[1-\frac{1}{\cos\alpha}\right]\cos\alpha_{opt}\right)}{\cos^f\alpha} \tag{1}$$

TRIM-SP simulations were done for a fixed surface binding energy of 3.38 eV, which is the heat of sublimation for beryllium. Since in the IIAX experiment, beryllium was saturated with deuterium at room temperature, a surface composed of a D/Be ratio of 0.33 was used based on saturation experiments [33]. VFTRIM-3D simulations use a vectorized version of TRIM-SP known as TRVMC, which uses a binding energy of 1 eV for hydrogen isotopes [32] and beryllium's heat of sublimation. This binding energy was also utilized by TRIM-SP for consistency.

The data shown in **Figure 3b** show a maximum between 300 and 500 eV, closely resembling BeO data taken by Roth et al. [20]. Beryllium has a high affinity for oxygen at room temperature, thus the surface binding energy is effectively increased, reducing the sputtering yield. In addition, deuterium-treated surfaces effectively decrease beryllium sputtering due to preferential sputtering of embedded deuterium atoms. As a consequence, the beryllium sputtering yield from deuterium-treated surfaces measured in IIAX is predicted well by VFTRIM-3D simulations. TRIM-SP simulations do not account for deuterium treatment, and thus their yields are higher than anticipated, coincidently matching VFTRIM-3D results. If deuterium saturation was used by TRIM-SP modeling, beryllium sputtering would be effectively decreased, thus not predicting the experimental data in IIAX. The ability for VFTRIM-3D to effectively model surface roughness also leads to the high predictability of experimental data both in IIAX and from Roth et al. [20].

3.1.2. High-Z material sputtering by light incident particles

High-Z material sputtering will be discussed for the cases of tin and tungsten sputtering. High-Z material sputtering for refractory materials such as tungsten is attractive due to its relatively low-sputtering yield and high-sputtering threshold. However, due to the plasma low tolerance for high-Z impurities due to radiation losses, impurity levels must remain low, $<10^{-4}$ (ratio of densities) in fusion plasmas [11]. Other high-Z materials such as tin are attractive from the standpoint of low-sputtering yield, relatively high-sputtering threshold, high thermal conductivity, and the potential for tin to be used as a liquid plasma-facing material due to its low melting point and low vapor pressure. For experimental data at normal incidence, the empirical formula (Eq. (1)) by Yamamura et al. was used as in the case for beryllium.

Figure 4a shows the results for TRIM-SP simulation of tin sputtering. The VFTRIM-3D simulations are done for a surface binding energy equal to the heat of sublimation of tin, 3.12 eV. The data presented are as expected with helium bombardment, leading to a larger tin sputtering than deuterium bombardment due to an effective energy transfer. A maximum for deuterium bombardment is reached at a slightly lower incident energy than for helium bombardment. The argument for when this maximum yield is reached is the same as for tungsten, noting that in addition for heavy materials, the penetration depths of deuterium and helium at low energies will be quite similar and thus their maxima remain close. Tin shows promise, in that its sputtering yield at energies ranging from 100 to 400 eV is about a factor of five less than beryllium sputtering. However, one would have to contend with radiation losses from tin's high Z equal to 50.

Figure 4b shows the experimental and VFTRIM-3D simulation results for tungsten self-sputtering as well as tungsten sputtering from deuterium and helium bombardment. The experimental data are for normal incidence taken by Eckstein et al. at low energy and has been

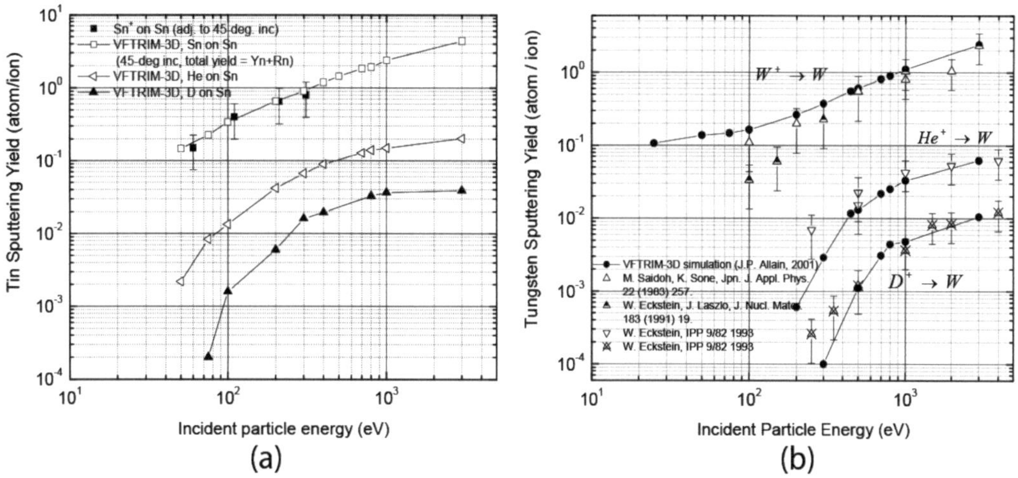

Figure 4. (a) Experimental and VFTRIM-3D simulation of tin sputtering by D, he, and Sn incident particles at 45° incidence. Normal incidence data are adjusted to 45° using Yamamura's formula (Eq. (1)) for oblique particle incidence. (b) Experimental and VFTRIM-3D simulation of tungsten sputtering by D, he, and self-ions at 45° incidence. Normal incidence data are adjusted to 45° using Yamamura's formula (Eq. (1)) for oblique particle incidence.

adjusted to 45° incidence for comparison [11]. TRIM-SP simulations were done for tungsten with a surface binding energy equal to its heat of sublimation of 8.68 eV. A mass density of 19.3 g/cm^3 was used in the simulation as well.

The VFTRIM-3D simulation predicts the experimental data reasonably well within the error bars. Helium bombardment shows a lower-sputtering threshold and larger yields compared to deuterium bombardment as expected from a lower mass ratio and a better energy transfer. Although the data are not shown for larger energies than 3 keV, the maximum tungsten-sputtering yield is expected at a lower incident energy for deuterium bombardment than for helium bombardment. This is due to the longer range of deuterium atoms in tungsten compared to that of incident helium, depositing less energy near the surface and thus turning the sputtering yield curve at a lower incident energy than helium.

3.1.3. *Effect of deuterium saturation on lithium and beryllium sputtering*

The sputtering yield of lithium and beryllium decreases with deuterium saturation of the surface. This is due to preferential sputtering of deuterium atoms over lithium or beryllium atoms when bombarded by incident energetic particles. In the case of deuterium treatment for beryllium target, an extensive review has been presented in previous work and is only referenced here [10, 24, 27]. The net effect of embedded deuterium atoms is the effective reduction of the beryllium and lithium-sputtering yield as demonstrated by VFTRIM-3D simulations, shown in **Figure 5a** for deuterium bombardment. The simulations maintained the surface binding energy fixed at 3.38 eV. The level of deuterium saturation is that described earlier with a D/Be ratio of 0.33. For lithium sputtering, deuterium saturation is modeled with a D/Li ratio of 0.5 as discussed earlier. The lithium surface binding energy is kept fixed at 1.68 eV.

The effect of deuterium saturation on beryllium sputtering is a bit stronger for helium bombardment (not shown here) than for deuterium bombardment. This is due to the effective energy transfer from the incoming deuterium to the embedded deuterium atoms in beryllium and lithium. This effect, however, is lessened with a lower amount of deuterium in the beryllium or lithium lattice. Therefore, the energy dependence for a given deuterium saturation level in a material cannot be simply determined from the energy dependence at 0% saturation by a constant multiplication. Similar results are found for lithium sputtering except for larger yields and a lower-sputtering threshold mostly due to a better energy transfer from deuterium and helium atoms to lithium target atoms.

Figure 5b further shows the importance of deuterium treatment on the *measured* absolute sputtering yield of solid lithium (i.e., IIAX experiments). **Figure 5b** shows experimental and VFTRIM-3D simulation results for He^+ bombardment on D-treated and non-D-treated lithium at 45° incidence. The figure plots the energy dependence of the absolute sputtering yield of lithium in atoms per incident ion. The lithium-sputtering yield functional behavior of the non-D-treated lithium target is shifted toward a maximum at higher energies (~1000 eV) for one of the experimental cases. The VFTRIM-3D results begin to diverge the experimental data at energies above 500 eV. In addition, the computational model used for the non-D-treated data is based on a mechanism for channeling energy from subsurface layers to the top layer [23]. The simulation model used to predict the D-treated data does not utilize this mechanism. This result implies that the absence of deuterium atoms at interstitial sites of the lithium bcc lattice allows for atoms from deeper in the sputtering cascade to transfer their momentum up to surface layer atoms, thus contributing to more sputtering.

The D-treated lithium-sputtering yield is measured to be significantly lower than bombardment with no deuterium treatment. As explained earlier, preferential sputtering is expected for

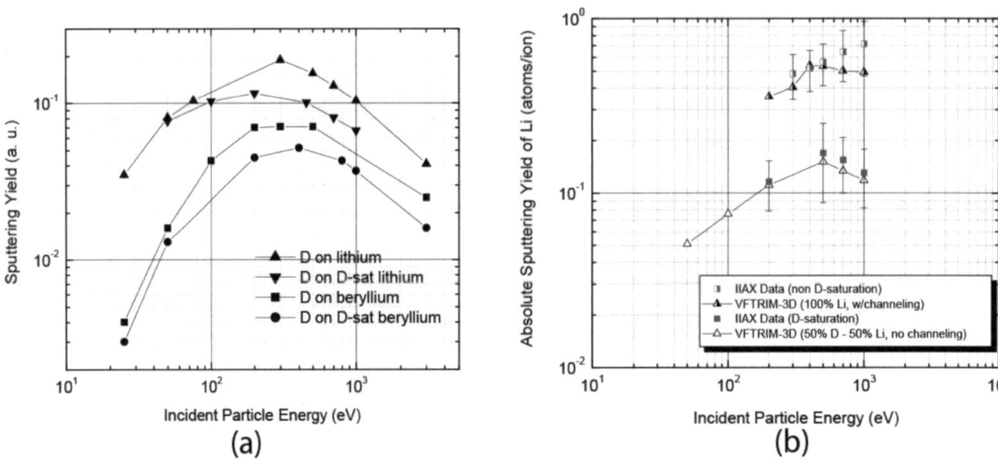

Figure 5. (a) VFTRIM-3D simulation of deuterium sputtering of D-treated and non-D-treated lithium and beryllium at 45° incidence. Deuterium saturation levels for lithium are 50% D/li, and for beryllium, 33% D/be. These are levels of saturation, which mimic those expected from PMI interactions in magnetic fusion devices at the wall boundary. (b) Energy dependence of 45° incidence he^+ bombardment on non-D-treated and D-treated solid lithium measurements and VFTRIM-3D simulation.

the lightest component and for the least bound species. The deuterium is sputtered preferentially and the surface, in time, is enriched in lithium. However, at doses in IIAX and doses found in typical plasma-facing conditions in tokamaks, the one-to-one ratio of lithium matrix atoms and saturating-deuteride species are kept over at least the depth of origin of sputtered species as a constant flux of deuterium atoms impinges on the lithium sample and a source of implanted deuterium atoms segregates to the surface over the time of dose [34–36].

Another factor is the competition between preferential sputtering on the one hand, and mixing or segregation on the other [37, 38]. These latter effects are less pronounced here since we have a surface that is "soaked" with deuterium atoms and not an alloy composed of deuterium and lithium constituents. Therefore, preferential sputtering mechanisms are justified as a viable interpretation. The binding of deuterium and lithium atoms is less likely than deuterium atoms penetrating and sitting at interstitial sites in bcc lithium.

3.1.4. *Self-sputtering*

Self-sputtering at the plasma boundary in fusion devices will occur for impurities, which have been injected into the plasma, are ionized, and return to strike solid surfaces at the wall or divertor regions. The momentum transfer between like masses is extremely effective due to maximum energy transfer, leading to increased sputtering and potentially diluting the plasma with impurities. There are two components in self-sputtering that must be considered. One is the erosion component due to physical sputtering and the other is the reflection of the incident particle into the plasma. Since both sources cannot be distinguished from each other, except their average particle energy, these sources must be added to obtain the total amount of particles injected into the plasma. **Figure 6a** shows the self-sputtering yield (sputtering +

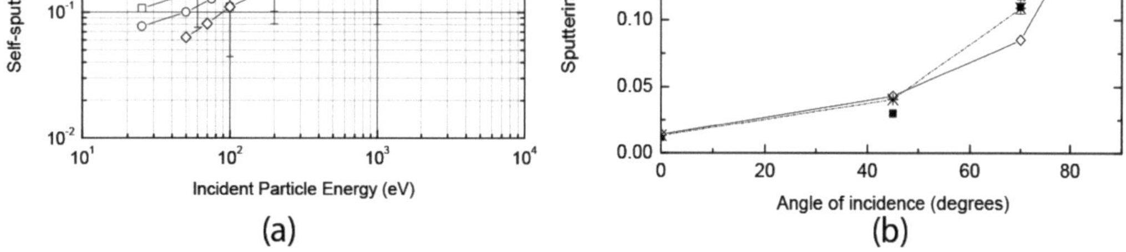

Figure 6. (a) Self-sputtering experimental and simulation yields (total = reflection and sputtering for TRIM-SP simulations) of tungsten, tin, beryllium, and lithium at 45° incidence. Normal incidence data are adjusted to 45° using the Yamamura formula for oblique particle incidence. (b) VFTRIM-3D, TRIM-SP and experimental data of 2 keV D^+ bombardment of graphite versus particle angle of incidence.

reflection) for the cases of tin, lithium, tungsten, and beryllium at 45° incidence. Normal incidence data are adjusted to 45° incidence by Yamamura's empirical formula (Eq. (1)) for oblique particle incidence.

Self-sputtering of tungsten and tin is simulated by VFTRIM-3D and compared with experimental data in **Figure 6a**. Due to the highly efficient transfer of energy process in self-sputtering, tungsten results in the largest yield still increasing at 3 keV. Tin self-sputtering [34–39] is shown to have a slightly lower-sputtering yield close to beryllium self-sputtering for energies up to 1 keV within experimental uncertainty. Beryllium self-sputtering is elaborated on in several papers [20–22]. The beryllium self-sputtering yield is found to have a maximum at about 1.2 keV. The data are predicted quite successfully as surface roughness is modeled by VFTRIM-3D. This is an important point that will be elaborated on in the next section regarding the dependence of the sputtering yield on the angle of incidence of the bombarding particle and the fact that in fusion devices PMI results in evolving surfaces that can roughen over the time scale of plasma-induced modification and operation.

Figure 6a also shows the experimental and computational results for lithium self-sputtering at 45° incidence. Computational data using TRIM-SP for incident-particle reflection are also included in the self-sputtering yield calculated. The experimental results are surprising. The self-sputtering yield of solid lithium maximizes at 700 eV to a value of 0.245 ± 0.100 atoms/ion. This is considerably lower than the values predicted by László and Eckstein [40]. There are two main reasons why the calculated values by László and Eckstein are significantly greater than our measured results. The computational model used by J. László et al. used the TRIM-SP, which assumes a smooth surface and neglects surface roughness. Second, the model does not utilize a compositional component to incorporate the effect of deuterium implantation at interstitial sites of the lithium sample as discussed in the previous section.

Within the experimental error, the simulation predicts the functional behavior of the sputter yield, except at higher energies ($E \geq 800$ eV) where the two begin to slightly diverge. This is due to a shorter mean range of the incident lithium ion in solid lithium compared to D or He. Thus, a large percentage of its kinetic energy is distributed among the top-most surface deuterium atoms, which lead to a preferentially larger deuterium erosion and a reduction of lithium sputtering. The fact that lithium self-sputtering is significantly reduced due to deuterium saturation of the surface is encouraging for its use as a potential plasma-facing component.

3.2. Sputtering yield dependence on angle of incidence

As discussed in the introductory sections of this review, the effect of roughness on the sputtering yield of materials is enhanced with an increase in the angle of incidence. **Figure 6b** shows the angle of incidence dependence of graphite sputtering by 2-keV incident deuterium ions. As the angle of incidence is increased, the effect of surface roughness is enhanced and thus simulation using TRIM-SP better predicts the data very well compared to using a smooth surface simulation with TRIM-SP [41]. **Figure 7a** demonstrates this effect for beryllium self-sputtering as well. Two sets of experimental data [10, 21] are plotted with VFTRIM-3D and TRIM-SP simulations for 1-keV beryllium self-sputtering. Again in this case, the experimental data are predicted successfully for VFTRIM-3D modeling roughness. The scatter in the experimental data at high angles of incidence is due to a variety of roughness levels in the materials as well as the creation of surface roughness in the course of ion bombardment.

The lithium self-sputtering yield for deuterium-treated and non-deuterium-treated surfaces is shown in **Figure 7b**. The dependence of the lithium-sputtering yield on the angle of incidence is also shown for two incident energies, 100 and 500 eV. Due to the importance of self-sputtering runaway, a thick black (red) line indicates unity lithium-sputtering yield. Unity self-sputtering means the sputtered flux is greater than the incident-particle flux and thus the potential for a runaway condition, which results in enhanced amounts of target material being eroded, re-ionized, and re-deposited with an impact on fusion device operation. The dependence of the lithium-sputtering yield on the angle of incidence is strong for higher oblique angles. This is due to the decrease in the penetration length of the incident bombarding atom and consequently greater energy deposition near the surface, increasing the probability to sputter. For deuterium-treated surfaces, the lithium-sputtering yield is lower than non-deuterium-treated surfaces at all angles of incidence shown and the given incident energies.

The lithium yield due to 500-eV incident lithium atoms on deuterium-treated lithium is slightly lower than the case for 100-eV lithium on non-deuterium-treated lithium. This is due to preferential sputtering of deuterium atoms playing a large role in reducing the lithium-sputtering yield even at higher energies. At 500 eV, lithium penetrates farther and thus the resulting lithium yield is lower. However, at larger angles of incidence, the energy deposition increases even for the case of deuterium-treated lithium, thus the yield crosses the low energy curve for non-deuterium-treated lithium. For the high-energy case with no deuterium coverage, we find that self-sputtering runaway is reached at a lower angle of incidence than for deuterium-treated lithium. This occurs at an angle of incidence of about 55°, while with

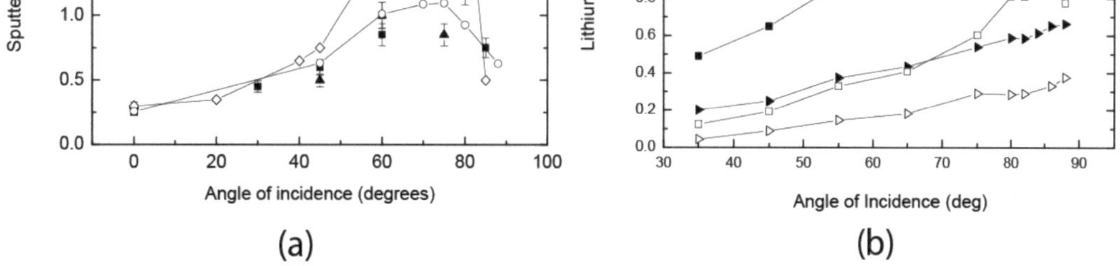

Figure 7. (a) VFTRIM-3D, TRIM-SP, and experimental data of 1 keV be$^+$ bombardment of beryllium against nominal particle angle of incidence. (b) Self-sputtering yield from lithium surface at about 200°C, for deuterium-treated and non-deuterium-treated lithium. TRIM-SP simulations are done for 45° incidence and 100 and 500 eV incident-particle energies.

deuterium-treated surfaces, the lithium yield remains under unity up to about 70° incidence at 500-eV incident-particle energy. This has some important implications. If one can maintain a one-to-one deuterium to lithium coverage, self-sputtering runaway of lithium could be dramatically reduced even for incident particles at high energies. This is important since deuterium-absorbed lithium plasma-facing surfaces give rise to low-recycling plasma regimes at the edge [42]. The comparison of the sputtering yield dependence on the incident angle *and* differences of incident-particle energy are illustrated in **Figure 8**. **Figure 8** shows the lithium sputter yield for both D-saturated and non-D-saturated surface conditions sputtered by 100 and 1000 eV D atoms. Notice that the enhancement with an incident angle decreases with a decreasing incident energy and in fact become equal at 60° incidence, which means that the energy deposition is predominant on the surface and only D saturation could significantly decrease the sputtering.

3.3. Secondary ion sputtering fraction in lithium sputtering

Another very important property in PMI is the surface charge density and the role of charge dynamics when sputtered atoms are released from the surface. The secondary ion sputtering fraction, defined as the fraction of ions to neutrals sputtered from the incident ions, has been measured for lithium sputtering by bombardment of D^+, He^+, and Li^+ at low energies and oblique incidence [13]. Such a measurement is important since in a fusion device, plasma-sputtered ions will immediately return to the surface due to the sheath potential and thus not

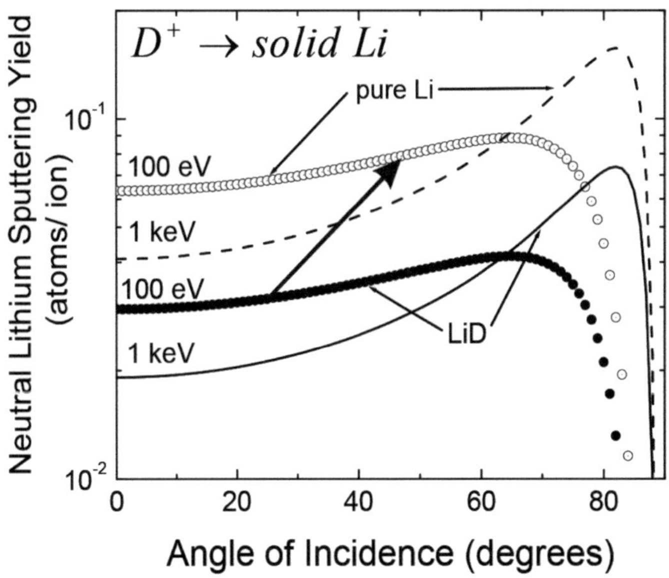

Figure 8. Lithium-sputtering yield versus angle of incidence using the Bohdansky-Sigmund-Yamamura (BSY) model. Open circles: 100 eV D on solid pure (100%) lithium; filled circles: 100 eV D on solid LiD (50% D-li); dashed line: 1 keV D on solid pure li; solid line: 1 keV D on solid LiD. As a deuterium-treated sample loses D near the li surface, the lithium-sputtering yield begins to increase and approaches the pure li yield. This is shown by the large arrow pointing in the direction of li-sputtering increase.

contribute to the sputtering yield. The secondary ion sputtering fraction does not vary significantly in the range of 500–1000 eV. The fraction of sputtered atoms in the ionic state is measured to be about 65% or two out of three sputtered atoms come out as ions. The dependence of the secondary ion fraction has been linked with the combination of chemical potential and work function of a surface with alkali metals having the highest yields.

4. Conclusion

An introduction to plasma-material interactions in fusion devices was provided in this chapter. The effects of varying surface roughness were described and the use of fractal dimensions as a viable model for simulating PMI of realistic surfaces. Physical sputtering and particle reflection were selected as primary mechanisms of PMI. Simulations and experimental data of low-Z and high-Z materials were provided. Fusion relevant ion-surface interactions for candidate materials were presented. These included combinations of D, T, and He on Li, Be, C, Sn, and W with and without D-saturation of the surfaces.

Results show that surface roughness is an important effect that must be accounted for in reflection and sputtering measurements, especially at low-incident-particle energies and oblique incidence. Low-Z materials such as lithium and beryllium suffer from low-sputtering thresholds, however, maintaining fairly low self-sputtering yields. High-sputtering thresholds on the other hand characterize high-Z materials but maintain high self-sputtering yields even at low bombarding energies. Oblique incidence is important to consider due to the strong dependence of sputtering on the incident-particle angle. Deuterium saturation of low-Z materials such as lithium or beryllium effectively reduces sputtering. Moreover, lithium has a high secondary ion-sputtering fraction, thus leading to an even lower-sputtering yield.

Acknowledgements

We acknowledge our funding sources DOE-OFES grant DE-SC0010717. To learn more about the Center for Plasma-Material Interactions and the Radiation Surface Science and Engineering Laboratory, visit our sites: (http://rssel.engineering.illinois.edu) and (http://cpmi.illinios.edu).

Author details

Jean Paul Allain* and David N. Ruzic

*Address all correspondence to: allain@illinois.edu

Department of Nuclear, Plasma and Radiological Engineering, Center for Plasma-Material Interactions, University of Illinois at Urbana-Champaign, Urbana, IL, USA

References

[1] Bringa EM. The Astrophysical Journal. 2008;**662**(1):372

[2] Greenberg JM. Surface Science. 2002;**500**:793

[3] Biersack J, Eckstein W. Applied Physics A. 1984;**34**:730

[4] Avnir D, Farin D, Pfeifer P. Nature. 1984;**308**:261

[5] Ruzic DN, Chiu HK. Journal of Nuclear Materials. 1989;**162-164**:904

[6] Ruzic DN. Nuclear Instruments and Methods B. 1990;**47**:118

[7] Haasz AA, Davis JW, Wu CH. Journal of Nuclear Materials. 1989;**162-164**:915

[8] Avnir D, Farin D, Pfeifer P. The Journal of Chemical Physics. 1983;**79**:3566

[9] Shaheen MA, Ruzic DN. Journal of Vacuum Science and Technology A. 1993;**11**:3085

[10] Eckstein W, Laszlo J. Journal of Nuclear Materials. 1991;**183**:19-24

[11] Federici G et al. Nuclear Fusion. 2001;**41**:12

[12] Stangeby PC. The plasma boundary of magnetic fusion devices. In: Scott P, Wilhelmsson H, editors. Plasma Physics Series. Boston: Institute of Physics Publishing; 2000

[13] Allain JP, Ruzic DN. Nuclear Fusion. 2001;**41**:12

[14] Chodura R. Physics of plasma wall interactions in controlled fusion. In: Post DE, Behrisch R, editors. NATO ASI Series B: Physics. Vol. 131. New York: Plenum Press; 1984. (DeWald AB, Bailey AW, Brooks JN, Physics Fluids 30, (1987) 267

[15] Sugai H. Vacuum. 1996;**47**(6-8):981-984

[16] Doerner RP et al. Journal of Nuclear Materials. 2001;**290-293**:166-172

[17] Myers SM, PM Richards WR. Wampler. Journal of Nuclear Materials. 1989;**165**:9

[18] Wampler WR. Journal of Nuclear Materials. 1984;**122-123**:1598

[19] Haasz AA, Davis JW. Journal of Nuclear Materials. 1997;**241-243**:1076

[20] Roth J, Eckstein W, Guseva MI. Fusion Engineering and Design. 1997;**37**:465-480

[21] Guseva MI et al. Journal of Nuclear Materials. 1997;**241-243**:1117-1121

[22] Hechtl E, Roth J, Eckstein W, Wu CH. Journal of Nuclear Materials. 1995;**220-222**:883-885

[23] Smith PC, Ruzic DN. Nuclear Fusion. 1998;**38**(5):673-679

[24] Ruzic DN, Smith PC, Turkot RB Jr. Journal of Nuclear Materials. 1997;**241-243**:1170-1174

[25] Guseva MI, Suvorov AL, Korshunov SN, Lazarev NE. Journal of Nuclear Materials. 1999; **266-269**:222-227

[26] Doerner RP et al. Journal of Nuclear Materials. 1998;**257**:51-58

[27] Alimov VK, Chernikov VN, Zakharov AP. Journal of Nuclear Materials. 1997;**241-243**: 1047-1052

[28] Roth J, Eckstein W, Bohdansky J. Journal of Nuclear Materials. 1989;**165**:199-204

[29] Roth J. Journal of Nuclear Materials. 1987;**145-147**:87-95

[30] Yamamura Y, Itikawa Y, Itoh N. Nagoya University Institute of Plasma Physics Report IPPJ-AM-26; 1983

[31] Kustner M, Eckstein W, Hecthl E, Roth J. Journal of Nuclear Materials. 1999;**265**:22

[32] Eckstein W. IPP Report 9/117. 1998

[33] Brooks JN et al. Journal of Nuclear Materials. 1997;**241-243**:294

[34] Sigmund P. Nuclear Instruments and Methods B. 1987;**27**:1-20

[35] Falcone G, Sigmund P. Applied Physics. 1981;**25**:307

[36] Sigmund P. Matematisk-Fysiske Meddelelser - Kongelige Danske Videnskabernes Selskab. 1993;**43**:255

[37] Sigmund P, Oliva A. Nuclear Instruments and Methods B. 1993;**82**:269

[38] Sigmund P, Glazov LG. Nuclear Instruments and Methods B. 2000;**164-165**:453

[39] Fontell A, Arminen E. Canadian Journal of Physics. 1969;**47**:2405

[40] László J, Eckstein W. Journal of Nuclear Materials. 1991;**184**:22-29

[41] Kustner M, Eckstein W, Dose V, Roth J. Nuclear Instruments and Methods B. 1998:320-331

[42] Brooks JN, Rognlien TD, Ruzic DN, Allain JP. Journal of Nuclear Materials. 2001;**290-293**: 185-190

Chapter 4

Stopping Power of Ions in a Magnetized Plasma: Binary Collision Formulation

Hrachya B. Nersisyan, Günter Zwicknagel and
Claude Deutsch

Additional information is available at the end of the chapter

http://dx.doi.org/10.5772/intechopen.77213

Abstract

In this chapter, we investigate the stopping power of an ion in a magnetized electron plasma in a model of binary collisions (BCs) between ions and magnetized electrons, in which the two-body interaction is treated up to the second order as a perturbation to the helical motion of the electrons. This improved BC theory is uniformly valid for any strength of the magnetic field and is derived for two-body forces which are treated in Fourier space without specifying the interaction potential. The stopping power is explicitly calculated for a regularized and screened potential which is both of finite range and less singular than the Coulomb interaction at the origin. Closed expressions for the stopping power are derived for monoenergetic electrons, which are then folded with an isotropic Maxwell velocity distribution of the electrons. The accuracy and validity of the present model have been studied by comparisons with the classical trajectory Monte Carlo numerical simulations.

Keywords: ion stopping, magnetized plasma target, binary collisions

1. Introduction

There is an ongoing in the theory of interaction of charged particle beams with plasmas. Although most theoretical works have reported on the energy loss of ions in a plasma without magnetic field, the strongly magnetized case has not yet received as much attention as the field-free case. The energy loss of ion beams and the related processes in magnetized plasmas are important in many areas of physics such as transport, heating, magnetic confinement of thermonuclear plasmas, and astrophysics. The range of the related topics includes ultracold plasmas [1, 2], the cooling of heavy ion beams by electrons [3–12], as well as many very dense systems involved in magnetized target fusions [11], or heavy ion inertial confinement fusion (ICF).

For a theoretical description of the energy loss of ions in a plasma, there exist some standard approaches. The dielectric linear response (LR) treatment considers the ion as a perturbation of the target plasma, and the stopping is caused by the polarization of the surrounding medium. It is generally valid if the ion couples weakly to the target. Since the early 1960s, a number of calculations of the stopping power (SP) within LR treatment in a magnetized plasma have been presented (see Refs. [13–37] and references therein). Alternatively, the stopping is calculated as a result of the energy transfers in successive binary collisions (BCs) between the ion and the electrons [37–45]. Here, it is necessary to consider appropriate approximations for the screening of the Coulomb potential by the plasma [8]. However, significant gaps between these approaches involve the ion stopping along magnetic field **B** and perpendicular to it. In particular, at high B values, the BC predicts a vanishingly parallel energy loss, which remains at variance with the nonzero LR one. Also, challenging BCLR discrepancies persist in the transverse direction, especially for vanishingly small ion projectile velocity v_i when the friction coefficient contains an anomalous term diverging logarithmically at $\mathbf{v}_\mathrm{i} = 0$ [23, 24]. For calculation of the energy loss of an ion, two new alternative approaches have been recently suggested. One of these methods is specifically aimed at a low-velocity energy loss, which is expressed in terms of velocity-velocity correlation and, hence, to a diffusion coefficient [34]. Next, in Ref. [27] using the Bhatnagar-Gross-Krook approach based on the Boltzmann-Poisson equations for a collisional and magnetized classical plasma, the energy loss of an ion is studied through a LR approach, which is constructed such that it conserves particle number locally.

An alternative approach, particularly in the absence of any relevant experimental data, is to test various theoretical methods against comprehensive numerical simulations. This can be achieved by a particle-in-cell (PIC) simulation of the underlying nonlinear Vlasov-Poisson Equation [10, 31]. While the LR requires cutoffs to exclude hard collisions of close particles, the collectivity of the excitation can be taken into account in both LR and PIC approaches. In the complementary BC treatment, the stopping force has been calculated numerically by scattering statistical ensembles of magnetized electrons from the ions in the classical trajectory Monte Carlo (CTMC) method [7, 10, 37–41]. For a review we refer to a recent monograph [8] which summarizes all theoretical and numerical methods and approaches also discussing the ranges of their validity.

The very recent upheaval of successful experiments involving hot and dense plasmas in the presence of kilotesla magnetic fields (e.g., at ILE (Osaka), CELIA (Bordeaux), LULI (Palaiseau), LLNL (Livermore)) remaining nearly steady during 10–15 ns strongly motivates the fusion as well as the warm dense matter (WDM) communities to investigate adequate diagnostics for their dynamic properties. This opens indeed a novel perspective by allowing magnetic fields to play a much larger if not a central role both in ICF and WDM plasmas. In this context proton or any nonrelativistic ion stopping is likely to provide an option of choice for investigating genuine magnetization features such as anisotropy, when the electron plasma frequency turns significantly lower than the cyclotron one [46]. In addition, an experimental test of proton or alpha particle stopping in a magnetized plasma is currently envisioned (see, e.g., Ref. [46] for a preliminary discussion). The parameters at hand are a fully ionized hydrogen plasma with a density up to 10^{20} cm^3 and temperature between 1 and 100 eV. The steady magnetic field can be up to 45 T strong. A preliminary examination based on comparing electron Debye length with corresponding Larmor radius indicates that to experience a strong influence of the magnetic field, the electron density should be comparable with a few 10^{16} cm^3. We expect these endeavors to lead to the very first unambiguous and genuine identification of an experimental magnetic signature for nonrelativistic ion stopping in plasmas.

Motivated by these recent developments, our purpose is to investigate the SP of an ion moving in a magnetized plasma in a wide range of the value of a steady magnetic field. The present paper is based on our earlier studies in Refs. [8, 24, 44, 45] where the second-order energy transfers for individual collisions of electron-ion [8, 24, 44] of any two identical particles, like electron-electron [44], and finally of two gyrating arbitrary charged particles [45] have been calculated with the help of an improved BC treatment. This treatment is—unlike earlier approaches of, e.g., Refs. [9, 42]—valid for any strength of the magnetic field. As the first application of the theoretical BC model developed in Refs. [8, 24, 44, 45], we have calculated in Ref. [47] the cooling forces on the heavy ion beam interacting with a strongly magnetized and temperature anisotropic electron beam. It has been shown that there is a quite good overall agreement with both the CTMC numerical simulations and the experiments performed at the ESR storage ring at GSI [48–50].

In Section 2 we introduce briefly a perturbative binary collision formulation in terms of the binary force acting between an ion and a magnetized electron and derive general expressions for the second-order (with respect to the interaction potential) stopping power. In contrast to the previous investigations in Refs. [8, 24, 44, 45], we here consider the (macroscopic) stopping force which is obtained by integrating the binary force of an individual electron-ion interaction with respect to the impact parameter and the velocity distribution function of electrons. That is, the stopping force for monoenergetic electrons is folded with a velocity distribution. The resulting expressions involve all cyclotron harmonics of the electrons' helical motion and are valid for any interaction potential and any strength of the magnetic field. In Section 2.4 we present explicit analytic expressions of this second-order stopping power for the specific case of a regularized and screened interaction potential [51, 52] which is both of finite range and less singular than the Coulomb interaction at the origin and which includes as limiting cases the Debye (i.e., screened) and the Coulomb potentials. For comparison of our expressions with previous approaches, we consider in Section 3 the corresponding asymptotic expressions for large and small ion velocities and strong and vanishing magnetic fields. The analytical expressions presented in Section 2.4 are evaluated numerically in Section 4 using parameters of the envisaged experiments on ion stopping [46]. In particular, we compare our approach with the CTMC simulations. The results are summarized and discussed in Section 5. The regularization parameter and the screening length involved in the interaction potential are briefly specified and discussed in Appendix A.

2. Theoretical model

2.1. Binary collision (BC) formulation

Let us consider two point charges with masses m, M and charges $-e$, Ze, respectively, moving in a homogeneous magnetic field $\mathbf{B} = B\mathbf{b}$. We assume that the particles interact with the potential $-Z\not{e}^2U(r)$ with $\not{e}^2 = e^2/4\pi\epsilon_0$, where ϵ_0 is the permittivity of the vacuum and

$\mathbf{r} = \mathbf{r}_1 - \mathbf{r}_2$ is the relative coordinate of the colliding particles. For two isolated charged particles, this interaction is given by the Coulomb potential, i.e., $U_C(\mathbf{r}) = 1/r$. In plasma applications U_C is modified by many-body effects and the related screening and turns into an effective interaction. In general, this effective interaction, which is related to the wake field induced by a moving ion, is non-spherically symmetric and depends also on the ion velocity. For any BC treatment, however, this complicated ion-plasma interaction must be approximated by an effective two-particle interaction $U(\mathbf{r})$. This effective interaction U may be modeled by a spherically symmetric Debye-like screened interaction $u_D(\mathbf{r}) = e^{-r/\lambda}/r$ with a screening length λ, given, e.g., by the Debye screening length λ_D (see, e.g., [16]), in case of low ion velocities and an effective velocity-dependent screening length $\lambda(v_i)$ for larger ion velocities v_i (see [53–55]). Further details on the choice of the effective interaction $U(r)$ are given in Ref. [47].

In the presence of an external magnetic field, the Lagrangian and the corresponding equations of particle motion cannot, in general, be separated into parts describing the relative motion and the motion of the center of mass (cm) [8]. However, in the case of heavy ions, i.e., $M \gg m$, the equations of motion can be simplified by treating the cm velocity $\mathbf{v}_{\rm cm}$ as constant and equal to the ion velocity $\mathbf{v}_i$, i.e., $\mathbf{v}_{\rm cm} = \mathbf{v}_i = const$. Then, introducing the velocity correction through relations $\delta v_{(t)} = \mathbf{v}_e(t) - \mathbf{v}_{e0}(t) = \mathbf{v}(t) - \mathbf{v}_0(t)$, where $\mathbf{v}(t) = r(t) = \mathbf{v}_e(t) - \mathbf{v}_i$ is the relative electron-ion velocity $\mathbf{v}_{e0}(t)$ and $\mathbf{v}_0(t) = \dot{\mathbf{r}}_0(t) = \mathbf{v}_{e0}(t) - \mathbf{v}_i$ are the unperturbed electron and relative velocities, respectively, the equation of relative motion turns into

$$r_0(t) = \mathbf{R}_0 + \mathbf{v}_r t + a[\mathbf{u} \sin(\omega_c t) - [\mathbf{b} \times \mathbf{u}] \cos(\omega_c t)], \tag{1}$$

$$\delta\dot{\mathbf{v}}(t) + \omega_c[\delta\mathbf{v}(t) \times \mathbf{b}] = -\frac{Z\not{e}^2}{m}\mathbf{f}[\mathbf{r}(t)]. \tag{2}$$

Here, $-Z\not{e}^2\mathbf{f}[\mathbf{r}(t)](\mathbf{f} = -\partial U/\partial\mathbf{r})$ is the force exerted by the ion on the electron, $\omega_c = eB/m$ is the electron cyclotron frequency, and $\delta\mathbf{v}(t) \to 0$ at $t \to -\infty$. In Eq. (1) $\mathbf{u} = (\cos\varphi, \sin\varphi)$ is the unit vector perpendicular to the magnetic field; the angle φ is the initial phase of the electron's helical motion; $\mathbf{v}_r = v_{e\parallel}\mathbf{b} - \mathbf{v}_i$ is the relative velocity of the guiding center of the electrons, where $v_{e\parallel}$ and $v_{e\perp}$ (with $v_{e\perp} \leq 0$) are the unperturbed components of the electron velocity parallel and perpendicular to $\mathbf{b}$, respectively; and $a = v_{e\perp}/\omega_c$ is the cyclotron radius. In Eq. (1), the quantities $\mathbf{u}$ and $\mathbf{R}_0$ are defined by the initial conditions. In Eq. (2) $\mathbf{r}(t) = \mathbf{r}_e(t) - \mathbf{v}_i t$ is the ion-electron relative coordinate.

2.2. The perturbative treatment

We seek an approximate solution of Eq. (2) in which the interaction force between the ion and electron is considered as a perturbation. Thus, we are looking for a solution of Eq. (2) for the variables $\mathbf{r}$ and $\mathbf{v}$ in a perturbative manner $\mathbf{r} = \mathbf{r}_0 + \mathbf{r}_1 + \ldots, \mathbf{v} = \mathbf{v}_0 + \mathbf{v}_1 + \cdots$, where $\mathbf{r}_0(t), \mathbf{v}_0(t)$ are the unperturbed ion-electron relative coordinate and velocity, respectively, and $\mathbf{r}_n(t), \mathbf{v}_n(t)(n = 1, 2, \cdots)$ are the nth-order perturbations of $\mathbf{r}(t)$ and $\mathbf{v}(t)$, which are proportional to Z^n.

The parameter of smallness which justifies such kind of expansion can be read off from a dimensionless form of the equation of motion Eq. (2) by scaling lengths in units of the screening length λ, velocities in units of the initial relative velocity v_0, and time in units of λ/v_0. Then, it is seen (see Ref. [47] for details) that the perturbative treatment is essentially applicable in cases where $|Z|\rlap{/}e^2/mv_0^2\lambda < 1$, that is, when the (initial) kinetic energy of relative motion $mv_0^2/2$, is large compared to the characteristic potential energy $|Z|\rlap{/}e^2/\lambda$ in a screened Coulomb potential. Or, expressed in velocities, the initial relative velocity v0 must exceed the characteristic velocity $v_d = \left(|Z|\rlap{/}e^2/m\lambda\right)^{1/2}$, that is, v_d here demarcates the perturbative from the non-perturbative regime. If this condition is met not only for a single ion-electron collision but in the average over the electron distribution, e.g., by replacing v0 with the averaged initial ion-electron relative velocity $\langle v_0\rangle$, i.e., $\langle v_0\rangle \gtrsim v_d$, we are in a regime of weak ion-target or, here, weak ion-electron coupling, which allows the use of perturbative treatments (besides BC also, e.g., linear response (LR)). For nonmagnetized electrons this is discussed in much detail in Refs. [53, 54]. Even though the particle trajectories are much more intricate in the presence of an external magnetic field, the given definitions and demarcations of coupling regimes are basically the same for magnetized electrons. That is, the applicability of a perturbative treatment is essentially related to the charge state Z of the ion and the typical range λ of the effective interaction, but not directly on the strength B of the magnetic field. The latter may affect the critical velocity v_d only implicitly via a possible change of the effective screening length λ with B.

The equation for the first-order velocity correction is obtained from Eq. (2) replacing on the right-hand side of the exact relative coordinate $\mathbf{r}(t)$ by $\mathbf{r}_0(t)$ with the solutions $\mathbf{v}_1(t) = \dot{\mathbf{r}}_1(t)$ and

$$r_1(t) = \frac{Z\rlap{/}e^2}{m}\{-\mathbf{b}\mathcal{Q}_{\|}(t) + Re[\mathbf{b}(\mathbf{b}\cdot\mathcal{Q}_{\perp}(t)) - \mathcal{Q}_{\perp}(t) + i[\mathbf{b}\times\mathcal{Q}_{\perp}(t)]]\}. \tag{3}$$

Here, we have introduced the following abbreviations:

$$\begin{aligned}\mathcal{Q}_{\|}(t) &= \int_{-\infty}^{t} \mathbf{b}\cdot\mathbf{f}[\mathbf{r}_0(\tau)](t-\tau)d\tau,\\ \mathcal{Q}_{\perp}(t) &= \frac{1}{i\omega_c}\int_{-\infty}^{t} \mathbf{f}[\mathbf{r}_0(\tau)]\left[e^{\omega_c(t-\tau)} - 1\right]d\tau\end{aligned} \tag{4}$$

and have assumed that all corrections vanish at $t \to -\infty$.

2.3. Second-order stopping power

We now consider the interaction process of an individual ion with a homogeneous electron plasma described by a velocity distribution function $f(\mathbf{v}_e)$ and a density n_e. We assume that the ion experiences independent binary collisions (BCs) with the electrons. The total stopping force, $\mathbf{F}(\mathbf{v}_i)$, acting on the ion is then obtained by multiplying the binary force $Z\rlap{/}e^2\mathbf{f}[\mathbf{r}(t)]$ by the element of the flux relative flux $n_e v_r d^2 s dt$, integrating with respect to time and folding with velocity distribution of the electrons. The impact parameter s introduced here in the electron flux is defined by $s = \mathbf{R}_{0\perp} = \mathbf{R}_0 - \mathbf{n}_r(\mathbf{n}_r\cdot\mathbf{R}_0)$ and is the component of $\mathbf{R}_0$ perpendicular to the relative velocity vector $\mathbf{v}_r$ with $\mathbf{n}_r = \mathbf{v}_r/v_r$. As can be inferred from Eq. (1), s represents the distance of the closest approach between the ion and the guiding center of the electron's helical motion.

The resulting stopping power, $S(\mathbf{v}_i) = -\frac{\mathbf{v}_i}{v_i} \cdot \mathbf{F}(\mathbf{v}_i)$, then reads

$$S(\mathbf{v}_i) = -\frac{Ze^2 n_e}{v_i} \int d\mathbf{v}_e f(\mathbf{v}_e) v_r \int d^2 s \int_{-\infty}^{\infty} \mathbf{v}_i \cdot \mathbf{f}[\mathbf{r}(t)] dt, \tag{5}$$

which is an exact relation for uncorrelated BCs of the ion with electrons. We evaluate this expression within a systematic perturbative treatment (see Ref. [47] for more details). First, we introduce the two-particle interaction potential $U(r)$, and the binary force $\mathbf{f}(\mathbf{r})$ is written using Fourier transformation in space. Furthermore, the factor $e^{i\mathbf{k}\cdot\mathbf{r}(t)}$ in the Fourier transformed binary force is expanded in a perturbative manner as $e^{i\mathbf{k}\cdot\mathbf{r}(t)} \simeq e^{i\mathbf{k}\cdot\mathbf{r}_0(t)}[1 + i(\mathbf{k}\cdot\mathbf{r}_1(t))]$, where $\mathbf{r}_0(t)$ and $\mathbf{r}_1(t)$ are the unperturbed and the first-order corrected relative coordinates (Eqs. (1) and (3)), respectively. Next, we consider only the second-order binary force $\mathbf{f}_2$ and the corresponding stopping force $\mathbf{F}_2$ with respect to the binary interaction since the averaged first-order force $\mathbf{F}_1$ (related to $\mathbf{f}_1$) vanishes due to symmetry reasons [8, 24, 44, 45, 47]. Within the second-order perturbative treatment, the stopping power can be represented as

$$S(\mathbf{v}_i) = -\frac{Ze^2 n_e}{v_i} \int d\mathbf{v}_e f(\mathbf{v}_e) v_r \int d^2 \mathbf{s} \int d\mathbf{k} U(\mathbf{k})(\mathbf{k}\cdot\mathbf{v}_i) \int_{-\infty}^{\infty} [\mathbf{k}\cdot\mathbf{r}_1(t)] e^{i\mathbf{k}\cdot\mathbf{r}_0(t)} dt. \tag{6}$$

From Eq. (6) it is seen that the second-order stopping power is proportional to Z^2. Inserting now Eqs. (1) and (3) into Eq. (6), assuming an axially symmetric velocity distribution $f(\mathbf{v}_e) = f(v_{e\parallel}, v_{e\perp})$, and performing the s integration, we then obtain

$$\begin{aligned} S = &-\frac{(2\pi)^4 Z^2 e^4 n_e}{m v_i} \int_{-\infty}^{\infty} dv_{e\parallel} \int_0^{\infty} f(v_{e\parallel}, v_{e\perp}) v_{e\perp} dv_{e\perp} \\ &\times \int d\mathbf{k} |U(\mathbf{k})|^2 (\mathbf{k}\cdot\mathbf{v}_i) \int_0^{\infty} \left[k_\parallel^2 + k_\perp^2 \frac{\sin(\omega_c t)}{\omega_c t} \right] \\ &\times J_0\left(2k_\perp a \sin\frac{\omega_c t}{2}\right) \sin(\mathbf{k}\cdot\mathbf{v}_r t) dt d, \end{aligned} \tag{7}$$

where J_n is the Bessel function of the nth order; $k_\parallel = (\mathbf{k}\cdot\mathbf{b})$ and $k_\perp$ are the components of $\mathbf{k}$ parallel and transverse to $\mathbf{b}$, respectively; and $v_{e\parallel}$ and $v_{e\perp}$ are the electron velocity components parallel and transverse to $\mathbf{b}$, respectively. This general expression (7) for the stopping power of an individual ion has been derived within second-order perturbation theory but without any restriction on the strength of the magnetic field B.

2.4. The SP for a regularized and screened coulomb potential

For an electron plasma with an isotropic Maxwell distribution, the velocity distribution relevant for the averaging in Eq. (7) is given by

$$f(v_e) = \frac{1}{(2\pi)^{3/2} v_{\text{th}}^3} e^{-v_e^2/2v_{\text{th}}^2}, \tag{8}$$

where the thermal velocity $v_t h$ is related to the electron temperature by $v_{\text{th}}^2 = T/m$ (here, the temperature is measured in energy units). Inserting Eq. (8) into expression (7) and assuming now a spherically symmetric potential $U = U(k)$ yields after performing the velocity integrations (see Ref. [56]), the stopping power

$$S(v_i) = \frac{8Z^2 e^4 n_e}{m\omega_c v_i} \frac{(2\pi)^4}{4} \int_0^\infty dk_\parallel \int_0^\infty U^2(k) k_\perp dk_\perp \int_0^\infty e^{-\frac{t^2}{2}k_\parallel^2 a^2} e^{-k_\perp^2 a^2 (1-\cos t)} \left(k_\parallel^2 + k_\perp^2 \frac{\sin t}{t}\right) t dt$$
$$\times \left[k_\perp a_{i\perp} \cos\left(k_\parallel a_{i\parallel} t\right) J_1(k_\perp a_{i\perp} t) + k_\parallel a_{i\parallel} \sin\left(k_\parallel a_{i\parallel} t\right) J_0(k_\perp a_{i\perp} t)\right]. \tag{9}$$

Here, we have introduced the thermal cyclotron radius of the electrons $a = v_{th}/\omega_c$, and $a_{i\perp} = v_{i\perp}/\omega_c$, $a_{i\parallel} = v_{i\parallel}/\omega_c$, where $v_{i\perp}$ and $v_{i\parallel}$ are the ion velocity components transverse and parallel to b, respectively. For the Coulomb interaction $U(k) = U_C(k)$, the full two-dimensional integration over the s-space results in a logarithmic divergence of the k integration in Eqs. (7) and (9). To cure this, cutoff parameters $k_{\min}$ and $k_{\max}$ must be introduced (see, e.g., Refs. [8, 24, 47] for details). These cutoffs are related to the screening of the interaction in a plasma target and the incorrect treatment of hard collisions in a classical perturbative approach. As an alternative implementation of this standard cutoff procedure, we here employ the regularized screened interaction $U(r) = U_R(r) = \left(1 - e^{-r/\lambda}\right) e^{-r/\lambda}/r$ with the Fourier transform

$$U_R(k) = \frac{2}{(2\pi)^2} \left(\frac{1}{k^2 + \lambda^{-2}} - \frac{1}{k^2 + d^{-2}}\right), \tag{10}$$

where $d^{-1} = \lambda^{-1} + \lambda\!\!\!^{-}{}^{-1}$. U_R represents a Debye-like screened interaction U_D (see Section 2.1) which is additionally regularized at the origin [51, 52] and thus removes the problems related to the Coulomb singularity in a classical picture and prevents particles (for $Z > 0$) from falling into the center of the potential. The parameter λ related to this regularization is here considered as a given constant or as a function of the classical collision diameter [47].

Substituting the interaction potential (10) into Eq. (9) and performing the $k_\parallel$ integration, we arrive, after lengthy but straightforward calculations, at

$$S(v_i) = \frac{4\sqrt{\pi} Z^2 e^4 n_e}{m v_{\text{th}}^2} v \int_0^\infty \frac{dt}{t} \int_0^1 d\zeta \exp\left[-v^2 \zeta^2 P(t,\zeta)\right] \Phi[\Psi(t,\zeta)]$$
$$\times \left[P_1(t,\zeta) + \frac{\sin(\alpha t)}{\alpha t} P_2(t,\zeta)\right] \frac{\zeta^2 (1-\zeta^2)}{G(t,\zeta)}, \tag{11}$$

where $P(t,\zeta) = \cos^2\vartheta + \sin^2\vartheta / G(t,\zeta)$ and

$$P_1(t,\zeta) = 2\cos^2\vartheta + P(t,\zeta)\left(1 - 2v^2\zeta^2\cos^2\vartheta\right), \tag{12}$$

$$P_2(t;\zeta) = \frac{2}{G(t,\zeta)} \left[\frac{\sin^2\vartheta}{G(t,\zeta)} + P(t,\zeta)\left(1 - \frac{v^2\zeta^2 \sin^2\vartheta}{G(t,\zeta)}\right)\right]. \tag{13}$$

Here, we have introduced the dimensionless quantities $v = v_i/\sqrt{2}v_{\text{th}}$, $\alpha = \omega_c\lambda/v_{\text{th}}$. ϑ is the angle between b and v_i, $\Psi(t,\zeta) = (t^2/2)(1-\zeta^2)/\zeta^2$, $G(t,\zeta) = \Theta(t)\zeta^2 + 1 - \zeta^2$, $\Theta(t) = \left(\frac{2}{\alpha t}\sin\frac{\alpha t}{2}\right)^2$, and

$$\Phi(z) = e^{-z} + e^{-\varkappa^2 z} - \frac{2}{\varkappa^2 - 1}\frac{1}{z}\left(e^{-z} - e^{-\varkappa^2 z}\right), \tag{14}$$

where $\varkappa = \lambda/d = 1 + \lambda/\lambda\!\!\!^{-}$.

Eq. (11) for the SP is the main result of the outlined BC treatment which will now be evaluated in the next sections.

3. Comparison with previous approaches

Previous theoretical expressions for the stopping power which have been extensively discussed by the plasma physics community (see, e.g., Refs. [3, 8] for reviews) basically concern the two limiting cases of vanishing and infinitely strong magnetic fields. We therefore investigate the present approach for these two cases, first for arbitrary interactions $U(\mathbf{k})$ and electron distributions $f(\mathbf{v}_e)$ as given by Eq. (7) and later for the more specific situation of the regularized interaction (10) and the velocity distribution (8) as given by Eq. (11).

3.1. General SP Eq. (7) at vanishing and infinitely strong magnetic fields

At vanishing magnetic field ($B \to 0$), $\sin(\omega_c t)/(\omega_c t) \to 1$ and the argument of the Bessel function in Eq. (7) should be replaced by $k_\perp v_{e\perp} t$. Then, denoting the second-order SP at vanishing magnetic field as S_0 and assuming spherically symmetric potential with $U = U(k)$, one obtains

$$S_0(\mathbf{v}_i) = \frac{4(2\pi)^2 Z^2 e^4 n_e}{m v_i^2} \mathcal{U} \int_0^{v_i} f(v_e) v_e^2 dv_e, \tag{15}$$

where $\mathcal{U}$ is the generalized Coulomb logarithm:

$$\mathcal{U} = \frac{(2\pi)^4}{4} \int_0^\infty U^2(k) k^3 dk. \tag{16}$$

Employing the regularized and screened potential $U(k)$ given by Eq. (10), the generalized Coulomb logarithm is $\mathcal{U} = \mathcal{U}_R = \Lambda(\varkappa)$ (see also Refs. [8, 24, 44, 45]), where

$$\Lambda(\varkappa) = \frac{\varkappa^2 + 1}{\varkappa^2 - 1} \ln \varkappa - 1. \tag{17}$$

Taking the bare Coulomb interaction with $U(k) = U_C(k) \sim 1/k^2$, Eq. (16) diverges logarithmically at $k \to 0$ and $k \to \infty$, and two cutoffs $k_{\min} = 1/r_{\max}$ and $k_{\max} = 1/r_{\min}$ must be introduced as discussed in Section 2.4. In this case the generalized Coulomb logarithm takes the standard form $\mathcal{U} = \mathcal{U}_C = \ln(k_{\max}/k_{\min}) = \ln(r_{\max}/r_{\min})$.

The asymptotic expression of Eq. (15) at high ion velocities can be easily derived using the normalization of the distribution function which results in

$$S_0(\mathbf{v}_i) \simeq \frac{4\pi Z^2 e^4 n_e}{m v_i^2}\ \mathcal{U}. \tag{18}$$

At an infinitely strong magnetic field ($B \to \infty$), the term in Eq. (7) proportional to $k_\perp^2$ and the argument of the Bessel function vanish since the cyclotron radius $a \to 0$. In this limit, denoting the SP as $S_\infty(v_i)$ and assuming a spherically symmetric interaction potential, we arrive at

$$S_\infty(\mathbf{v}_i) = \frac{2\pi Z^2 e^4 n_e}{m}\ \mathcal{U} v_i \sin^2\vartheta \int \frac{1}{v^5}\left(v_{i\|}v_{e\|} - 2v_{e\|}^2 + v_i^2\right) f_e(\mathbf{v}_e) d\mathbf{v}_e. \tag{19}$$

The corresponding high-velocity asymptotic expression is given by

$$S_\infty(\mathbf{v}_i) = \frac{2\pi Z^2 e^4 n_e}{m v_i^2}\ \mathcal{U} \sin^2\vartheta. \tag{20}$$

Eqs. (15) and (19) and their asymptotic expressions for high velocities in Eqs. (18) and (20), respectively, agree with the results derived by Derbenev and Skrinsky in Ref. [57] in case of the Coulomb interaction potential, i.e., with $\mathcal{U} = \mathcal{U}_C$. Using instead a regularized interaction potential and thus the Coulomb logarithm, $\mathcal{U}_R$ allows closed analytic expressions and converging integrals and avoids any introduction of lower and upper cutoffs "by hand" in order to restrict the domains of integration. Moreover, employing the bare Coulomb interaction may, as pointed out by Parkhomchuk [58], result in asymptotic expressions which essentially different from Eqs. (19) and (20), which is related to the divergent nature of the bare Coulomb interaction (see Ref. [47]).

3.2. Some limiting cases of Eq. (11)

Next, we discuss some asymptotic regimes of the SP (Eq. (11)) where the regularized interaction (Eq. (10)) and the isotropic velocity distribution (Eq. (8)) have been assumed. In the high-velocity limit where $v_i > (\omega_c\lambda, v_{\text{th}})$, only small t contributes to the SP (Eq. (11)) due to the short time response of the electrons to the moving fast ion. In this limit we have $\sin(\alpha t)/\alpha t \to 1$. The remaining t integration can be performed explicitly. This integral is given by [47].

$$\int_0^\infty \frac{dt}{t}\Phi[\Psi(t,\zeta)] \equiv \Lambda(\varkappa). \tag{21}$$

Here, the function $\Phi(z)$ is determined by Eq. (14), and $\Lambda(\varkappa)$ is the generalized Coulomb logarithm (Eq. (17)). The remaining expressions do not depend on the magnetic field, i.e., ω_c, as a natural consequence of the short time response of the magnetized electrons. In fact,

$\sin(\alpha t)/\alpha t \to 1$ and $G(t,\zeta) \to 1$ and the related t integration (Eq. (21)) are also valid for vanishing magnetic field $\alpha \to 0$. Integration by parts turns Eq. (11) into

$$S_0 = \frac{4\pi Z^2 e^4 n_e}{m v_i^2} \Lambda(\varkappa) \left[\mathrm{erf}(v) - \frac{2}{\sqrt{\pi}} v e^{-v^2} \right], \tag{22}$$

where $\mathrm{erf}(z)$ is the error function and $v = v_i/\sqrt{2}v_{\mathrm{th}}$ is again the scaled ion velocity. The SP (Eq. (22)) is isotropic with respect to the ion velocity $\mathbf{v}_i$ and represents the two limiting cases of high velocities at arbitrary magnetic field and arbitrary velocities at vanishing field. Of course, expression (22) can be also obtained by performing the remaining integration in the nonmagnetized SP (Eq. (15)) using the isotropic velocity distribution (Eq. (8)) and $\mathcal{U} = \Lambda(\varkappa)$.

A further increase of the ion velocity finally yields

$$S_0 \simeq \frac{4\pi Z^2 e^4 n_e}{m v_i^2} \Lambda(\varkappa), \tag{23}$$

which completely agrees with the asymptotic expression (18) in case of $\mathcal{U} = \Lambda(\varkappa)$. Inspecting Eq. (23) shows that the SP does not depend explicitly on the electron temperature T at sufficiently high velocities, while T may still be involved in the generalized Coulomb logarithm $\Lambda(\varkappa)$.

At $B \to 0$ and small velocities $(v_i < v_{th})$, the SP (Eq. (22)) becomes

$$S_0 \simeq \frac{4\pi\sqrt{2\pi} Z^2 e^4 n_e}{3 m v_{\mathrm{th}}^3} v_i \Lambda(\varkappa). \tag{24}$$

Now, we consider the situation when the magnetic field is very strong and the electron cyclotron radius is the smallest length scale, $\omega_c \lambda \gg (v_i, v_{\mathrm{th}})$, and the SP is only weakly sensitive to the transverse electron velocities and, hence, is affected only by their longitudinal velocity spread. In this limit $\sin(\alpha t)/\alpha t \to 0$ and $G(t,\zeta) \to 1 - \zeta^2$ are obtained from Eq. (11) after straightforward calculations:

$$S_\infty = \frac{4\pi\sqrt{\pi} Z^2 e^4 n_e}{m v_{\mathrm{th}}^2} v \Lambda(\varkappa) \int_0^1 e^{-v^2\zeta^2 P(\zeta)} \zeta^2 d\zeta \left[2\cos^2\vartheta + P(\zeta)\left(1 - 2v^2\zeta^2\cos^2\vartheta\right) \right], \tag{25}$$

where $P(\zeta) = \cos^2\vartheta + \sin^2\vartheta/(1-\zeta^2)$.

After changing the variable ζ in Eq. (25) to $x = \zeta[P(\zeta)]^{1/2}$ and some subsequent rearrangement, Eq. (25) can be expressed alternatively as

$$S_\infty = \frac{2\sqrt{\pi} Z^2 e^4 n_e}{m v_{\mathrm{th}}^2} v \Lambda(\varkappa) \sin^2\vartheta \int_{-\infty}^{\infty} \frac{e^{-v^2 x^2} x^2 dx}{1 + x^2 - 2x\cos\vartheta)^{3/2}}. \tag{26}$$

Up to the definition of the Coulomb logarithm (i.e., $\mathcal{U} = \Lambda(\varkappa)$ versus $\mathcal{U} = \mathcal{U}_C$), the expressions are identical to those obtained by Pestrikov [59].

In particular, at $\vartheta = 0$ and $\vartheta = \pi/2$ (i.e., when ion moves parallel or transverse to the magnetic field, respectively), Eq. (25) (or Eq. (26)) yields

$$S_\infty = \frac{4\sqrt{\pi}Z^2 e^4 n_e}{m v_{\text{th}}^2} \Lambda(\varkappa) v e^{-v^2}, \tag{27}$$

$$S_\infty = \frac{2\sqrt{\pi}Z^2 e^4 n_e}{m v_{\text{th}}^2} \Lambda(\varkappa) v e^{-v^2/2} \left[(1+v^2) K_0\left(\frac{v^2}{2}\right) - v^2 K_1\left(\frac{v^2}{2}\right) \right]. \tag{28}$$

respectively, where $K_n(z)$ (with $n = 0, 1$) is the modified Bessel function. It is also constructive to obtain the angular averaged stopping power. From Eq. (25) one finds

$$\overline{S}_\infty(v) = \frac{1}{2}\int_0^\pi S_\infty(v,\vartheta)\sin\vartheta d\vartheta = \frac{4\pi Z^2 e^4 n_e}{3 m v_i^2} \Lambda(\varkappa) \left\{ \text{erf}(v) + \frac{2}{\sqrt{\pi}} v \left[v^2 E_1(v^2) - e^{-v^2} \right] \right\}, \tag{29}$$

where $E_1(z)$ is the exponential integral function.

In the high-velocity limit with $\omega_c \lambda \gg v_i \gg v_t h$, the SP (Eq. (25)) becomes

$$S_\infty \simeq \frac{2\pi Z^2 e^4 n_e}{m v_i^2} \Lambda(\varkappa) \left\{ \sin^2\vartheta \left[\text{erf}(v) - \frac{2}{\sqrt{\pi}} v e^{-v^2} \right] + \frac{4}{\sqrt{\pi}} v^3 e^{-v^2} \cos^2\vartheta \right\}. \tag{30}$$

With further increase of the ion velocity, we can then neglect the exponential term in Eq. (30), while $\text{erf}(v) \to 1$ yields the asymptotic expression (Eq. (20)) (for $\mathcal{U} = \Lambda(\varkappa)$).

The SP given by Eq. (30) (or Eq. (20) with $\mathcal{U} = \Lambda(\varkappa)$) decays as the corresponding SP (Eq. (23)) like $\sim v_i^{-2}$ with the ion velocity. But here, the parallel SP (Eq. (27)) vanishes exponentially at $\vartheta = 0$ which is a consequence of the presence of a strong magnetic field, where the electrons move parallel to the magnetic field. If the ion moves also parallel to the field (i.e., $\vartheta = 0$), the averaged stopping force must vanish within the BC treatment for symmetry reasons.

Finally, we also investigate the case of small velocities at strong magnetic fields. Considering a small ion velocity $(v \ll 1)$ in Eq. (25), we arrive at

$$S_\infty = \frac{4\pi Z^2 e^4 n_e}{m v_{\text{th}}^2} \Lambda(\varkappa) v \left\{ \sin^2\vartheta \left[\ln\left(\frac{2}{v\sin\vartheta}\right) - \frac{\gamma}{2} - 1 \right] + \cos^2\vartheta \right\}, \tag{31}$$

where $\gamma \simeq 0.5772$ is Euler's constant. Now, it is seen that the SP, S^∞, leads at low ion velocities $v \ll 1$ and for a nonzero ϑ to a term which behaves as $\sim v\ln(1/v)$. Thus, the corresponding friction coefficient diverges logarithmically at small v. This is a quite unexpected behavior compared to the well-known linear velocity dependence without magnetic field (see asymptotic expressions above). Finally, at $\vartheta = 0$ the logarithmic term vanishes and the SP behaves as $S^\infty \sim v$.

4. Features of the SP (Eq. (11)) and comparison with CTMC simulations

In this section we study some general properties of the SP of individual ions resulting from the BC approach by evaluating Eq. (11) numerically. We consider the effect of the magnetic field on the SP at various temperatures of the plasma. The density $n_e \simeq 10^{16}\text{cm}^{-3}$ and the temperatures $T \simeq 1\,\text{eV}$, 10 or 100 eV of the electron plasma, are in the expected range of the envisaged experiments on proton or alpha particles stopping in a magnetized target plasma [46] (see corresponding **Figures 1–3**). As an example we choose proton projectile for our calculations. In all examples considered below, the regularization parameter $\lambda_0 = 10^{-10}\,\text{mm}$ thereby meets the condition $\lambda_0 \gg b_0(0)$, i.e., λ_0, and does not affect noticeably the SP (Eq. (11)) at low and medium velocities as shown in Appendix A (see also Ref. [47] for more details).

For a BC description beyond the perturbative regime, a fully numerical treatment is required. In the present cases of interest, such a numerical evaluation of the SP is rather intricate but can be successfully implemented by classical trajectory Monte Carlo (CTMC) simulations [37–40]. In the CTMC method, the trajectories for the ion-electron relative motion are calculated by a numerical integration of the equations of motion (Eq. (2)). The stopping force is then deduced by averaging over a large number (typically 10^5–10^6) of trajectories employing a Monte Carlo

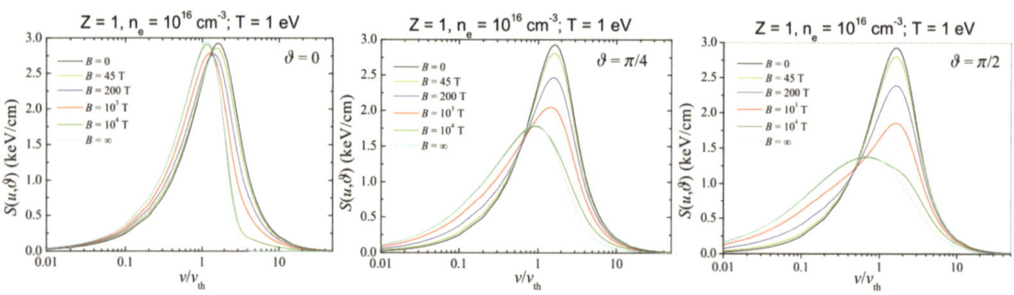

Figure 1. The SP [in keV/cm] for protons as a function of the ion velocity v_i [in units of v_{th}] and for fixed plasma temperature $T = 1\,\text{eV}$. The theoretical stopping power (Eq. (11)) is calculated for $\lambda_0 = 10^{-10}\text{m}$ (see appendix a for details) and for an electron plasma with $n_e = 10^{16}cm^{-3}$ in a magnetic field of $B = 0$ (black), 45 T (green), 200 T (blue), 10^3 T (red), 10^4 T (green), and $B = \infty$ (cyan). The angle ϑ between **B** and $\mathbf{v_i}$ is $\vartheta = 0$ (left), $\vartheta = \pi/4$ (center), and $\vartheta = \pi/2$ (right). The CTMC results for $B = \infty$ case are shown by the filled circles.

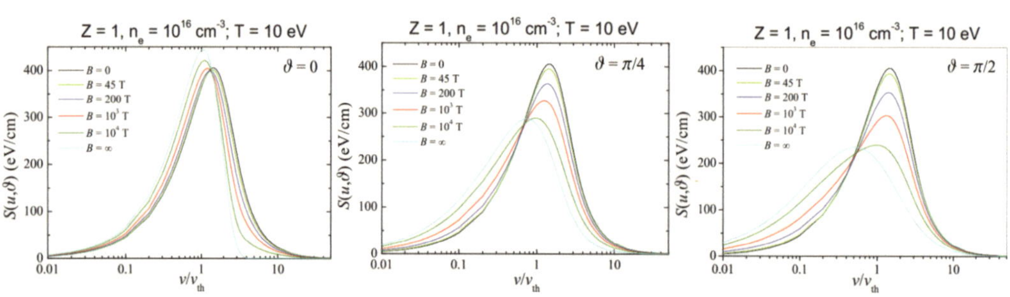

Figure 2. Same as in **Figure 1** but for $T = 10\,\text{eV}$. The SP is given in units eV/cm.

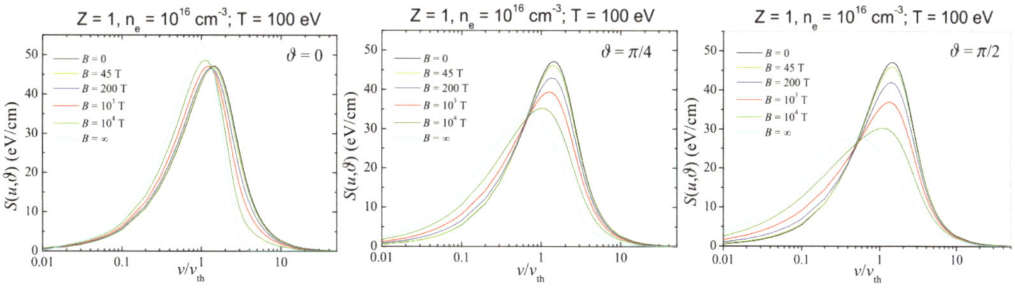

Figure 3. Same as in **Figure 1** but for $T = 100\,\text{eV}$. The SP is given in units eV/cm.

sampling for the related initial conditions. For a more detailed description of the method, we refer to Refs. [8, 44, 45]. Both the analytic perturbative treatment and the non-perturbative numerical CTMC simulations are based on the same BC picture and use the same effective spherical screened interaction $U(r)$. The following comparison of these both approaches thus essentially intends to check the validity and range of applicability of the perturbative approach as it has been outlined in the preceding sections.

5. Stopping profiles and ranges

5.1. General trends

The parameter analysis initiated on **Figures 1–3** at $n_e = 10^{16}\text{cm}^{-3}$ and $T = 1 - 10 - 100\,\text{eV}$ is implemented for monitoring a possible experimental vindication through a fully ionized hydrogen plasma out of high-power laser beams available on facilities such as ELFIE (Ecole Polytechnique) or TITAN (Lawrence Livermore) [62]. The given adequately magnetized targets (in the 20–45 T range) would then be exposed to TNSA laser-produced proton beams out of the same facilities, in the hundred keV-MeV energy range [62].

Therefore, we are looking for the most conspicuous effect of the applied magnetized intensity B on the proton stopping.

Fixing n_e and varying T (see **Figures 1–3**) display an ubiquitous and increasing anisotropy shared by the stopping profiles (SP) with increasing B and θ and angle between $\vec{\text{B}}$ and initial projectile velocity $\vec{\text{V}}$.

Moreover, that anisotropy evolves only moderately between $\theta = \frac{\pi}{4}$ and $\frac{\pi}{2}$.

Another significant feature is the extension to any $\text{B} \neq 0$ of the $\text{B} = 0$ scaling $\frac{n_e}{T}$. For instance, SP at $n_e = 10^{12}\ \text{cm}^{-3}$ and $\text{T} = 1$ eV, at a given θ, is equivalent to that for $n_e = 10^{14}\ \text{cm}^{-3}$ and $\text{T} = 100$ eV.

As expected, B effects impact essentially the low-velocity section ($\frac{\text{V}}{\text{V}_{\text{th}}}$, V_{th} = target electron thermal velocity) of the ion stopping profile. One can observe, increasing with B, a shift to the left of SP maxima, as shown in **Figure 4** at B = 45 tesla, for the profiles displayed in **Figure 3**, with θ-averaged SP remaining close to $\theta = \frac{\pi}{2}$.

Switching now attention to corresponding ranges, down to projectile at rest $\left(E_p = 0\right)$, one witnesses on **Figure 5** the counterpart of the above-noticed SP behavior.

In a low projectile velocity $\left(\frac{V}{V_{th}} \leq 1\right)$, one gets the largest B effects and the smallest proton ranges attributed to the highest B. The fan of B ranges then merges on a given point, located between 10 keV and 100 keV at $n_e = 10^{16} cm^{-3}$, and then inverts itself with increasing B featuring now increasing ranges. Moreover, the aperture of the fan of ranges increases steadily with θ.

Finally, it can be observed that for $\theta = 0$, the infinite magnetized range looks rather peculiar and reminiscent of the ion projectile gliding on $\vec{B} \parallel \vec{V}$ [8, 34].

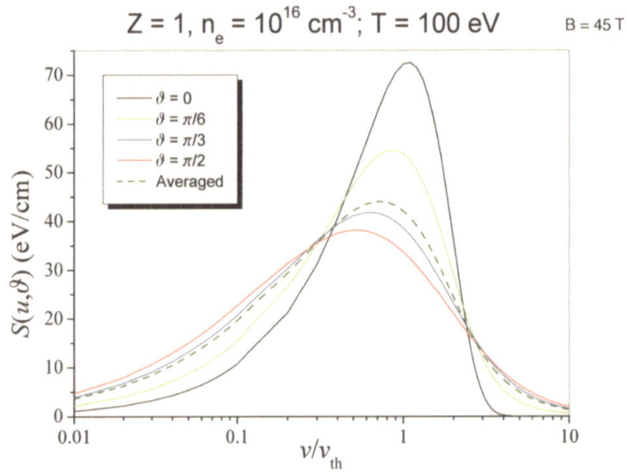

Figure 4. Same as in **Figure 3** restricted to B = 45 T, featuring θ-dependent and θ-averaged SP in eV/cm.

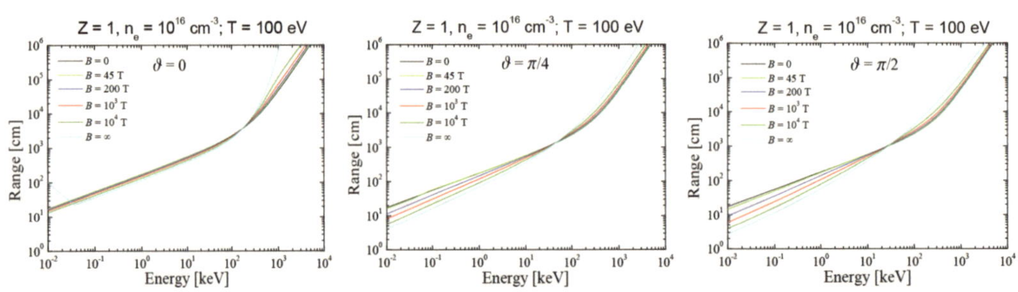

Figure 5. Ranges, down to zero energy pertaining to SP in **Figure 3**.

5.2. Specific trends

The projected experimental setup [62] could manage constant, static, and homogeneous B values up to 45 T. So, we are let to investigate n_e range limits within which significant B effects can be observed.

Obviously, $n_e = 10^{12}\ \text{cm}^{-3}$ is expected to show quantitatively larger B impact than $10^{18}\ \text{cm}^{-3}$.

Giving attention to proton ranges of T dependence in a low-density plasma $\left(n_e = 10^{12}\text{cm}^{-3}\right)$ at $T = 1$ and 100 eV, respectively, (**Figure 6**), one witnesses the smallest ranges for $\frac{V}{V_{th}} \ll 1$, increased by four orders of magnitude between 1 and 100 eV while remaining essentially unchanged for $\frac{V}{V_{th}} \geq 1$. Turning now to $n_e = 10^{18}\text{cm}^{-3}$ at $T = 1$ eV, one can see that the given SP remains quasi-isotropic, hardly θ-dependent, except at extreme magnetization ($B = \infty$). Discrepancies between B = 0 and 20 T remain visible only for $\frac{V}{V_{th}} \leq 2$. $B = \infty$ does not feature anymore the highest stopping when $\frac{V}{V_{th}} \leq 1$. Also, $B = 10^3$ SP exhibits a few top wigglings. Upshifting T at 10 eV yields back $n_e = 10^{18}\text{cm}^{-3}$ SPs very similar to these displayed on **Figure 2** ($n_e = 10^{16}\ \text{cm}^{-3}$, $T = 10$ eV) **Figure 7**.

Corresponding proton ranges ($n_e = 10^{18}\ \text{cm}^{-3}$, $T = 1$ eV) are shown in **Figure 8**.

Experimentally, accessible and very small ranges are thus documented for $\frac{V}{V_{th}} \leq 1$. Here, $B = 0$ and 20 T data remain everywhere distinguishable.

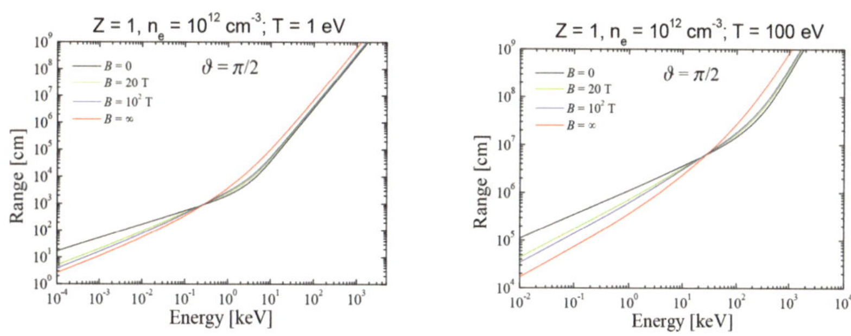

Figure 6. Proton ranges down to the rest of the target with $n_e = 10^{12}\ \text{cm}^{-3}$, $T = 1$, and 100 eV at $\theta = \frac{\pi}{2}$.

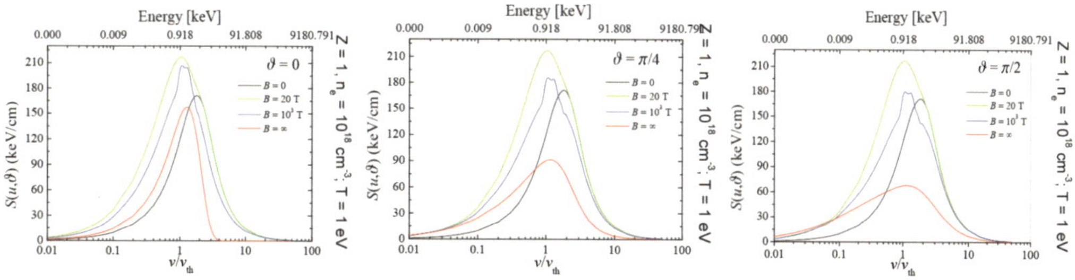

Figure 7. Same as in **Figure 1** for $n_e = 10^{18}\ \text{cm}^{-3}$ and $T = 1$ eV with $\theta = 0, \frac{\pi}{4}$, and $\frac{\pi}{2}$.

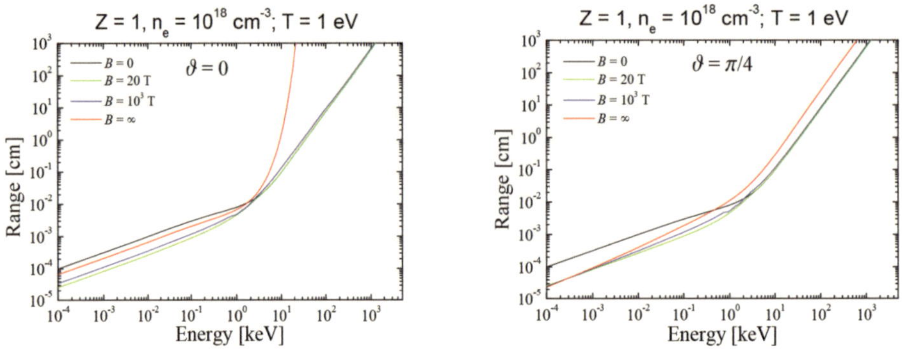

Figure 8. Proton ranges in electron target $n_e = 10^{12}$ cm^{-3} and T = 1 eV with $\theta = 0$ and $\frac{\pi}{4}$.

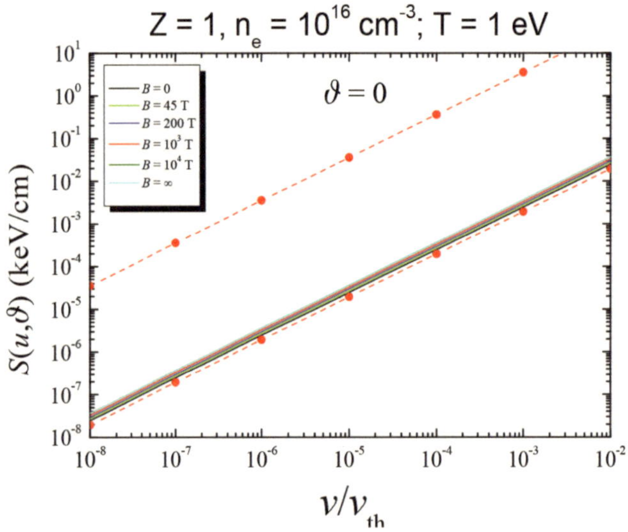

Figure 9. Very-low-velocity proton slowing down on target protons at B = 10^3T (upper straight line) contrasted to target electron stopping (any B, lower straight lines).

5.3. Very-low-velocity proton slowing down

Up to now we limited our investigation to proton stopping by target electrons. In the very-low-velocity regime $V \leq V_{thi}$, the target protons can also contribute significantly as evidenced on **Figure 9**. This topic will be more thoroughly addressed in a separate presentation.

6. Summary

We developed and extensively used a kinetic approach based on a binary collision formulation and suitably regularized Coulomb interaction, to numerically document for any value of the applied magnetization B, the stopping of a proton projectile in a fully ionized hydrogen plasma target. Both ion projectile and target plasma parameters have been selected in order to fit a planned ion-plasma interaction experiment in the presence of an applied magnetic field $\tilde{B}$. It should be pointed out that we restricted the target plasma to its electron component. It therefore remains to include the target ion contribution to proton stopping [63], thus featuring a complete low-velocity ion slowing down.

More generally, we expect that the present investigation, experimentally geared as it is, could help to bridge a long-standing and persisting gap between theoretical speculations and experimental facts in the field of nonrelativistic ion stopping in magnetized target plasmas.

Acknowledgements

The work of H.B.N. has been supported by the State Committee of Science of the Armenian Ministry of Education and Science (Project No. 13-1C200). This work was supported by the Bundesministerium für Bildung und Forschung (BMBF) under contract 06ER9064.

Appendix A: Adjustment of the effective interaction

Our results (Eq. (11)) were derived by using the screened interaction $U_R(r)$. As already mentioned, the use and the modeling of such an effective two-body interaction are a major but indispensable approximation for a BC treatment where the full ion-target interaction is replaced by an accumulation of isolated ion-electron collisions. The replacement of the complicated real non-spherically symmetric potential, like the wake fields as shown and discussed in Ref. [60], with a spherically symmetric one is, however, well motivated by earlier studies on a BC treatment at vanishing magnetic field (see Refs. [53–55]). It was shown by comparison with 3D self-consistent PIC simulations that the drag force from the real nonsymmetric potential induced by the moving ion can be well approximated by an BC treatment employing a symmetric Debye-like potential with an effective velocity-dependent screening length $\lambda(v_i)$. In these studies also a recipe was given how to derive the explicit form of $\lambda(v_i)$, which turned out to be not too much different from a dynamic screening length of the simple form $\lambda(v_i) = \lambda_D\left[1 + (v_i/v_{th})^2\right]^{1/2}$. Here, $\lambda_D = v_{th}/\omega_p$ is the Debye screening length at $v_i = 0$, ω_p is the electron plasma frequency, and v_{th} is a thermal velocity of electrons. Although no systematic studies about the use of such an effective interaction with a screening length $\lambda(v_i)$ have been made for ion stopping in a magnetized electron plasma, the replacement of the real interaction by a velocity-dependent spherical one should be a reasonable approximation also in this case. The introduced dynamical screening length $\lambda(v_i)$ also implies the assumption of a weak perturbation of the electrons by the ion and linear screening where the screening length is independent of the ion charge Ze, which coincide with the regimes of perturbative BC (see, e.g., Ref. [54]). Therefore, we do not consider here possible nonlinear screening effects.

Next, we specify the parameter λ which is a measure of the softening of the interaction potential at short distances. As we discussed in the preceding sections, the regularization of the potential (Eq. (10)) guarantees the existence of the s integrations, but there remains the problem of treating accurately hard collisions. For a perturbative treatment, the change in relative velocity of the particles must be small compared to v_r, and this condition is increasingly difficult to fulfill in the regime $v_r \to 0$. This suggests to enhance the softening of the potential near the origin of the smaller v_r. Within the present perturbative treatment, we employ a dynamical regularization parameter $\lambda(v_i)$ [44, 45], where $\lambda^2(v_i) = Cb_0^2(v_i) + \lambda_0^2$ and $b_0(v_i) = |Z|e^2/m(v_i^2 + v_{\rm th}^2)$. Here, b_0 is the averaged distance of the closest approach of two charged particles in the absence of a magnetic field, and λ_0 is some free parameter. In addition we also introduced $C \simeq 0.292$ in $\lambda(v_i)$. In Refs. [44, 45], this parameter is deduced from the comparison of the second-order scattering cross sections with an exact asymptotic expression derived in Ref. [61] for the Yukawa type (i.e., with $\lambda \to 0$) interaction potential. As we have shown in Refs. [44, 45] employing the dynamical parameter (v_i), the second-order cross sections for electron-electron and electron-ion collisions excellently agree with CTMC simulations at high velocities. Also, the free parameter λ_0 is chosen such that $\lambda_0 \ll b_0(0)$, where $b_0(0)$ is the distance $b_0(v_i)$ at $v_i = 0$. From the definition of (v_i), it can be directly inferred that λ_0 does not play any role at low velocities, while it somewhat affects the size of the stopping force at high velocities when $b_0(v_i) \lesssim \lambda_0$. More details on the parameter λ_0 and its influence on the cooling force are discussed in Ref. [47].

Author details

Hrachya B. Nersisyan[1,2]*, Günter Zwicknagel[3] and Claude Deutsch[4]

*Address all correspondence to: hrachya@irphe.am

1 Plasma Theory Group, Institute of Radiophysics and Electronics, Ashtarak, Armenia

2 Centre of Strong Fields Physics, Yerevan State University, Yerevan, Armenia

3 Institut für Theoretische Physik II, Universität Erlangen-Nürnberg, Erlangen, Germany

4 LPGP, Université Paris-Saclay, Orsay, France

References

[1] Killian TC. Ultracold neutral plasmas. Science. 2007;**316**:705

[2] Deutsch C, Nersisyan HB, Zwicknagel G. Equilibrium and transport in strongly coupled and magnetized ultracold plasmas. In: Shukla PK, Mendonça JT, Eliasson B, Resedes D, editors. Strongly Coupled Ultra-Cold and Quantum Plasmas. Vol. 1421. AIP Conf. Proc; 2012. p. 3

[3] Sørensen AH, Bonderup E. Electron cooling. Nuclear Instruments & Methods. 1983;**215**:27

[4] Poth H. Electron cooling, Theory, Experiment, Application. Physics Reports. 1990;**196**:135

[5] Meshkov IN. Electron cooling, status and perspectives. Physics of Particles and Nuclei. 1994;**25**:631

[6] Men'shikov LI. Electron cooling. Physics-Uspekhi. 2008;**51**:645

[7] Zwicknagel G. Electron cooling of ions and antiprotons in traps. In: Nagaitsev S, Pasquinelli RJ, editors. Beam Cooling and Related Topics. Vol. 821. AIP Conf. Proc.; 2006. p. 513

[8] Nersisyan HB, Toepffer C, Zwicknagel G. Interactions between Charged Particles in a Magnetic Field: A Theoretical Approach to Ion Stopping in Magnetized Plasmas. Heidelberg: Springer; 2007

[9] Möllers B, Walter M, Zwicknagel G, Carli C, Toepffer C. Drag force on ions in magnetized electron plasmas. Nuclear Instruments and Methods in Physics Research Section B: Beam Interactions with Materials and Atoms. 2003;**207**:462

[10] Mollers B, Toepffer C, Walter M, Zwicknagel G, Carli C, Nersisyan HB. Cooling of ions and antiprotons with magnetized electrons. Nuclear Instruments and Methods in Physics Research Section A: Accelerators, Spectrometers, Detectors and Associated Equipment. 2004;**532**:279

[11] Cereceda C, Deutsch C, DePeretti M, Sabatier M, Nersisyan HB. Dielectric response function and stopping power of dens magnetized plasmas. Physics of Plasmas. 2000;**7**:2884

[12] Cereceda C, DePeretti M, Deutsch C. Stopping power for arbitrary angle between test particle velocity and magnetic field. Physics of Plasmas. 2005;**12**:022102

[13] Rostoker N, Rosenbluth MN. Test particle in a completely ionized plasma. Physics of Fluids. 1960;**3**:1

[14] Rostoker N. Kinetic equation with a constant magnetic field. Physics of Fluids. 1960;**3**:922

[15] Akhiezer IA, Eksp Z. Theory of the interaction of a charged particle with a plasma in a magnetic field. Teor. Fiz. 1961;**40**:954; Sov. Phys. JETP 1961;**13**:667

[16] Akhiezer AI, Akhiezer IA, Polovin RV, Sitenko AG, Stepanov KN. Plasma Electrodynamics. 1st ed. Vol. 1. Pergamon: Oxford; 1975

[17] Honda N, Aona O, Kihara T. Fluctuations in a plasma 3: Effect of a magnetic field on stopping power. Journal of the Physical Society of Japan. 1963;**18**:256

[18] May RM, Cramer NF. Test ion energy loss in a plasma with a magnetic field. Physics of Fluids. 1970;**13**:1766

[19] Boine-Frankenheim O, DAvanzo J. Stopping power of ions in a strongly magnetized plasma. Physics of Plasmas. 1996;**3**:792

[20] Nersisyan HB. Stopping of charged particles in a magnetized classical plasma. Physical Review E. 1998;**58**:3686

[21] Nersisyan HB. Stopping of heavy ions in a magnetized strongly coupled plasmas: high velocity limit. Contributions to Plasma Physics. 2005;**45**:46

[22] Nersisyan HB, Deutsch C. Correlated fast ion stopping in magntized classical plasma. Physics Letters A. 1998;**246**:325

[23] Nersisyan HB, Walter M, Zwicknagel G. Stopping power of ions in a magnetized two-temperature plasma. Physical Review E. 2000;**61**:7022

[24] Nersisyan HB, Zwicknagel G, Toepffer C. Energy loss of ions in a magntized plasma: Conformity between linear response and binary collision treatment. Physical Review E. 2003;**67**:026411

[25] Nersisyan HB. Ion stopping in a dense magnetized plasmas. Nuclear Instruments and Methods in Physics Research Section B: Beam Interactions with Materials and Atoms. 2003;**205**:276

[26] Nersisyan HB, Zwicknagel G, Toepffer C. Conformity between linear response and binary collision treatment of an ion energy loss in a magnetized quantum plasma. European Physical Journal D: Atomic, Molecular, Optical and Plasma Physics. 2004;**28**:235

[27] Nersisyan HB, Deutsch C, Das AK. Number-conserving linear response study of low velocity ion stopping in a collisional magnetized classical plasma. Physical Review E. 2011;**83**. DOI: 036403

[28] Nersisyan HB, Deutsch C. Low velocity ion stopping in a a collisional plasma. European Physical Journal: Web of Conferences. 2013;**59**:05019

[29] Nersisyan HB, Deutsch C. Energy loss of ions by electric field fluctuations in a magnetized plasma. Physical Review E. 2011;**83**:066409

[30] Seele C, Zwicknagel G, Toepffer C, Reinhard P-G. Time-dependent stopping power and influence of an infinite magnetic field. Physical Review E. 1998;**57**:3368

[31] Walter M, Toepffer C, Zwicknagel G. Stopping power of ions in a magnetized two-temperature plasmas. Nuclear Instruments and Methods in Physics Research Section B: Beam Interactions with Materials and Atoms. 2000;**168**:347

[32] Walter M. Ph.D. thesis. Ion stopping in a magnetized plasma. University of Erlangen; 2002

[33] Steinberg M, Ortner J. Energy loss of a charged particle in magnetized quantum plasma. Physical Review E. 2001;**63**:046401

[34] Deutsch C, Popoff R. Low ion-velocity slowing down in a strongly magnetized plasma target. Physical Review E. 2008;**78**:056405

[35] Deutsch C, Popoff R. Low ion-velocity slowing down in a classical plasma. Nuclear Instruments and Methods in Physics Research Section A: Accelerators, Spectrometers, Detectors and Associated Equipment. 2009;**606**:212

[36] Deutsch C, Zwicknagel G, Bret A. Ultracold plasmas: A paradigm for strongly coupled and classical electron fluids. Journal of Plasma Physics. 2009;**75**:799

[37] Zwicknagel G. Theory and Simulation of the Interaction of Ions with Plasmas: Nonlinear Stopping, Ion-Ion Correlation Effects and Collisions of Ions with Magnetized Electrons, Habilitation Treatise. University of Erlangen; 2000. [http://www.opus.ub.uni-erlangen.de/opus/volltexte/2008/913/]

[38] Zwicknagel G. Ion-electron collisions in a homogeneous magnetic field. In: Bollinger JB, Spencer RL, Davidson RC, editors. Non-Neutral Plasma Physics III. Vol. 498. AIP Conf. Proc.: 1999. p. 469

[39] Zwicknagel G. Nonlinear energy loss of highly charged heavy ions. Nuclear Instruments and Methods in Physics Research Section A: Accelerators, Spectrometers, Detectors and Associated Equipment. 2000;**441**:44

[40] Zwicknagel G, Toepffer C. Energy loss of ions by collisions with magnetized electrons. In: Anderegg F, Schweikhard L, Driscoll CF, editors. Non-Neutral Plasma Physics IV. Vol. 606. AIP Conf. Proc.; 2002. p. 499

[41] Zwicknagel G. Trapped charged particles and fundamental interactions. In: Blaum K, Herfurth F, editors. Trapped Charged Particles and Fundamental Interactions, Lecture Notes in Physics. Vol. 749. Berlin: Springer-Verlag; 2008. pp. 69-96

[42] Toepffer C. plaScattering of magnetized electrons by ions. Physical Review A. 2002;**66**: 022714

[43] Nersisyan HB, Zwicknagel G. Conformity between linear response and binary collision treatments of an ion energy loss in a magnetized quantum plasma. Journal of Physics A: Mathematical and General. 2004;**37**:11073

[44] Nersisyan HB, Zwicknagel G. Binary collisions of charged particles in a magnetic field. Physical Review E. 2009;**79**:066405

[45] Nersisyan HB, Zwicknagel G. Energy transfer in binary collisions of two gyrating charged particles in a magnetic field. Physics of Plasmas. 2010;**17**:082314

[46] Deutsch C, Nersisyan HB, Bendib A. Diagnostics of arbitrary magnetized and dense plasmas of ICF/WDM concern. In: Proceedings of the 20th International Symposium on Heavy- Ion Inertial Fusion, HIF 2014, Lanzhou, China, 11–15 August, 2014. DOI: http://dx.doi.org/10.1017/S0263034615000919 in Laser Part. Beams 2015

[47] Nersisyan HB, Zwicknagel G. Cooling force on ions in a magnetized electron plasma. Physical Review Accelerators and Beams. 2013;**16**:074201

[48] Winkler T. Ph.D. thesis; Electron cooling of highly charged ions. University Heidelberg; 1996

[49] Winkler T et al. Electron cooling force measurements for highly charged ions in the ESR. Hyperfine Interactions. 1996;**99**:277

[50] Winkler T, Beckert K, Bosch F, Eickhoff H, Franzke B, Nolden F, Reich H, Schlitt B, Steck M. Electron cooling of highly chargd ions in the ESR. Nuclear Instruments and Methods in Physics Research Section A: Accelerators, Spectrometers, Detectors and Associated Equipment. 1997;**391**:12

[51] Kelbg G. Theorie des quanten plasmas. Annalen der Physik (Berlin). 1963;**467**:219

[52] Deutsch C. Nodal expansion in a real matter plasma. Physics Letters A. 1977;**60**:317

[53] Zwicknagel G. Theory and simulation of heavy ion stopping in plasma. Laser and Particle Beams. 2009;**27**:399

[54] Zwicknagel G. Nonlinear energy loss of heavy ions in plasmas. Nuclear Instruments and Methods in Physics Research Section B: Beam Interactions with Materials and Atoms. 2002;**197**:22

[55] Zwicknagel G, Toepffer C, Reinhard P-G. Stopping of heavy ions in plasmas at strong coupling. Physics Reports. 1999;**309**:117; 1999;**314**:671(E)

[56] Gradshteyn IS, Ryzhik IM. Tables of Integrals, Series and Products. 2nd ed. New York: Academic; 1980

[57] Derbenev YS, Skrinsky AN. The effect of a magnetic field on electron cooling. Particle Accelerator. 1978;**8**:235

[58] Parkhomchuk VV. In: Poth H, editor. Proceedings of the Workshop on Electron Cooling and Related Applications (ECOOL84, 1984). Karlsruhe: KfK; 1985. p. 71; KfK Report No. 3846

[59] Pestrikov DV. Longitudinal cooling force due to magnetized electrons. Nuclear Instruments and Methods in Physics Research Section A: Accelerators, Spectrometers, Detectors and Associated Equipment. 2005;**554**:13

[60] Peter T, Meyer-ter-Vehn J. Energy loss of heavy ions in dense plasmas 1. Linear and nonlinear Vlasov theory for the stopping power. Physical Review A. 1991;**43**:1998

[61] Hahn H, Mason EA, Smith FJ. Quantum transport cross sections for ionized gases. Physics of Fluids. 1971;**14**:278

[62] Chen S, Fuchs J. Ecole Polytechnique. Private Communication; July 2014

[63] Tashev B, Baimbetov F, Deutsch C, Fromy P. Low velocity ion stopping in binary ionic mixtures. Physics of Plasmas. 2008;**15**:102701

Chapter 5

Mode Transition and Hysteresis in Inductively Coupled Plasma Sources

Shu-Xia Zhao

Additional information is available at the end of the chapter

http://dx.doi.org/10.5772/intechopen.76654

Abstract

In this chapter, the characteristics of low-temperature inductively coupled plasma sources, that is, non-equilibrium, weakly ionized and bounded plasma, are described. The phenomenon of mode transition and hysteresis is one of the main physics aspects that happens in this source. Via a hybrid model, the behaviors of plasma parameters, electron kinetics and neutral species during mode transition are presented. Still, the role of metastables and multistep ionization on triggering the hysteresis is investigated. Using a fluid model that couples the equivalent circuit module, the discontinuity of mode transition and hysteresis are observed by tuning the matching network impedance. The work indicates the mutual interaction between the plasma and the circuit excites hysteresis. Besides these findings, the other important aspects of this phenomenon are briefly discussed. To the author, the exploration on the precursors that trigger hysteresis is the most attractive topic. The investigations advance the improvement of analytical theory, numerical modeling and experimental diagnostics of low-temperature plasma physics.

Keywords: low-temperature plasma, inductively coupled plasma, mode transition and hysteresis, hybrid model, fluid model, equivalent circuit, multistep ionizations

1. Introduction

The inductively coupled plasma (ICP) source is one of the most important low-temperature plasma sources that find widespread applications in many fields [1], such as plasma photonic crystals, synthesis of nanomaterials and nanostructured materials, atomic layer processing, agriculture and innovative food cycles, medicines, environments, plasma-assisted combustion and chemical conversion and aerospace application (propulsion and flow control) and so on. Driven within the domains of radio frequency electromagnetic and rather low-pressure

(~mTorrs) ranges, the ICP sources present several advantages, such as high-plasma density, high anisotropy in the sheath, independent control of incident flux density and energy and simple low-cost reactor configuration (unwanted for the static magnetic devices) over some other plasma sources, such as capacitively coupled plasma and electron cyclotron resonance reactor [2]. As compared to the atmospheric discharges, this sort of low-pressure radio frequency plasma sources are known for their non-equilibrium properties, that is, $T_e \gg T_i > T_n$, where T_e, T_i and T_n are temperatures of electrons, ions and neutrals, respectively [3], due to the low-temperature peculiarity of this type of plasma source. Another essential feature is its weak ionization degree that ensures the abundance of collisions and reactions between charged species and neutrals, which is quite different with the high-temperature fully ionized plasmas where only the Coulomb interactions between charged species are important [4]. Of great importance is the diversity in the mutual interactions among charged and neutral species, which are classified into elastic and inelastic collisions with respect to the principle of kinetic energy balance. Regarding species specialty and colliding outcomes, the inelastic collisions can be described as type (1) ionization, dissociation, electronic, rotational, vibrational excitation, attachment, detachment and de–excitation, which mainly occur between electrons and neutrals; type (2) recombination, associations, charge exchange, excitation transfer and penning ionizations, which mainly happen among heavy species (meant to all species except for electrons); and type (3) the spontaneous radiation from excited state species (without a trigger) [5]. The elastic scattering to some extent determines plasma transport process and hence spatial characteristics of plasma via the parameter of momentum transfer collision frequency, while the inelastic collisions that sustain the weakly ionized plasma mainly determine the energy loss mechanism and give steady-state plasma components optical emission. Finally, all low-temperature plasma sources are generated in chambers with their respective fixed configurations and more importantly with limit space dimension. This means that all the plasmas are bounded plasmas, as compared with the space plasma; therefore, the sheath, produced on all bound surfaces, forms one important constituent of low-temperature plasma physics [6]. In a word, non-equilibrium, weak ionization and plasma bounds characterize the low-pressure radio frequency plasma source as complicated and multi-disciplinary.

Even with the above complexity, rich and fruitful interesting physics phenomena and mechanisms are already revealed in these low-pressure and radio frequency plasma sources via present efforts. In particular, in the ICP sources, pulsed radio frequency power source [7], standing wave effects [8], nonlinear harmonics [9], double coil discharges [10], anomalous skin effects [11], nonlocal electron kinetics [12], mode transition and hysteresis [13] and so on are still or have been hot research frontiers that draw attention. In this chapter, the mode transitions and hysteresis topic is focused upon. This topic has been historically studied well and continually occupies people's attention due to its complexity of the multi-factor interactions and potential application in achieving stable plasma sources for the processing technique. The ICP source is famous for its capacity of operating at two different modes, that is, capacitive and inductive modes. The capacitive mode is sustained by radial and axial electromagnetic fields, analogous to conventional capacitively coupled plasma source that is excited by the electrostatic field and hence is abbreviated as the E mode. The inductive mode is sustained by the azimuthal electromagnetic field caused by coil current and is abbreviated as H mode. Remember that the power source applied to the coil is temporally varied in the range of radio frequency. At low-coil power, the ICP source is maintained at E mode, where the plasma density and optical emissions are weak, and the glow area of discharge is more localized under the coil. As we increase the coil power, the discharge transfers abruptly or smoothly toward H mode, where the plasma density and current are significantly increased and the optical emission is strengthened. Moreover, the discharge is more uniform. Interestingly, at certain circumstances, when cycling the power source, the trajectories of plasma parameters versus upward and downward powers don't coincide; hence, hysteresis is formed and the ICP source is therefore famous for its other feature, that is, the existence of two stable states at one fixed power value. In labs of academia or enterprise, the ICP sources are triggered from the E mode at the beginning and then transferred to H mode. Most of the plasma processing techniques prefer to be conducted in the H mode due to its better plasma properties. Therefore, understanding the E–H mode transition and hysteresis is very meaningful to the related industry.

This chapter is outlined as follows. In Section 2, the major achievements of the author on this topic are presented. Three subtopics and the used methodology are discussed and described, aimed at demonstrating to readers the characteristics of plasma parameters, electron kinetics and neutral species during mode transition and excitation of discontinuous mode transition and hysteresis by the external circuit. Finally, the conclusion and further remarks are given in Section 3.

2. Theoretical and experimental investigations of mode transitions and hysteresis: An overview

2.1. Characteristics of basic plasma parameters

In this part, the characteristics of electron parameters, density, temperature and energy distribution function and plasma potential at two modes are presented via the two-dimensional hybrid model [14]. The hybrid model consists of three parts, that is, fluid module, electron Monte Carlo module and electromagnetic module. Species density and momentum, together with the electrostatic field generated by net charge density (analogous to ambipolar diffusion field), are given by the fluid module. Electron transport and collision coefficients, and the effective electron temperature, are calculated through the Monte Carlo method and then transferred to the fluid module. The electromagnetic module calculates the electromagnetic field generated via the coil current and voltage through the Maxwellian's equations, based on the electron conductivity from the fluid module. Both the electrostatic and electromagnetic fields are sent to the Monte Carlo module to push the electrons via Newton's law. The interactions of three modules are illustrated by the model flowchart in **Figure 1**. The three modules are iterated with each other until a final steady state is achieved. In this chapter, a cylindrical inductively coupled plasma reactor with planar coil is used, as shown in **Figure 2**.

In **Figures 3** and **4**, the calculated electron density and temperature profiles versus coil current at the pressure of 20 mTorr are presented. In **Figure 3**, at low-coil current, 10 A, the density

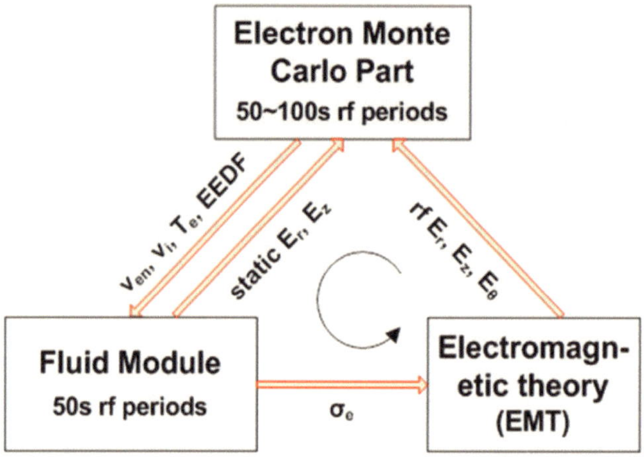

Figure 1. Flowchart of the hybrid model.

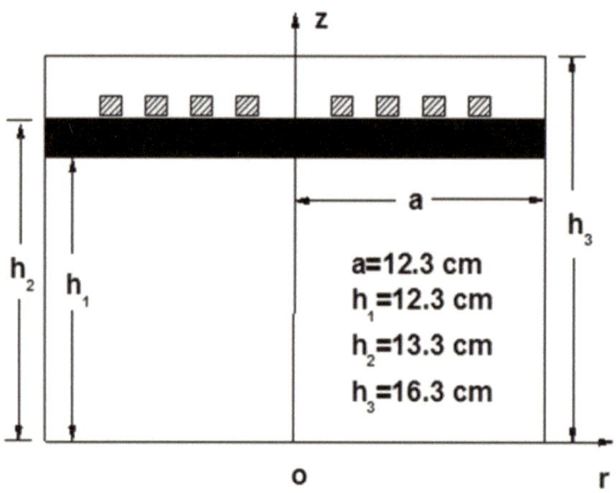

Figure 2. Schematic of the cylindrical inductively coupled discharge configuration.

magnitude is low and the profile is smooth. At high coil current, 40 A, the density magnitude is high, more or less four factors higher than the 10 A case. Meanwhile, the density is peaked under the coil, as referred to the reactor in **Figure 2**. The E–H mode transition happened along with increase in the coil current. In **Figure 4**, at E mode, that is, 10 A, the electron temperature is high around the plasma chamber bound but sinks at the discharge center region. This is because the ambipolar diffusion potential barrier suppresses the electrons from entering the sheath for heating due to the lack of elastic collisions at low pressure. As known, this is a representative feature for the capacitive discharges [15]. At the H mode, that is, 40 A, the temperature profile is substantially changed. It peaks under the coil and more or less decreases toward the center, bottom and sidewall. Besides, the sink area of the temperature profile is significantly shrunk and moves toward the coil, as compared to the E mode. The appearance of temperature sink at different coil currents and its alteration with coil current is related to the

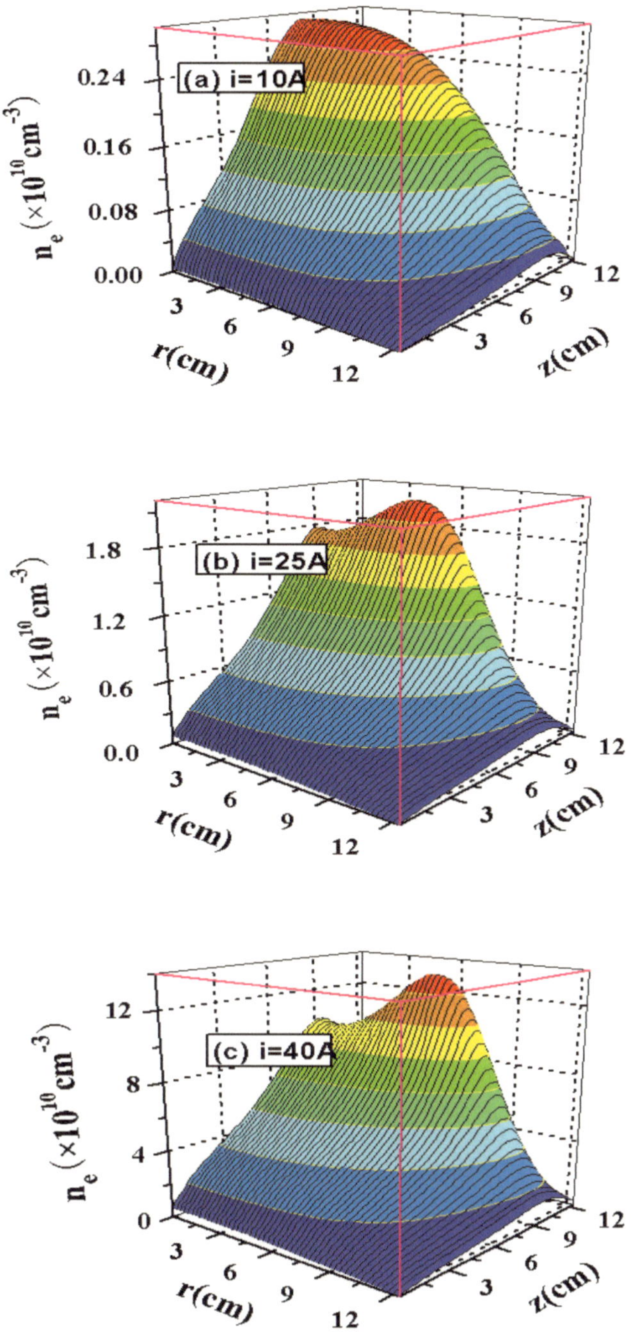

Figure 3. Electron density n_e profile versus coil current at the pressure of 20 mTorr.

spatial potential distribution in **Figure 5**, where the potential barrier is shifted from the discharge center to the coil with the coil current and meanwhile the area of potential barrier is decreased.

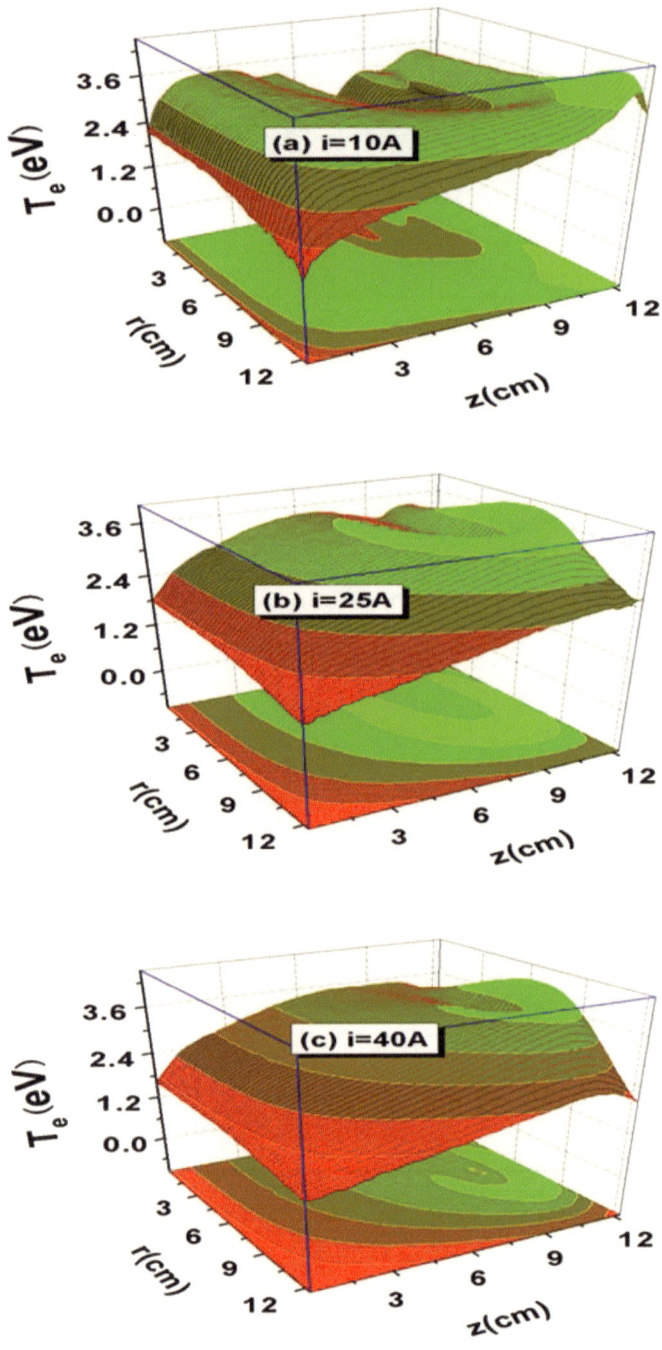

Figure 4. Electron temperature T_e profile versus coil current at the pressure of 20 mTorr.

In **Figure 6**, the electron energy distribution functions (EEDFs) of E and H modes, sampled at the discharge center, are compared at different pressures. At low pressure, that is, 20 mTorr, a prominent low-energy peak is found in the EEDF of the E mode due to the suppression of

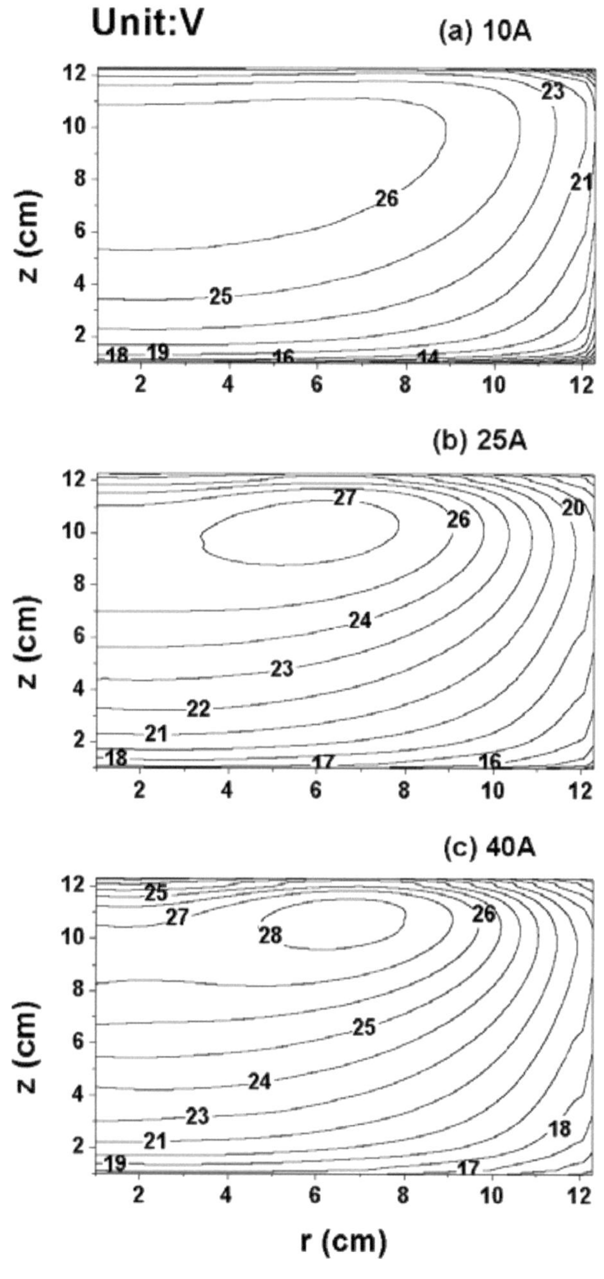

Figure 5. Plasma potential profile versus coil current at the pressure of 20 mTorr.

potential barrier, and it disappears at H mode because the barrier shifts toward the coil. At high pressures, that is, 50 and 100 mTorr, the EEDFs evolve to an opposite way, that is, low-energy electrons' amount of the H mode is higher than the E mode. This is because at high pressures the suppression of the potential barrier is not important anymore due to the frequent elastic collisions between electrons and neutrals at high electron densities. Hence in the electron

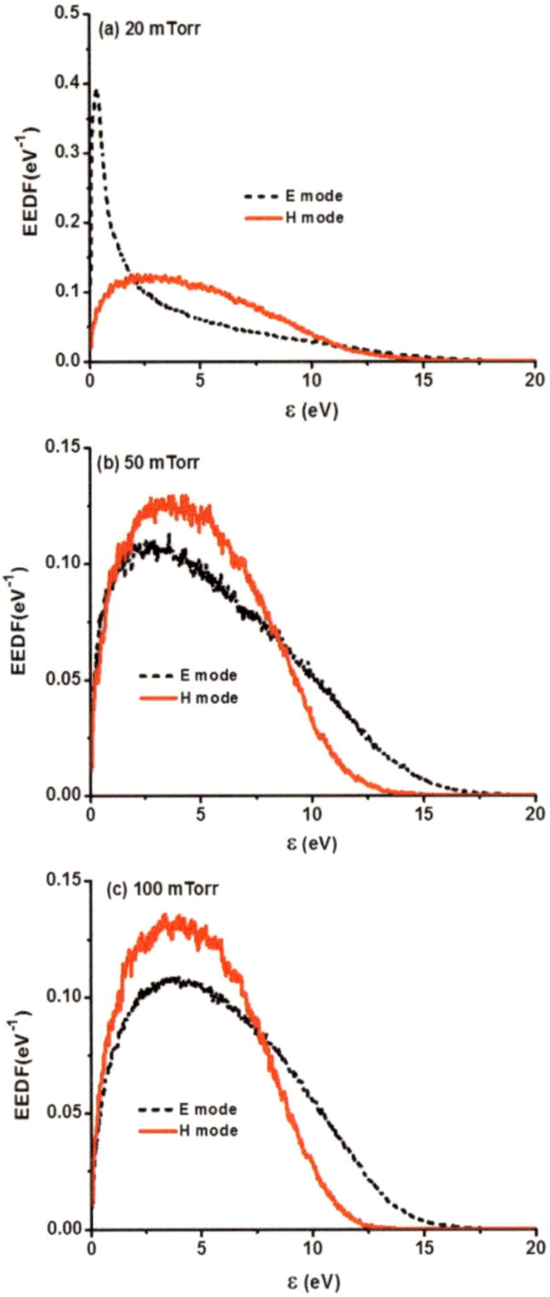

Figure 6. Comparisons of electron energy distribution function (EEDF) of E and H modes at different pressures. The EEDFs are sampled at the discharge center.

temperature profile (see **Figure 7**), the sink region disappears. The temperature profiles of E and H mode are representative of the capacitive and inductive discharges, and, as is well known, the temperature value in the E mode is higher than in the H mode [16]. To demonstrate the electron kinetics better, in **Figure 8**, the electron energy probability function (EEPF) variation with coil

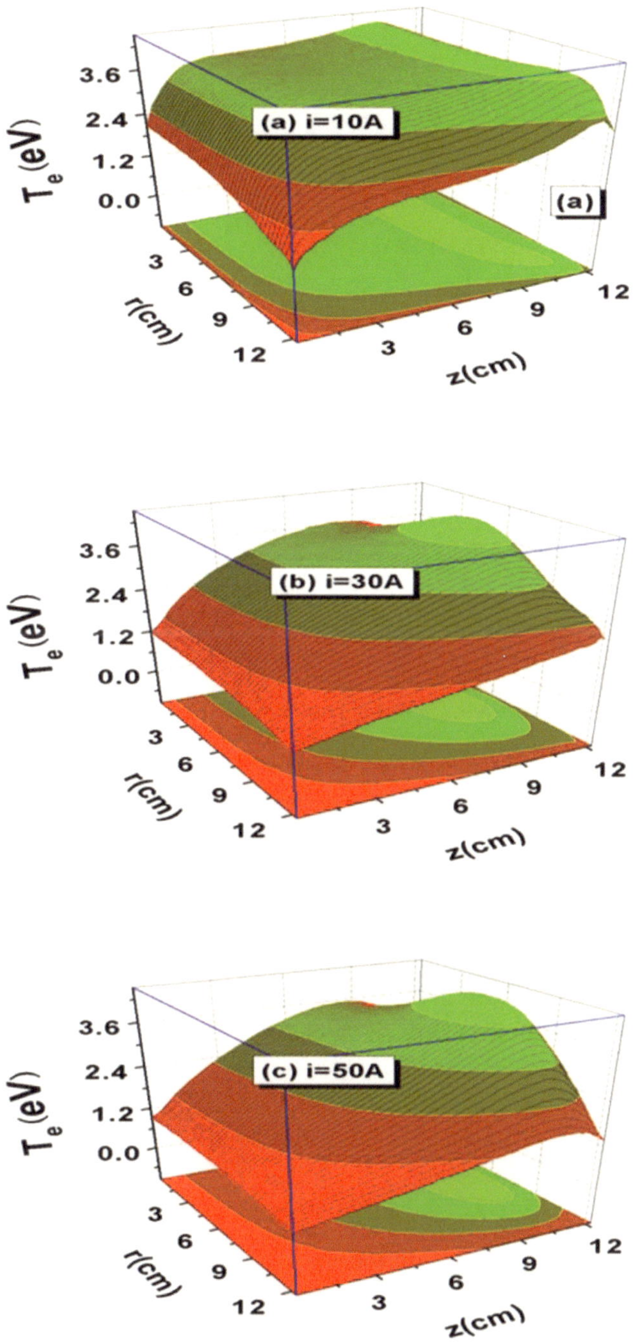

Figure 7. Electron temperature T_e profile versus coil current at the pressure of 100 mTorr.

current at low pressure of 20 mTorr is shown. Clearly, at the E mode, that is, with the coil current less than and equal to 20 A, the obvious three-temperature Maxwellian distribution is observed. The low-energy electron peak, as mentioned before, is formed by the suppression of the potential

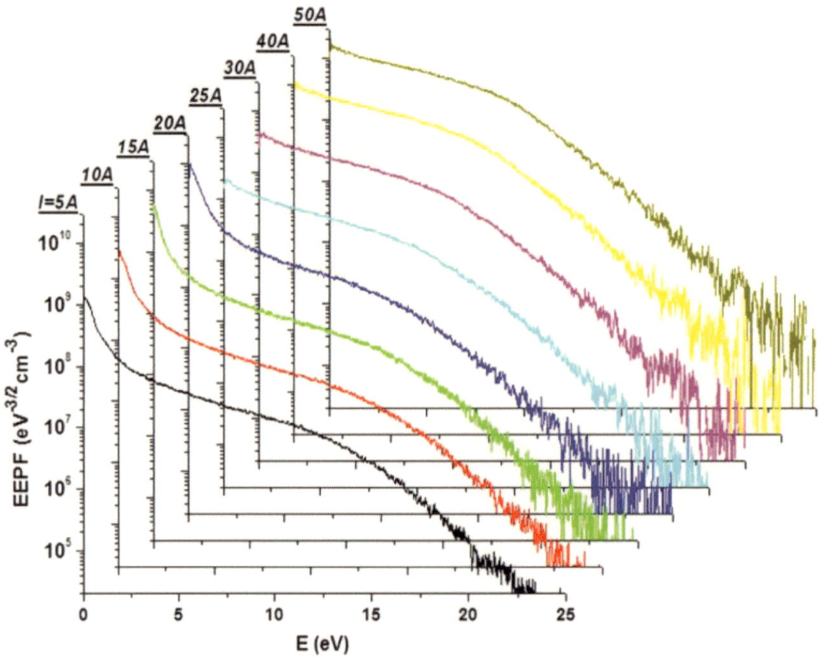

Figure 8. Evolution of electron energy probability (EEPF) against coil current at the pressure of 20 mTorr.

barrier, while the depletion of high-energy electrons tail is caused by the inelastic electron-neutral collisions, such as excitations and ionizations. At H mode, that is, with coil current equal to and larger than 25 A, the two-temperature Maxwellian EEDF in the elastic collision energy range, that is, less than 11.56 eV (the excitation threshold), now disappears due to high electron density and frequent collisions, and the high-energy electrons' depletion via inelastic collisions still exists because the electron density is not high enough for the e–e Coulomb collisions thermalizing these two electron swarms [17, 18].

In a word, the hybrid model successfully captures the main characteristics of plasma parameters during the mode transition, including both the macroscopic plasma properties and microscopic electron kinetics, and all these predictions presented here agree well with the experimental measurements.

2.2. Behavior of metastable neutrals

The behavior of metastable neutrals during the mode transition is investigated by the above hybrid model, with the advanced reaction set that includes the metastables and all relevant reactions [19]. In **Figure 9**, the metastable densities, sampled at the discharge center, versus applied power at different pressures are presented. The densities at low and high pressures, that is, 30 and 300 mTorr, both first increase and then decrease with the power, and the decreasing trend at high pressure is more obvious. Hence, the metastables density increases with power at E mode while decreases with power at H mode, which is different with the electron density trend in Section 2.1. In **Figure 10**, the metastables density profiles at different

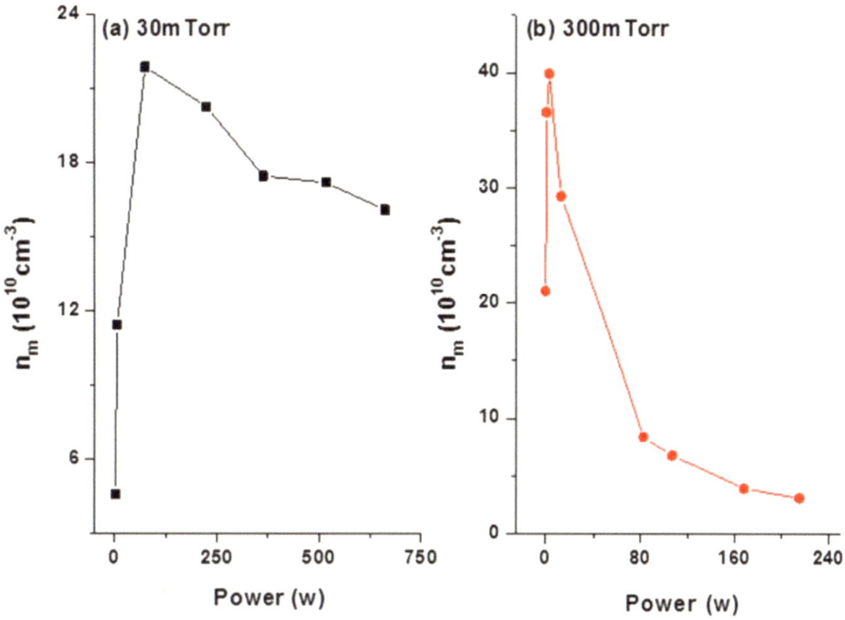

Figure 9. Metastables densities, sampled at the discharge center, versus power at different pressures.

coil currents are presented. It is shown that the peak density keeps increasing with coil current, however, the peak location basically shifts from the discharge center at E mode toward the coil at H mode, thus leading to the non-monotonic varying trends of metastables' densities at the discharge center in **Figure 9**. The localizing trend of metastables density to the coil with coil current is caused by the fact that multistep ionization becomes more and more important as the plasma density is increased, to an extent when the negative source, that is, multistep ionization rate larger than the excitation rate, is formed. The stationary metastables continuity equation with a negative source can be characterized as Bessel's equation with imaginary argument that shows spatial characteristics analogous to the localized profile. This localizing effect is more important at high pressure due to the prevalence of multistep ionizations; hence, the decreasing trend of metastables density with power is more obvious at high pressures. Last, the model predicted a non-monotonic variation of metastables density during mode transition which agrees well with the experiment [20].

Besides the exploration of metastables evolution along with mode transition, the role of metastables in determining the hysteresis is still investigated through the hybrid model. The behind mechanisms that generate hysteresis are difficultly identified since it is a process that is interplayed by so many elements. In the literature, many papers ascribed the hysteresis to the multistep ionizations [13, 21]. To assess this argument, in **Figure 11**, the influence of metastables on electron density and temperature variations versus power is presented. Inclusion of metastables and multistep ionization overall elevates electron densities and meanwhile reduces electron temperatures against the power; however, it does not trigger hysteresis. Besides, the metastables change the trend of electron temperature with power at the H mode.

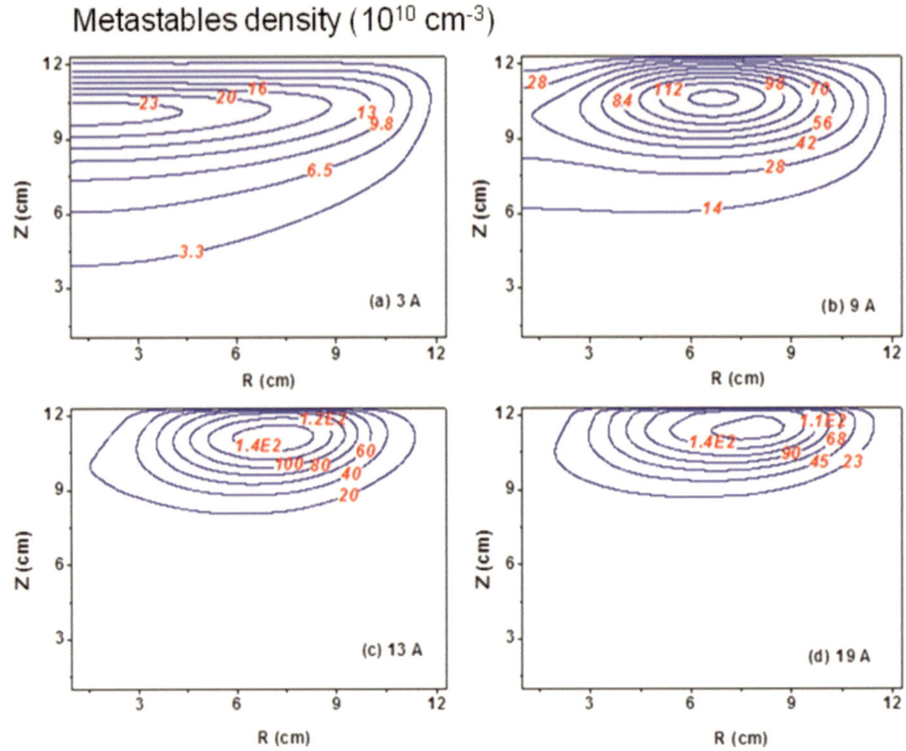

Figure 10. Metastables density profiles at different coil currents at the pressure of 100 mTorr.

The decrease of temperature with power when including metastables is caused by the fact that ionizations consume electron kinetic energy more effectively than excitations, as revealed by a novel electron mean energy Equation [22].

2.3. Discontinuous mode transition and hysteresis excited by matching network

The discontinuity feature of mode transition and interesting hysteresis phenomena have attracted people's attention for years. They are easily observed in the experiments [16, 23] and can be analytically predicted by stationary zero-dimensional global model [24, 25]. However, it is difficult for the self-consistent multidimensional models to capture the discontinuity and hysteresis unless the external circuit module is taken into account. In this chapter, the conventional fluid model that describes the pure inductive mode is extended by including the capacitive mode and advanced by introducing an equivalent circuit module [26, 27]. The diagram of equivalent circuit is illustrated in **Figure 12**. It consists of radio frequency (RF) power source, matching network that consists of parallel and series capacitors and capacitive and inductive coupling branches. The capacitive coupling components include dielectric window capacitor, sheath transferred capacitor and bulk plasma-transferred resistance. The inductive coupling branch is actually based on a transformer model [13], where the coil itself and plasma-transferred inductor and resistor are included and the relations between the coil and plasma-transferred

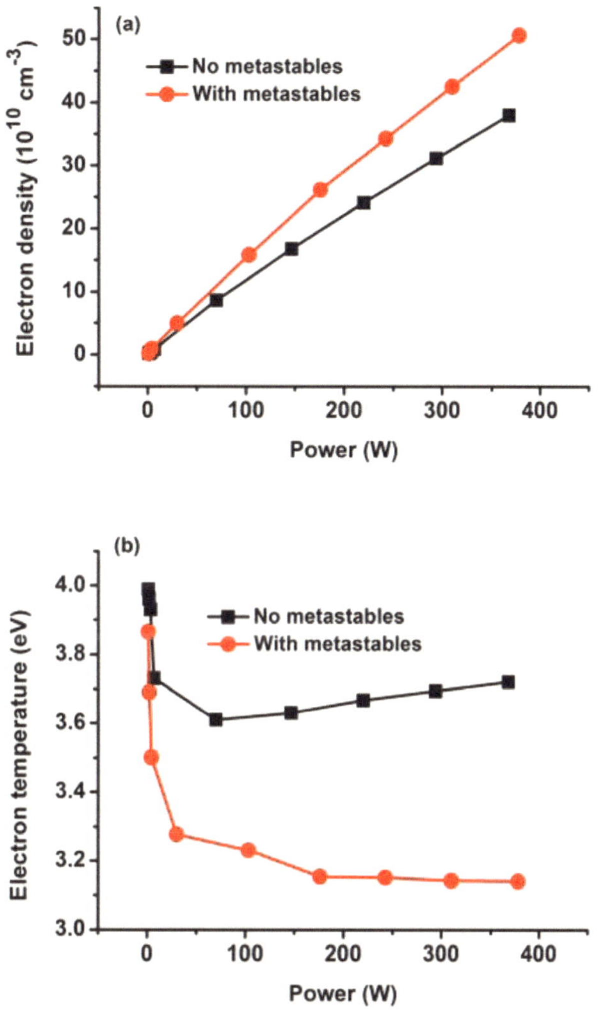

Figure 11. Electron densities (a) and temperatures (b) versus power at two cases, i.e., (1) no metastables and (2) with metastables in the model.

impedances are illustrated in the square of **Figure 12**. The plasma resistances in the capacitive and inductive branches are both transferred through the Ohm's heating mechanism but the capacitive resistance is based on radial and axial plasma current components [3] and the inductive resistance on the azimuthal current component [28].

Via the circuit module, the coil current and voltage, boundary conditions for the electromagnetic module to calculate fields can be given through Kirchhoff's law. More importantly, after considering the circuit module the discontinuity of mode transition and hysteresis can be captured by a fluid model since the mutual interacting details between the circuit and plasma, probably nonlinear, are contained. Of more significance is that two excitation means of mode transitions that have widely been seen in experiments, that is, by means of varying power

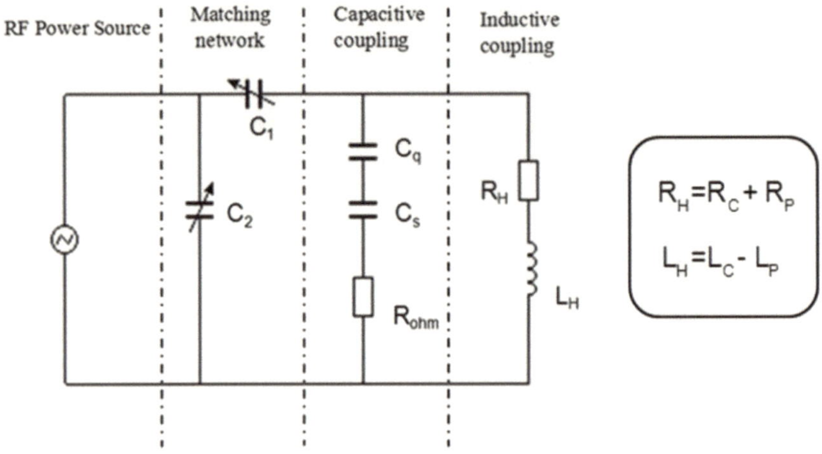

Figure 12. Components of equivalent circuit module.

[16, 23] and matching network [29], can be both captured by this advanced fluid model that couples an external circuit module.

In **Figure 13**, the discontinuous mode transition at a low pressure of 20 mTorr and hysteresis at high pressure of 100 mTorr is perfectly generated by the fluid model, via the alteration of electron density versus series capacitance of matching network. This prediction agrees well with the experimental observations that hysteresis mostly appears at relatively high pressures [21, 23]. Accordingly, the electron temperature just displays mode transition at low pressure, but hysteresis at high pressure, as shown in **Figure 14**. The variations of electron density and temperature with E–H mode transition predicted by the fluid model are in agreement with the hybrid model in Section 2.1.

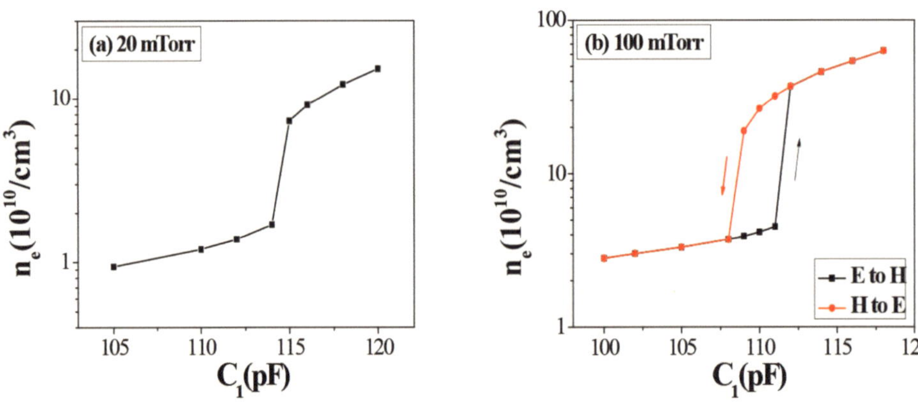

Figure 13. Discontinuous electron density variation versus the series capacitance of matching network at low pressure of 20 mTorr (a) and hysteresis loop of electron density against the series capacitance at high pressure of 100 mTorr.

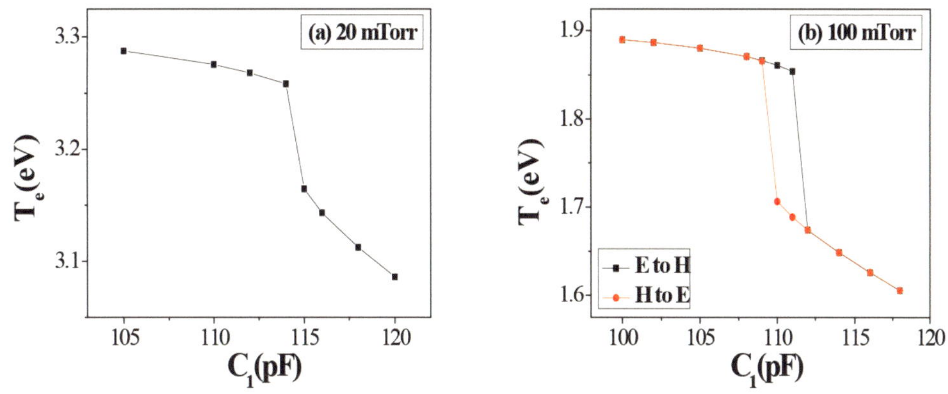

Figure 14. Discontinuous electron temperature variation versus the series capacitance of matching network at low pressure of 20 mTorr (a) and hysteresis loop of electron temperature against the series capacitance at high pressure of 100 mTorr.

Interestingly, the plasma-transferred impedance evolves similarly to the plasma parameters, that is, discontinuously jumping at low pressure and displaying hysteresis at high pressure. In **Figures 15** and **16**, the plasma resistance and inductance of inductive branch and sheath width and capacitance of capacitive branch are plotted against the series capacitance, respectively, at the high pressure of 100 mTorr. In **Figure 15**, at the E–H mode transition, the plasma resistance and inductance both increase because of high-plasma density and strong azimuthal current density. The high-plasma inductance at the H mode weakens the system inductance according to the formula in **Figure 12**, as determined by the law of electromagnetic induction. In **Figure 16**, the sheath width significantly decreases with E–H mode transition due to the scaling law [6] and the sheath transferred capacitance, inversely proportional to mean sheath width, increases substantially. At the H–E mode transition of the hysteresis loop, the opposite cases happen.

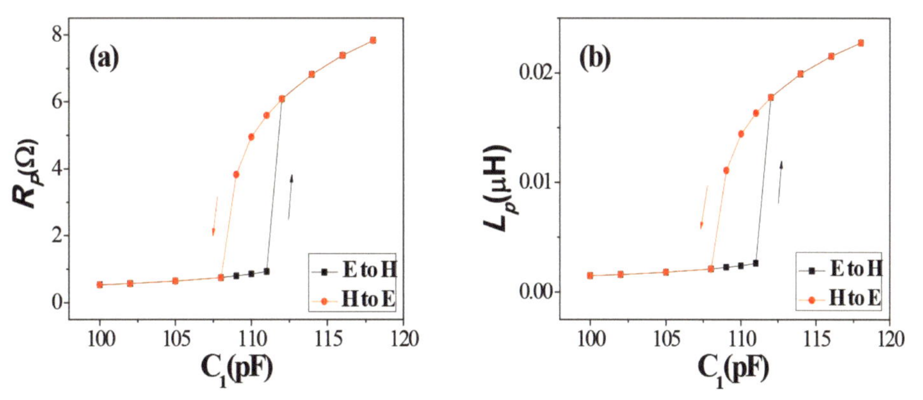

Figure 15. Variations of plasma-transferred resistance (a) and inductance (b) of the inductive branch of equivalent circuit in the hysteresis loop at high pressure of 100 mTorr.

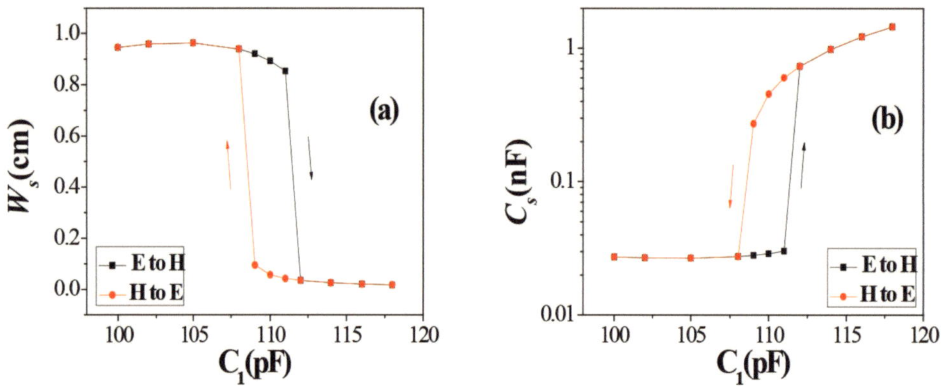

Figure 16. Variations of sheath width (a) and transferred sheath capacitance (b) in the hysteresis loop at high pressure of 100 mTorr.

3. Conclusion and further remarks

The low-pressure radio frequency ICP source is characterized as non-equilibrium, weakly ionized and bounded plasma and finds wide applications in many fields. It holds many interesting physical phenomena and mechanisms. One is the mode transition and hysteresis that happen at two operating modes, that is, inductive and capacitive modes. In this chapter, the characteristics of plasma parameters and neutrals during mode transition are presented by a hybrid model. Moreover, the discontinuity feature of mode transition and hysteresis excited by adjusting the matching network are predicted by a fluid model that couples an external equivalent circuit module. Still, the role of metastables on triggering hysteresis is discussed and the interesting hysteresis loop formed by plasma-transferred impedance is analyzed. The present chapter indicates that the mutual interaction of plasma with circuit is the reason which excites the hysteresis.

Note that the mode transitions and hysteresis of ICP sources are very complicated. Besides the above representative features, it still exhibits research values in the topics of reactive gas mixtures, such as O_2 [30], CF_4/Ar [31], SO_2 [32], ammonia [33] and so on and double hysteresis loop [29], inverse hysteresis [34], spatial characteristics [35], optical emission [36], electrical diagnostics [37], instability of electronegative plasma source [38] and so on. To the author, the exploration of precursors that triggers hysteresis, for instance, metastables and multistep ionizations [13, 21], electron energy distribution function [39], power coupling efficiency [40], sheath [24, 41], external circuit [26, 27] and nonlinear mechanisms [13] and so on, is the most attractive topic. The investigations greatly advance the improvements of analytical theory, numerical modeling, and experimental diagnostics of low-temperature plasma physics.

Acknowledgements

Gratitude is given due to the support by National Natural Science Foundations of China (Grant No. 11305023), and the useful discussions with Hui-Jing Xu are especially acknowledged.

Author details

Shu-Xia Zhao

Address all correspondence to: zhaonie@dlut.edu.cn

Dalian University of Technology, Dalian, China

References

[1] Adamovich I, Baalrud SD, Bogaerts A, Bruggeman PJ, Cappelli M, Colombo V, Czarnetzki U, Ebert U, Eden JG, Favia P, Graves DB, Hamaguchi S, Heiftje G, Hori M, Kaganovich ID, Kortshagen U, Kushner MJ, Mason NJ, Mazouffre S, Mededovic Thagard S, Metelmann HR, Mizuno A, Moreau E, Murphy AB, Neimira BA, Oehrlein GS, Petrovic ZL, Pitchford LC, Pu YK, Rauf S, Sakai O, Samukawa S, Starikovskaia S, Tennyson J, Terashima K, Turner MM, van de Sanden MCM, Vardelle A: The 2017 plasma roadmap: Low temperature plasma science and technology. Journal of Physics D: Applied Physics 2017;**50**:323001. DOI: 10.1088/1361-6463/aa76f5

[2] Zhao SX, Xu X, Li XC, Wang YN. Fluid simulation of the E-H mode transition in inductively coupled plasma. Journal of Applied Physics. 2009;**105**:083306. DOI: 10.1063/1.311 2009

[3] Chabert P, Braithwite N. Physics of Radio-Frequency Plasmas. New York: Cambridge University Press; 2011.17p

[4] Bittencourt JA. Fundamentals of Plasma Physics. 3rd ed. Berlin/Heidelberg/New York: Springer-Verlag; 2004.580p

[5] Tennyson J, Rahimi S, Hill C, Tse L, Vibhakar A, Akello-Egwel D, Brown DB, Dzarasova A, Hamilton JR, Kaksch D, Mohr S, Wren-Little K, Bruckneier J, Agarwal A, Bartschat K, Bogaerts A, Booth JP, Goeckner MJ, Hassouni K, Itikawa Y, Graams BJ, Krishnakumar E, Laricchiuta A, Mason NJ, Pandey S, Petrovic ZL, Pu YK, Ranjan A, Rauf S, Schulze J, Turner MM, Ventzek P, Whitehead JC, Yoon JS. QDB: A new database of plasma chemistries and reactions. Plasma Sources and Science Technology. 2017;**26**:055014. DOI: 10.1088/1361-6595/aa6669

[6] Lieberman MA, Lichtenberg AJ. Principles of Plasma Discharges and Materials Processing. 2nd ed. New York: Wiley-Interscience; 2005.387p

[7] Economou DJ. Pulsed plasma etching for semiconductor manufacturing. Journal of Physics D: Applied Physics. 2014;**47**:303001. DOI: 10.1088/0022-3727/47/30/303001

[8] Lee SH, Cho JH, Huh SR, Kim GH. Standing wave effect on plasma distribution in an inductively coupled plasma source with a short antenna. Journal of Physics D: Applied Physics. 2014;**47**:015205. DOI: 10.1088/0022-3727/47/1/015205

[9] Si XJ, Zhao SX, Xu X, Bogaerts, Wang YN. Fluid simulations of frequency effects on nonlinear harmonics in inductively coupled plasma. Physics of Plasmas. 2011;**18**:033504. DOI: 10.1063/1.3566007

[10] Mishra A, Kim TH, Kim KN, Yeom GY. Mass spectrometric study of discharges produced by a large-area dual-frequency-dual-antenna inductively coupled plasma source. Journal of Physics D: Applied Physics. 2012;**45**:475201. DOI: 10.1088/0022-3727/45/47/475201

[11] Tyshetskiy YO, Smolyakov AI, Godyak VA. Reduction of electron heating in the low-frequency anomalous-skin-effect regime. Physical Review Letters. 2003;**90**:255002. DOI: 10.1103/PhysRevLett.90.255002

[12] Hagelaar GJM. Fluid description of non-local electron kinetics in inductively coupled plasmas. Plasma Sources Science and Technology. 2008;**17**:025017. DOI: 10.1088/0963-0252/17/2/025017

[13] Turner MM, Lieberman MA. Hysteresis and the E-to-H mode transition in radiofrequency inductive discharges. Plasma Sources Science and Technology. 1999;**8**:313-324. DOI: 10.1088/0963-0252/17/2/025017

[14] Zhao SX, Gao F, Wang YN. Dynamic investigation of mode transition in inductively coupled plasma with a hybrid mode. Journal of Physics D: Applied Physics. 2009;**42**:225203. DOI: 10.1088/0022-3727/42/22/225203

[15] Godyak VA, Piejak RB. Abnormally low electron energy and heating-mode transition in a low-pressure argon rf discharge at 13.56MHz. Physical Review Letters. 1990;**65**:996-999

[16] Ostrikov KN, Xu S, Yu MY. Power transfer and mode transitions in low-frequency inductively coupled plasmas. Journal of Applied Physics. 2000;**88**:2268-2271

[17] Gao F, Zhao SX, Li XS, Wang YN. Comparison between experiment and simulation for argon inductively coupled plasma. Journal of Applied Physics. 2000;**88**:2268-2271

[18] Vasenkov A, Kushner MJ. Electron energy distributions and anomalous skin effects in high-plasma-density inductively coupled discharges. Physical Review E. 2002;**66**:066411

[19] Zhao SX, Wang YN. Investigation of the effect of metastable atoms on mode transition in argon inductive discharge via a hybrid model. Journal of Physics D: Applied Physics. 2010;**43**:275203. DOI: 10.1088/0022-3727/43/27/275203

[20] Daltrini AM, Moshkalev SA, Morgan TJ, Piejak RB, Graham WG. Plasma power measurement and hysteresis in the E-H transition of a rf inductively coupled plasma system. Applied Physics Letters. 2008;**92**:061504. DOI: 10.1063/1.2844885

[21] Lee MH, Lee KH, Hyun DS, Chung CW. On the hysteresis in E to H and H to E transitions and the multistep ionization in inductively coupled plasma. Applied Physics Letters. 2007;**90**:191502. DOI: 10.1063/1.2734501

[22] Zhao SX. Non-monotonic behavior of electron temperature in argon inductively coupled plasma and its analysis via novel electron mean energy equation. Physics of Plasmas. 2018;**25**(3):033516

[23] Daltrini AM, Moshkalev SA, Monteiro MJR, Besseler E, Kostryukov A, Machia M. Mode transitions and hysteresis in inductively coupled plasmas. Journal of Applied Physics. 2007;**101**:073309. DOI: 10.1063/1.2715845

[24] El-Fayoumi IM, Jones IR, Turner MM. Hysteresis in the E- to H-mode transition in a planar coil, inductively coupled RF argon discharge. Journal of Physics D: Applied Physics. 1998; **31**:3082-3094

[25] Lee MH, Chung CW. On the E to H mode and H to E transition mechanisms in inductively coupled plasma. Physics of Plasmas. 2006;**13**:063510. DOI: 10.1063/1.2212387

[26] Xu HJ, Zhao SX, Zhang YR, Gao F, Li XC, Wang YN. Equivalent circuit effects on mode transitions in H_2 inductively coupled plasmas. Physics of Plasmas. 2015;**22**:043508. DOI: 10.1063/1.4917335

[27] Xu HJ, Zhao SX, Gao F, Zhang YR, Li XC, Wang YN. Discontinuity of mode transition and hysteresis in hydrogen inductively coupled plasma via a fluid model. Chinese Physics B. 2015;**24**:115201. DOI: 10.1088/1674-1056/24/11/115201

[28] El-Fayoumi IM, Jones IR. The electromagnetic basis of the transformer model for an inductively coupled RF plasma source. Plasma Sources Science and Technology. 1998;**7**: 179-185

[29] Gao F, Zhao SX, Li XS, Wang YN. Effects of matching network on the hysteresis during E and H mode transitions in argon inductively coupled plasma. Physics of Plasmas. 2010;**17**: 103507. DOI: 10.1063/1.3496385

[30] Wegner T, Küllig C, Meichsner J. On the E-H transition in inductively coupled radio frequency oxygen plasmas: I. Density and temperature of electrons, ground state and singlet metastable molecular oxygen. Plasma Source Science and Technology. 2017;**26**: 025006. DOI: 10.1088/1361-6595/26/2/025006

[31] Liu W, Gao F, Zhao SX, Li XC, Wang YN. Mode transition in CF_4+Ar inductively coupled plasma. Physics of Plasmas. 2013;**20**:123513. DOI: 10.1063/1.4858900

[32] Zaplotnic R, Vesel A, Mozetic M. Investigation of reactive plasma species created in SO_2 by an inductively coupled RF discharge in E- and H- mode. Journal of Applied Physics. 2016;**120**:163302. DOI: 10.1063/1.4966171

[33] Draskovic-Bracun A, Mozetic M, Zaplotnik R. E- and H-mode transition in a low pressure inductively coupled ammonia plasma. Plasma Processes and Polymers. 2018;**15**:e1700105. DOI: 10.1002/ppap.201700105

[34] Lee MH, Chung CW. Observation of inverse hysteresis in the E to H mode transitions in inductively coupled plasmas. Plasma Source Science and Technology. 2010;**19**:015011. DOI: 10.1088/0963-0252/19/1/015001

[35] Lee HC, Kim YC, Chung CW. Evolution of two-dimensional plasma density on the E-H heating mode transition in planar type inductively coupled plasma. IEEE Transactions on Plasma Science. 2011;**39**:2536-2537. DOI: 10.1109/TPS.2011.2135382

[36] Morishita S, Hayashi Y, Makabe T. Spatiotemporal optical structure of plasmas in the E-to-H transition in an inductively coupled plasma in Ar. Plasma Source Science and Technology. 2010;**19**:055007. DOI: 10.1088/0963-0252/19/5/055007

[37] Lee HC, Chung CW. Comparisons of the electrical characteristics by impedance matching conditions on the E-H and H-E transition and the hysteresis of inductively coupled plasma. Thin Solid Films. 2012;**521**:185-188. DOI: 10.1016/j.tsf2011.12.015

[38] Corr SC, Steen PG, Graham WG. Instabilities in an inductively coupled oxygen plasma. Plasma Source Science and Technology. 2003;**12**:265-272

[39] Lee HC, Chung CW. Effect of electron energy distribution on the hysteresis of plasma discharge: Theory, experiment and modeling. Scientific Reports. 2015;**5**:15254. DOI: 10.1038/srep15254

[40] Suzuki K, Nakamura K, Ohkubo H, Sugai H. Power transfer efficiency and mode jump in an inductive RF discharge. Plasma Source Science and Technology. 1998;**7**:13-20

[41] Cunge G, Crowley B, Vender D, Turner MM. Characterization of the E to H transition in a pulsed inductively coupled plasma discharge with internal coil geometry: Bi-stability and hysteresis. Plasma Source Science and Technology. 1999;**8**:576-586

Chapter 6

Numerical Modeling of Partial Discharge Development Process

Cheng Pan, Ju Tang and Fuping Zeng

Additional information is available at the end of the chapter

http://dx.doi.org/10.5772/intechopen.79215

Abstract

Partial discharge (PD), a type of low-temperature plasma, indicates a discharge event that does not bridge the electrodes of an electrical insulation system under high voltage stress. It is common in power equipment, such as transformers, cables, gas-insulated switchgears, and so on. The occurrence of PD could deteriorate the insulation performance of the equipment, but, meanwhile, it is often used to diagnose the insulation status. Therefore, it is very necessary to clarify the PD mechanism, and through modeling the PD process, a better understanding of the phenomenon could be attained. Although PD is essentially a gas discharge phenomenon, it possesses some distinctive features, for example, very narrow discharge channel, short time duration, and stochastic behavior, which determine the simulation method of PD different from that for the other types of plasmas. This chapter seeks to propose a simulation method that could reflect the physical processes of PD development after introducing some background knowledge about PD and analyzing the shortcomings of existent models.

Keywords: simulation model, partial discharge, streamer, fluid equations, discharge time lag

1. Introduction

Partial discharge (PD) is usually observed in power equipment, such as transformers, cables, gas insulated switchgears, and so on, which indicates a gas breakdown in essence induced by a local electric field distortion. It should be noted that it does not bridge the electrodes, differing from the gas breakdown across conductors. The remaining component of the whole insulation which does not suffer from PD could be oil, solid, or gas. On one hand, during the PD process, the heat energy, the charges with high velocity and chemical-active substances are released to erode and change the composite of the remaining component, leading to the deterioration of insulation performance and even the insulation failure. For example, as for high voltage power cable, PD resulting from the insulation defects could induce degradation of the solid dielectric due to chemical effect and physical attack by charge bombardment, and electric trees will be present after long-term service [1]. When the solid dielectric is across by the trees, an insulation fault takes place. On the other hand, PD parameters, such as discharge magnitude, discharge time, and so on, are determined by the characters of the gas and the remaining insulation. In terms of this, the PD measurement is often employed to diagnose the insulation status of power equipment. Whether understanding the negative effect of PD on insulation or equipment condition maintenance in the usage of PD measurement, it is based on the clear PD mechanism.

In essence, PD is a gas breakdown phenomenon. Similar to the other types of low-temperature plasmas, the temperature of electrons during a PD is much higher than that of ions, which is equivalent to the neutral gas molecules. However, PD also shows some distinctive features. For example, because PD always results from the local defect with a high electric field, the discharge channel is very narrow (the radius may be 100 μm) and the duration time is very short (several to tens of nanoseconds). During a PD sequence, once previous PD is terminated, and the subsequent one may take place after several milliseconds or even several days [2]. This phenomenon indicates that PD has a stochastic behavior, due to not only the effect of gas itself but also the interaction between gas breakdown and the remaining insulation. Therefore, as for the PD, the mere investigation of gas breakdown is meaningless. On the contrary, the interaction between PD and the remaining insulation should be considered. More importantly, a large number of PD data should be obtained to seek for its statistical characters because of its stochastic behavior.

According to the type of the remaining insulation and electrode configuration, PD could be divided into three categories [3]: internal discharge, surface discharge, and corona, as in **Figure 1**. Internal discharge indicates a gas breakdown taking place in a cavity embedded in solid or liquid dielectric. Generally, the former is more common. It consists of the streamer development and the interaction between streamer and cavity walls. A surface charge usually occurs along the solid dielectric surface due to a large tangential component of electric field,

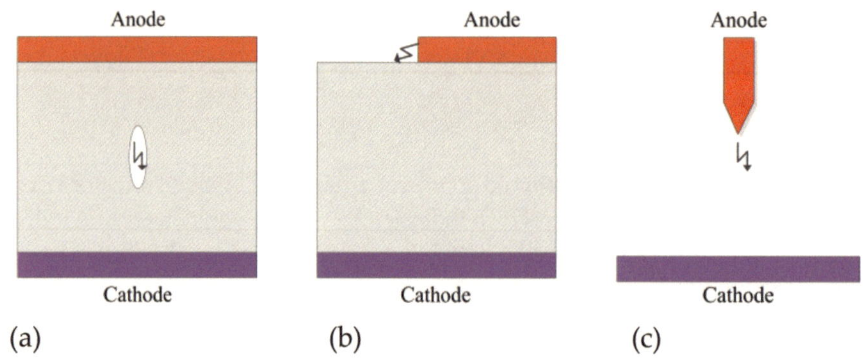

Figure 1. Three categories of PD: (a) internal discharge, (b) surface discharge, and (c) corona.

during which the interaction between streamer development and dielectric dominates. Corona often takes places in the local region around a conductor, which mainly involves the streamer development. Therefore, internal discharge could best represent PD, because it includes the two processes. In fact, the majority of PD simulations are concentrated on the internal discharge (also called cavity discharge) [4–7]. And in this chapter, we also focus on it.

There are many factors that could affect PD characters, such as the applied voltage (voltage waveform, amplitude, and frequency), electrode configuration, cavity (transportation parameters of gas, location, and size), remaining dielectric (permittivity, conductivity, and surficial parameters, e.g., morphology, surface trap distribution), and so on. To sum it up, two intrinsic factors behind them determine the evolution of PD behavior, that is, electric field and seed electrons. Generally speaking, two conditions must be simultaneously satisfied in order that a gas breakdown can take place: there must be at least one free electron in the gas, and the electric field must be of sufficient strength and duration time to ensure that this electron generates a sequence of avalanches [8]. Based on the conditions, it is inferred that the supply of free electrons and electric field affect not only the occurrence of PD but also its characters. Actually, the electric field is related to the applied voltage, electrode configuration, residual charges within the cavity, the cavity size, and the permittivity of remaining dielectric, while the supply of free electron depends on the gas status and surficial conditions of dielectric, corresponding to the volume generation and surface emission, respectively [9].

Looking back at the evolution of PD simulation methods, the a-b-c model was initially proposed [10–12], in which the discharge process was considered as charging-discharging of capacitors. Subsequently, some researchers held that the discharge could be represented by the increase of gas conductivity, and the current continuity equation was used to calculate discharge parameters [13–15]. On the contrary, others thought that a discharge was actually the deployment of charges in the cavity, and Poisson's equation was enough [16–18]. Obviously, these models could represent the transient phenomenon of a discharge, but not reflect its physical processes. In recent years, a plasma model was employed to simulate single PD [19–21], in which the impact ionization, drift, diffusion, recombination, and other processes were quantitatively described by fluid equations. This model successfully obtained microscopic physical processes of a PD, but did not take the stochastic characters into account.

In this chapter, we firstly reviewed PD simulation models in brief, which consisted of the a-b-c model, Pedersen's model, conductance model, Niemeyer's model and plasma model, and analyzed their merits and drawbacks. Then, an advanced model was constructed to obtain physical processes, including the streamer propagation and surface charge dynamics, and macroscopic parameters, for example, discharge magnitude and moment of continuous PDs, so that a comprehensive analysis was available.

2. Review of PD simulation models

Since a-b-c model was proposed, numerical modeling of PD has been developed for decades of years. During this period, many kinds of simulation models have been constructed, which could be roughly divided into two categories: based on the point of view of circuit and based on the point of view of field. The former indicates a-b-c model and the latter consists of Pedersen's model, conductance model, and Niemeyer's model.

2.1. a-b-c model

The a-b-c model or the three-capacitor model is the original one to interpret the PD mechanism [3], and then it is usually employed to simulate the stochastic characters of PD [10, 11]. In the model, the dielectrics between electrodes, including the gas and solid insulation, are considered as capacitors, as in **Figure 2**. In detail, C_1 indicates cavity capacitance, C_2 is the capacitance of dielectric in series with the cavity, and C_3 is the capacitance of solid dielectric in parallel with the cavity. Besides, R_1, R_2, and R_3 indicate the resistance of corresponding part, respectively.

The occurrence and termination of PD depend on the potential difference across the cavity, U_1. When U_1 exceeds the inception voltage, a discharge will take place and will stop when it is less than the extinction voltage. If a discharge occurs, C_1 is short-circuited, leading to a fast transient current to flow in the circuit due to a voltage difference between the voltage source and across C_2. Based on the analysis of capacitor charging-discharging processes, the apparent charge magnitude, which reflects PD intensity, could be calculated.

It could be found that this model is very simple, but it can represent the transient process related to a discharge event and is often used to explain some experimental results. However, it could not describe the discharge process physically, and the concept, capacitor, is not strictly valid, because the interface between the cavity and the solid dielectric is not equipotential when a discharge takes place [22].

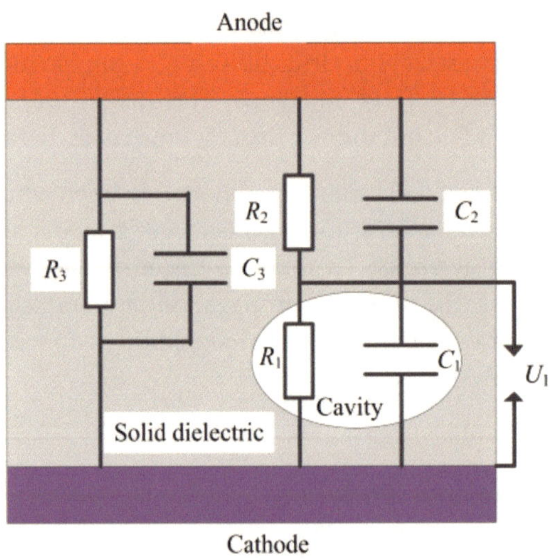

Figure 2. a-b-c model.

2.2. Pedersen's model

There are two important parameters of PD, that is, physical charges and apparent charges. The former indicate the charges generated during a discharge process, while the latter are measured charges through external circuit. In order to establish the link between physical charges and apparent charges, Pedersen proposed a model to describe the transient process [23]. Without considering the charge exchange between solid dielectric and the adjacent electrode, the amount of apparent charges equals the induced charges at an electrode surface due to charge generation, recombination, and movement during a discharge process. Therefore, if the physical charge distribution is known, the apparent charges could be calculated [24]

$$Q_{\text{app}} = -\iiint \lambda \rho \mathrm{d}V - \iint \lambda \sigma \mathrm{d}s \tag{1}$$

where ρ and σ indicate volume and surface charge density within the cavity, respectively. λ, a dimensionless function, depends on the charge location, which satisfies Laplace equation

$$\nabla \cdot (\varepsilon_0 \varepsilon_{\mathrm{r}} \nabla \lambda) = 0 \tag{2}$$

where ε_0 is the vacuum permittivity, and ε_{r} the relative permittivity.

Pedersen's model is helpful to understand the measured results by using the pulse current method. However, the apparent charges depend on physical charge distribution which results from the discharge process and keeps unknown in this model.

2.3. Conductance model

When PD takes place, a plasma region with a high charge concentration in the cavity is formed, so the gas conductivity largely increases in comparison with the initial state. Based on this fact, the discharge process is simplified by the variation of gas conductivity [13], which can be described by the following equations:

$$\nabla \cdot \boldsymbol{D} = \rho \tag{3}$$

$$\nabla \cdot \boldsymbol{J} + \frac{\partial \rho}{\partial t} = 0 \tag{4}$$

where $\boldsymbol{D}$ is the electric displacement field, $\boldsymbol{J}$ the free current density. At the initial state, the gas conductivity is set to be zero. When a discharge takes place, it is set to be γ_{gd} and hence the electric field distribution within the cavity changes. In terms of the electric field evolution, some PD parameters are obtained, for example, apparent charges and physical charges.

Forssen compared the simulation results with the experimental data, and they were in general agreement but with a slight difference. Furthermore, Illias developed the simulation model by taking the surface emission and temperature variation during the discharge into account [14]. However, in any case, the increment of gas conductivity could not represent the PD process.

2.4. Niemeyer's model

Niemeyer considered PD within the cavity as a streamer-type discharge, because only this type could be detected and has engineering significance [9]. After analyzing the physical processes of PD, he proposed several equations to describe PD, as follows:

$$\int_{0}^{x_{cr}} \overline{\alpha}[E(x)]dx \geq K_{cr} \tag{5}$$

$$\Delta U_{res} \approx E_{ch} l_{str} \tag{6}$$

$$q = \pm g\pi\varepsilon_0 l \Delta U_{PD} \tag{7}$$

Eq. (5) is actually the well-known critical avalanche criterion, in which $\overline{\alpha}$, the function of electric field, indicates the effective ionization coefficient, K_{cr} the logarithm of a critical number of electrons that has to accumulate in the avalanche head to make the avalanche self-propagating by its own space charge field, and x_{cr} the distance within $\overline{\alpha}$ which exceeds zero. In terms of it, the inception field of PD occurrence could be obtained. Eq. (6) simply describes the streamer propagation, where E_{ch} is the electric field in the discharge channel, U_{res} the residual voltage instantaneously after discharge, and l_{str} the distance to which streamer could propagate. Eq. (7) establishes the relationship between physical charges and potential difference before and after a PD, in which g is a dimensionless proportionality factor and l the cavity scale.

Based on the model, Niemeyer simulated PD behaviors within a spherical cavity by considering the stochastic supply of free electrons, which agreed with experimental data qualitatively and quantitatively although there was a slight disagreement in the phase and magnitude distributions of PD. However, there is a significant shortcoming that the electric field distribution was assumed to be uniform within the cavity. Considering this point, Illias developed the simulation model in which the deployed charges were not uniform and Poisson's equation was employed to calculate PD parameters [16, 17].

2.5. Plasma model

In terms of physical processes, a cavity PD is similar to the filamentary dielectric barrier discharge (DBD) [25]. As for the latter, fluid equations are widely used to simulate gas discharge process [26, 27], which describe the impact ionization, charge drift, diffusion, recombination, and some secondary effects. In recent years, several researchers employed them to simulate the PD occurring in a cavity [18–20]. For example, Novak and Bartnikas established a two-dimensional breakdown model based on the continuity equations for electrons and ions to examine the influence of surface charges upon the partial discharge behavior [19]. In terms of it, the evolution of electric field and charge concentration distribution within the cavity during the discharge process was obtained, as well as the discharge current pulse.

However, the behaviors of single PD could not represent that of continuous PDs due to the memory effect. On one hand, residual charges generated by previous discharge land on the cavity surface and affect the electric field distribution within the cavity, leading to the change of subsequent PD characters. On the other hand, the accumulated surface charges may provide free electrons for the next PD occurrence. The interaction between adjacent PDs could not be represented by singe PD. Therefore, it is necessary to establish a simulation model which could present the discharge development process and take the memory effect into account to obtain the stochastic characters of PD sequences.

3. Numerical modeling of PD sequences using fluid equations

As for the PD simulation, on one hand, the model should reflect physical processes as much as possible, and on the other hand, a large number of data should be obtained to get the statistical parameters of repetitive PDs due to the stochastic characters. There is a contraction that taking too much physical processes into account must result in the model complexity and a large calculation consumption which is not beneficial to statistical analysis. Therefore, some important processes should be considered in the simulation model, while others are abandoned.

By reviewing the PD simulation models, it is found that two processes are crucial to cavity PD characters, that is, streamer development and surface process. Obviously, the apparent charges that could be detected by pulse current method are determined by streamer development in the cavity. Surface process mainly consists of charge accumulation on the interface and surface emission of charge. After the streamer lands on the dielectric surface, charges accumulate and will affect the subsequent PD behavior. Besides, surface emission could provide free electrons for the next PD. It should be noted that the distribution of surface charges generated by previous discharge does not keep unchanged until subsequent one takes place. Due to the surface or bulk conductivity of dielectric, the accumulated charges may decay. To sum it up, the streamer development and surface charge accumulation reflect a single PD process, while surface charge accumulation, decay, and emission represent the interaction of adjacent discharges during a PD sequence, which should be considered in the simulation model.

3.1. Simulation model construction

Because sandwich-type samples are widely used in the experimental researches on PD, a cylindrical cavity with a diameter of 2 mm and a height of 0.25 mm is employed in our simulation model, as in **Figure 3**. The cavity, full of atmospheric pressure air, is embedded within the solid dielectric, of which the relative permittivity equals 2.3. The thickness of dielectric barriers is set to be identical to the cavity height. Although during the discharge process, the temperature of cavity may slightly increase due to the joule heating from discharges, the temperature variation is neglected in our model, which means that the pressure in the cavity keeps unchanged.

The streamer development is quantitatively described by fluid equations, as follows:

$$\frac{\partial N_e}{\partial t} = N_e\alpha|\boldsymbol{W}_e| - N_e\eta|\boldsymbol{W}_e| - N_eN_p\beta - \nabla\cdot(N_e\boldsymbol{W}_e - D\nabla N_e) \tag{8}$$

$$\frac{\partial N_p}{\partial t} = N_e\alpha|\boldsymbol{W}_e| - N_eN_p\beta - N_nN_p\beta - \nabla\cdot\left(N_p\boldsymbol{W}_p\right) \tag{9}$$

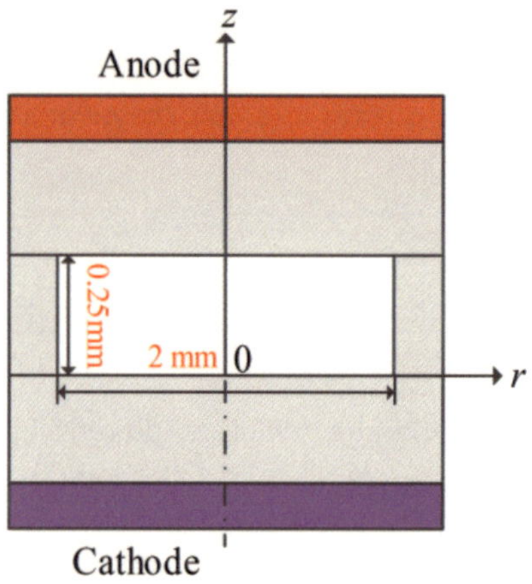

Figure 3. Configuration of simulation model.

$$\frac{\partial N_n}{\partial t} = N_e \eta |\mathbf{W}_e| - N_n N_p \beta - \nabla \cdot (N_n \mathbf{W}_n) \tag{10}$$

where N indicates the bulk charge concentration within the cavity, e, p, and n the symbols for electron, positive ion, and negative ion, respectively, t discharge time, α, η, β, and D denote the ionization, attachment, recombination, and electron diffusion coefficients, respectively, and $\mathbf{W}$ the drift velocity. Eqs. (8)–(10) reflect the transportation processes of electrons, positive and negative ions, which includes impact ionization, drift, diffusion, attachment, and recombination. However, the secondary processes, for example, photoionization, are neglected due to two reasons: (1) photoionization is crucial to the streamer development in long gaps but not so important for short gaps [28] and (2) the calculation of the secondary effect is extremely complicated, especially for the photoionization [29], which would bring about great difficulties of the PD sequence simulation. The detailed expressions of the above transport parameters come from Morrow's paper [30], and we list them in Appendix A.

After the streamer arrives at the interface between the cavity and the dielectric, the charges will accumulate on the dielectric surface. We use the following equation to describe the transition from volume charges to surface charges:

$$\sigma \Delta S = (N_p - N_e - N_n) e \Delta V \tag{11}$$

where ΔS and ΔV represent the area and volume of unit grid after meshing, respectively. Surface charge distribution is assumed to keep unchanged during the discharge process.

During the streamer development, the influence of space charges on the electric field should not be neglected, so Poisson's equation is employed to obtain the electric field within the cavity:

$$\nabla^2 \varphi = -\frac{e}{\varepsilon_r \varepsilon_0} (N_p - N_e - N_n) \tag{12}$$

At the upper and lower surfaces of cavity, the boundary conditions for Poisson's equation are

$$\varepsilon_0 E_z\left(d_g^-\right) - \varepsilon_r \varepsilon_0 E_z\left(d_g^+\right) = \sigma_u \tag{13}$$

$$\varepsilon_r \varepsilon_0 E_z(0^-) - \varepsilon_0 E_z(0^+) = \sigma_d \tag{14}$$

where d_g is the cavity height, $E_z\left(d_g^-\right)$ and $E_z\left(d_g^+\right)$ indicate the z-component of electric field at both sides of the upper surface, while $E_z(0^-)$ and $E_z(0^+)$ represent the z-component of electric field at both sides of the lower surface, σ_u and σ_d denote the surface charge density at the upper and lower surfaces.

An initial electron-positive ion pair with a concentration of 10^{13} cm^{-3} is placed near the upper or lower surface to induce the streamer and avoid Townsend phase of gas discharge [28]. It should be noted that this assumption differs from the consideration of free electrons, which will be described in the later text. During the streamer development, charge concentration varies quickly, and an area with a steep concentration gradient appears at the head of the streamer. Meanwhile, the value of charge concentration should maintain positive, which cannot be guaranteed by the traditional finite difference method. So, the flux-corrected transport (FCT) algorithm is used to solve the convection term of charge continuity equations to overcome the two problems [31–33], which is listed in Appendix B.

In general, the time step for FCT is chosen based on the electron drift velocity, however, which may not apply to the circumstance in our simulation model. It is because apart from the streamer development, its extinguishment process also needs to be obtained which is responsible for the accumulation of electrons and ions. However, the drift velocity of electrons is about 100 higher than that of ions, and the choice of time step must lead to the large increase of calculation consumption at the later stage of discharge when ion drift dominates. Instead, if it is chosen based on the ion drift velocity, the accuracy of the calculation cannot be guaranteed at the initial stage of discharge. Therefore, as a compromise, the time step is set according to whether there are any electrons within the cavity volume. In detail, during the initial stage of streamer development, it is determined by the electron drift velocity. After electrons completely accumulate at the interface, it depends on the drift velocity of a positive ion or a negative one (both are the same). The expression for the time step is

$$\Delta t_{e,p} = 0.1 \frac{\Delta z}{\left|W_{e,p}\right|^{max}} \tag{15}$$

where Δt is the time step, Δz the grid length along z-direction, and $|W|^{max}$ the maximum value of charge drift velocity within the cavity.

According to Pedersen's model, the apparent charges are determined by charge transportation within the cavity, which could be detected by pulse current method. However, due to the effect of dielectric barriers, the pulse obtained at the external circuit may not reflect the streamer propagation. So, we use Sato's equation to calculate the current due to free charge movement [34], as follows:

$$I = \frac{e}{U_a} \iiint_V (N_p \mathbf{W}_p - N_e \mathbf{W}_e - N_n \mathbf{W}_e) \cdot \mathbf{E}_a dV \tag{16}$$

where U_a indicates the applied voltage, $\mathbf{E}_a$ the applied field, and V the discharging volume.

On one hand, the field within the cavity should exceed a critical value so that a discharge may take place. Based on the ignition condition of streamer, the critical field is expressed as follows [35]:

$$E_b = \frac{24410\left(\frac{P}{760} d_g\right) + 6730\sqrt{\frac{P}{760} d_g}}{d_g} \tag{17}$$

where P is in Torr. After a discharge takes place, electrons and ions accumulate on the dielectric surface. Due to the recombination of charges from gas, surface, and bulk conduction of dielectric, the accumulated surface charges will decay until the next discharge occurs. It is found from our previous experiments that the decaying discipline of surface charges could be expressed as [36].

$$\frac{\sigma_p}{\sigma_{p0}} = e^{\frac{-t}{\eta_p}} \tag{18}$$

$$\frac{\sigma_e}{\sigma_{e0}} = e^{\frac{-t}{\eta_e}} \tag{19}$$

where σ_{p0} and σ_{e0} indicate initial positive charge and electron density at dielectric surfaces, respectively. η_p and η_e equal 312.5 and 568.8 ms, both of which represent the surface charge decay time for positive ions and electrons. The negative ion is neglected because its concentration is much lower in comparison with electron and positive ions.

On the other hand, although free electrons from the volume ionization and surface emission are formulated, their supply shows a strong scholastic behavior. Hence, there is usually a time delay between the instant of application of an electric field in excess of the critical field and the onset of breakdown, which is called a discharge time lag (strictly speaking, it is a statistical time lag, but the formative time lag is very short for cavity discharge and could be neglected). In order to simplify the physical process of free electron production, the discharge time lag is introduced to our model. Some experimental and simulation results show that the discharge time lag is not completely random, but is subject to exponential distribution [37, 38], which is expressed as

$$\tau = \begin{cases} \infty & E_z < E_b \\ -\ln(1 - P_d)/\zeta & E_z \geq E_b \end{cases} \tag{20}$$

where P_d indicates the discharge probability, which belongs to [0, 1) and is random, ζ the rate parameter of exponential distribution.

In terms of Eq. (17), the critical field for gas breakdown is calculated, and it equals 67,000 V/cm. In this case, the potential difference across the electrodes is 3130 V. Because the PD mechanism at AC voltage has been studied by many authors [2, 4–7], and a comprehensive understanding about it has been obtained, the PD mechanism under DC voltage needs to be clarified. In this chapter, the DC voltage with an amplitude of 3200 V is applied to the anode, and the cathode is grounded all the time. Of course, this model is also applied to the circumstance of AC voltage application.

3.2. Simulation results

3.2.1. A PD development process

The process of PD development in the cavity consists of two stages: the streamer propagation and surface charge accumulation. **Figures 4** and **5** show the temporal and spatial distribution of electrons and positive ions during this process, respectively. After discharge conditions are satisfied, the streamer is initiated near the lower surface of dielectric. With the help of applied field, electrons propagate toward the anode. At 0.72 ns, the head of streamer arrives at the upper surface of dielectric. Based on this, the streamer development velocity could be

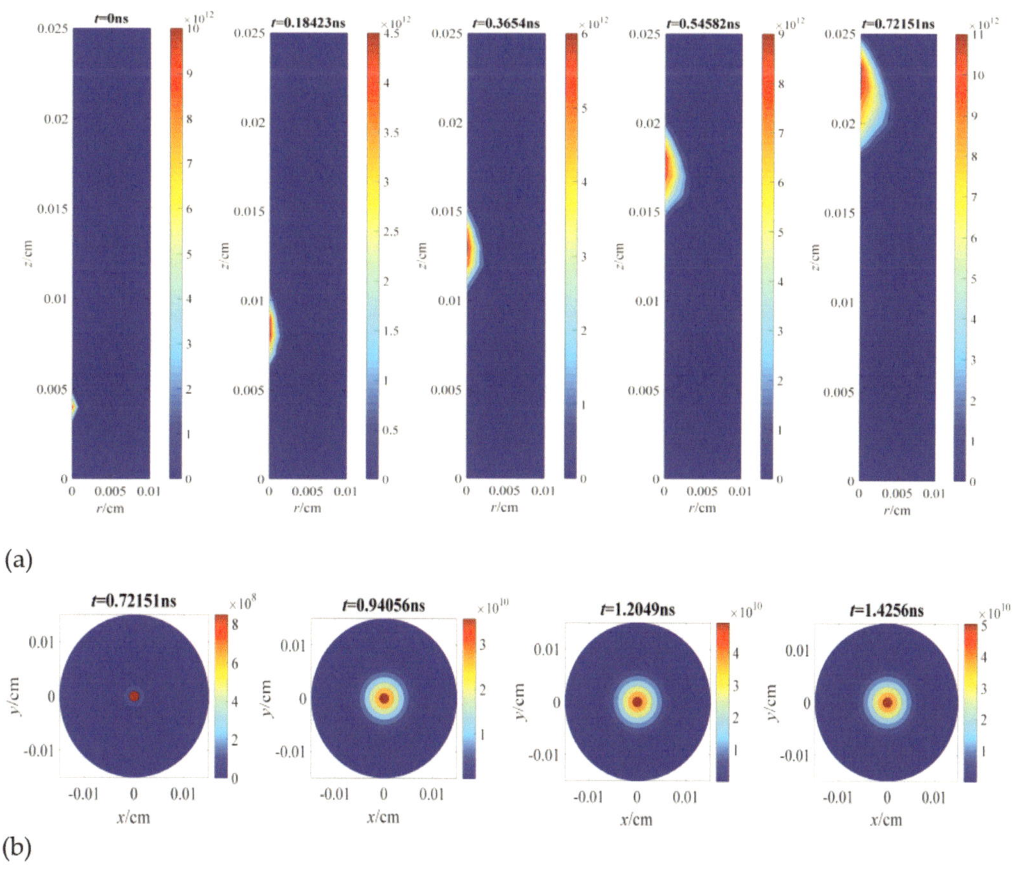

Figure 4. Evolution of electron concentration distribution during the first PD (a) within the cavity volume (unit: cm^{-3}) and (b) on the upper surface of the cavity (unit: cm^{-2}).

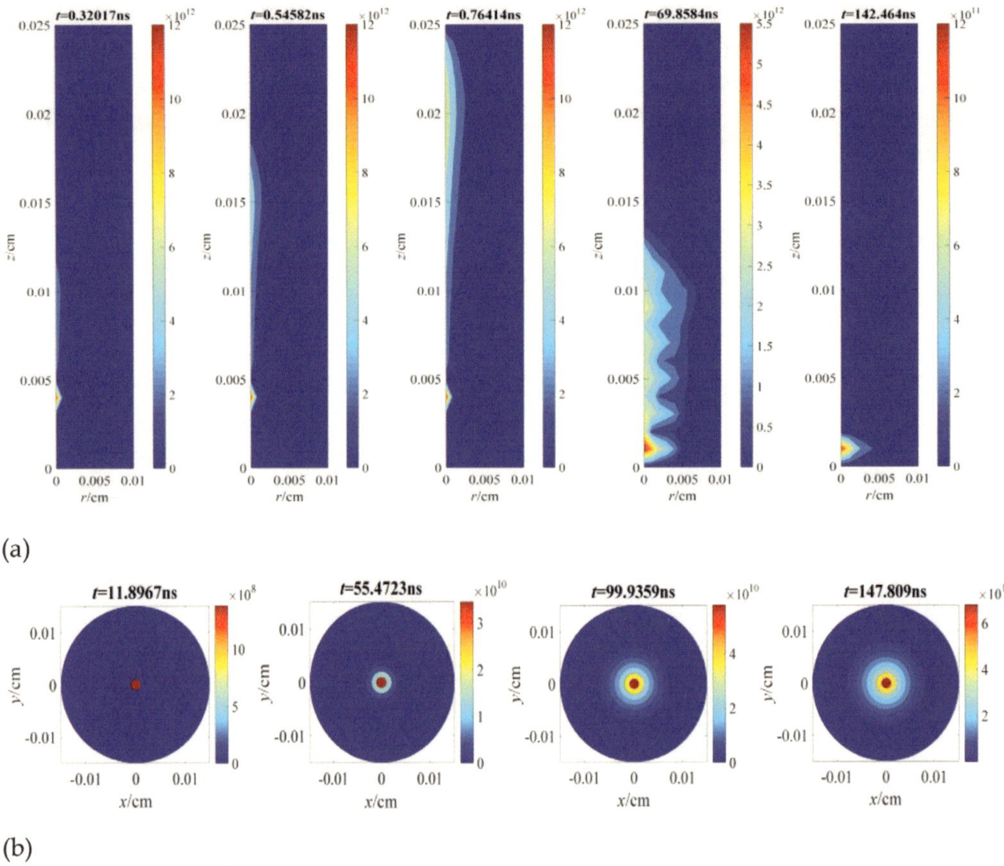

Figure 5. Evolution of positive ion concentration distribution during the first PD (a) within the cavity volume (unit: cm^{-3}), (b) on the lower surface of cavity (unit: cm^{-2}).

calculated, which equals 3.5×10^7 cm/s and the order is in accordance with other researcher's simulation result [39]. Then, electrons begin to accumulate on the upper surface of dielectric, and the density of surface charges reaches a saturation value after 1.4 ns. During this period, positive ions almost maintain stationary because the drift velocity is approximately 1/100 of the electron. However, positive ions seem to move according to **Figure 5**, and the distribution appearance looks like a ladle, which are attributed to the impact ionization of electrons. At 11.9 ns, a large number of positive ions land on the lower surface of dielectric, and the accumulation is terminated at 147.8 ns. Therefore, the accumulation time of electrons is much shorter than that of positive ions.

Based on the simulation results, it is found that the distribution of surface charges appears as a spot, and the maximum charge density locates at the middle of a spot. Compared with the experimental results [36], the distribution shape and surface density level (0.1 nC/mm^2) are identical, which show that the simulation results are reasonable. However, there are some slight differences due to the simplification of model.

Charge transportation within the cavity will induce a current pulse, as in **Figure 6**, which could reflect the streamer development. The peak value of pulse appears at 0.72 ns; at this moment, the streamer head arrives at the upper surface of the cavity. The pulse width lasts for 1.4 ns; during this period, the accumulation of electrons is terminated. On the contrary, positive ions still move in the cavity volume. It is inferred that positive ions have a minor contribution to the current pulse because of their low drift velocity. A low-inductance resistor connected to the cathode is usually employed to detect a current, but this current slightly differs from that in **Figure 6** [40].

3.2.2. A PD sequence

A PD sequence consisting of 100 continuous discharges is obtained by the simulation (**Figures 4–6** show the first discharge development process). **Figure 7** shows the discharge time and the peak value of current of each discharge. In terms of this information, some statistical parameters of PDs, for example, discharge frequency and average discharge magnitude, could be calculated, and discharge patterns could be depicted.

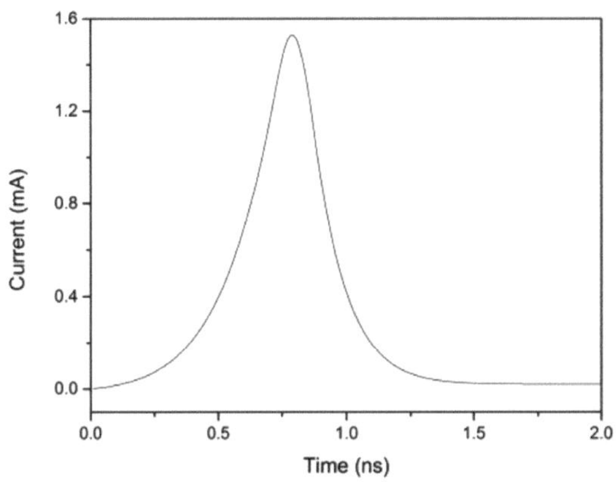

Figure 6. Current pulse waveform of the first PD obtained by simulation.

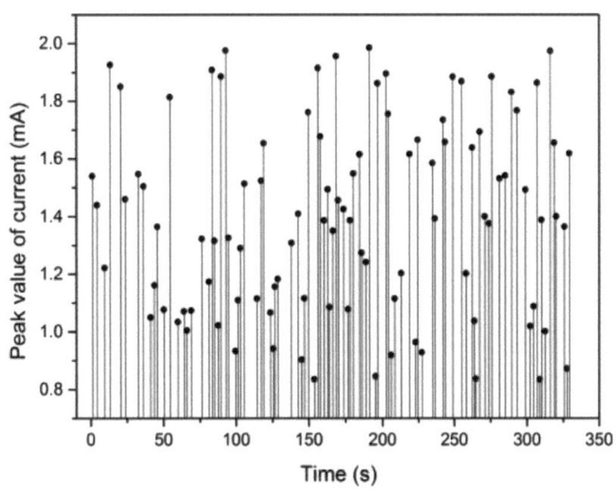

Figure 7. A PD sequence with 100 continuous discharges.

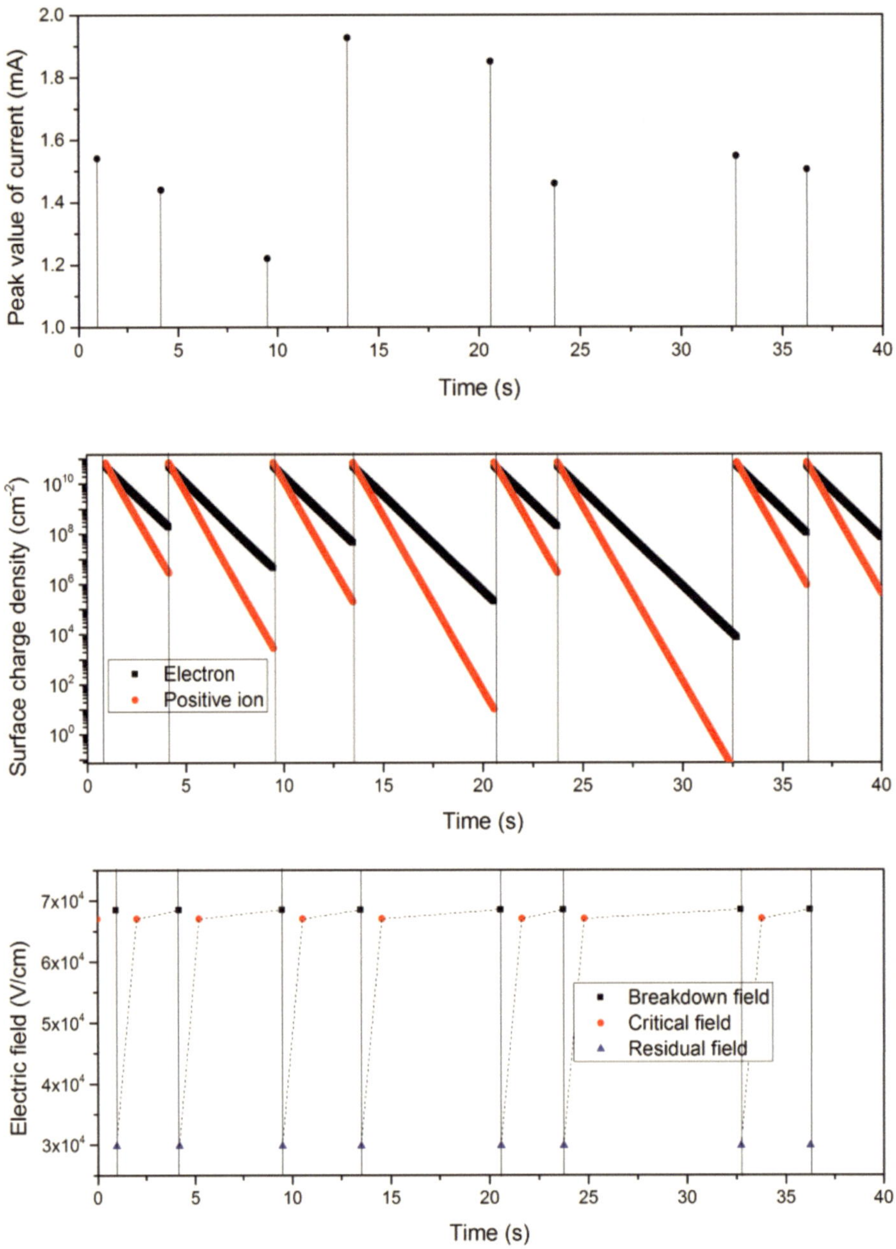

(a)

(b)

(c)

Figure 8. The first eight discharges during the PD sequence: (a) discharge time and magnitude, (b) surface charge decaying process, and (c) evolution of electric field within the cavity.

Besides, by analyzing the PD sequence, the interaction between adjacent discharges is obtained. **Figure 8** shows the temporal evolution of surface charges and electric field within the cavity of first eight discharges. The first PD does not take place immediately after the voltage application due to the existence of a discharge time lag. After the discharge is terminated, the electric field within the cavity is dramatically reduced (as in **Figure 8c**), which is attributed to the effect of surface charge accumulation. Then, the surface charges begin to decay, and the electric field within the cavity gradually recovers. After it exceeds the critical value, and the condition for discharge time lag is satisfied, the next PD takes place.

During the process of surface charge decaying (as in **Figure 8b**), the initial concentration of electrons and positive ions is approximately identical, but residual charges are completely distinct at the moment when a next discharge occurs. Due to the decay rate of positive charges faster than that of electrons, the concentration of residual negative surface charges is much higher. Therefore, compared with positive ions, residual electrons resulting from previous discharge have a larger influence on the subsequent one during a PD sequence.

4. Conclusions

PD, a type of low-temperature plasma, has some distinctive features, which determines its simulation method different from that of other types. In detail, as for the most representative PD type, cavity PD, it is necessary to take the streamer propagation, surface charge accumulation and decay, free electron supply into account so that the PD mechanism could be clarified. Besides, due to the stochastic character of PD, a large number of PD data must be obtained with the help of simulation.

Traditional simulation models about PD could be mainly divided into two categories: based on the point of view of circuit and based on the point of view of field. The former indicates a-b-c model, in which the discharge process is replaced by capacitor charging and discharging. The latter consists of Pedersen's model, conductance model, and Niemeyer's model, in which the discharge process is modeled by the variation of gas volume conductivity or significant simplification of discharge process. Anyway, these models could not reflect the PD development process physically.

Based on the simulation method for a single PD, we develop it by using fluid equations combined with Poisson's equation. In terms of the model, microscopic physical processes, that is, streamer development and surface charge accumulation, could be obtained, as well as macroscopic parameters, that is, discharge current and discharge time, and the interaction between adjacent discharges. It is found that electrons and positive ions, respectively, land on the two surfaces of the cavity, and the accumulation time of positive ions is much longer than that of electrons. During a PD sequence, the decay of surface charges resulting from previous discharge could be considered to be the key factor, contributing to the occurrence of the subsequent one.

Acknowledgements

We would like to thank Prof. George Chen from the University of Southampton for the inspiring discussion and appreciate the financial support from the National Natural Science Foundation of China (51607128).

A. Appendix

The transportation parameters for air are expressed by the following equations:

$N = 2.69 \times 10^{19}$ cm^{-3} indicates the number of gas molecules per unit volume, and E is the local field in V/cm

$$W_e = \begin{cases} -(E/|E|)(7.4 \times 10^{21}|E|/N + 7.1 \times 10^6) & |E|/N > 2.0 \times 10^{-15} \\ -(E/|E|)(1.03 \times 10^{22}|E|/N + 1.3 \times 10^6) & 10^{-16} \le |E|/N \le 2.0 \times 10^{-15} \\ -(E/|E|)(7.2973 \times 10^{21}|E|/N + 1.63 \times 10^6) & 2.6 \times 10^{-17} \le |E|/N < 10^{-16} \\ -(E/|E|)(6.87 \times 10^{22}|E|/N + 3.38 \times 10^4) & |E|/N < 2.6 \times 10^{-17} \end{cases} \quad \text{(A1)}$$

$$W_n = \begin{cases} -2.7E & |E|/N > 5.0 \times 10^{-16} \\ -1.86E & |E|/N \le 5.0 \times 10^{-16} \end{cases} \quad \text{(A2)}$$

$$W_p = 2.34E \quad \text{(A3)}$$

$$\frac{\alpha}{N} = \begin{cases} 2.0 \times 10^{-16} e^{\frac{-7.248 \times 10^{-15}}{|E|/N}} & |E|/N > 1.5 \times 10^{-15} \\ 6.619 \times 10^{-17} e^{\frac{-5.593 \times 10^{-15}}{|E|/N}} & |E|/N \le 1.5 \times 10^{-15} \end{cases} \quad \text{(A4)}$$

$$\frac{\eta_2}{N} = \begin{cases} 8.889 \times 10^{-5}|E|/N + 2.567 \times 10^{-19} & |E|/N > 1.05 \times 10^{-15} \\ 6.089 \times 10^{-4}|E|/N - 2.893 \times 10^{-19} & |E|/N \le 1.05 \times 10^{-15} \end{cases} \quad \text{(A5)}$$

$$\eta_3/N^2 = 4.7778 \times 10^{-59}(|E|/N)^{-1.2749} \quad \text{(A6)}$$

$$\eta = \eta_2 + \eta_3 \quad \text{(A7)}$$

where η_2 and η_3 are the two-body and three-body attachment coefficients, respectively.

$$\beta = 2.0 \times 10^{-7} \quad \text{(A8)}$$

$$D = 0.3341 \times 10^9 (|E|/N)^{0.54069} |W_e/E| \quad \text{(A9)}$$

B. Appendix

Based on the axisymmetric character of sample configuration in our model, the cylindrical coordinate system is employed, so the convection term could be rewritten as

$$\frac{\partial w}{\partial t} = -\frac{\partial f}{\partial r} - \frac{\partial g}{\partial z} \quad \text{(A10)}$$

where $w = rN$, $f = rNW_r$, $g = rNW_z$, $\mathbf{W} = W_r\mathbf{e}_r + W_z\mathbf{e}_z$, $\mathbf{e}_r$, and $\mathbf{e}_z$ are the unit vectors along r and z directions, respectively. To solve this equation, six steps are needed:

(1) to obtain the low-order flux ${F^L}_{i+\frac{1}{2},j}$, ${G^L}_{i,j+\frac{1}{2}}$

$$ {F^L}_{i+\frac{1}{2},j} = (2i+1)\pi\Delta z\Delta r\Delta t(W_r)_{i+\frac{1}{2},j}w\begin{cases} w = i\Delta rN_{i,j} & (W_r)_{i+\frac{1}{2},j}\geq 0 \\ w = (i+1)\Delta rN_{i+1,j} & (W_r)_{i+\frac{1}{2},j} < 0 \end{cases} \tag{A11} $$

$$ {G^L}_{i,j+\frac{1}{2}} = \Delta tS_i(W_z)_{i,j+\frac{1}{2}}w\begin{cases} w = i\Delta rN_{i,j} & (W_z)_{i,j+\frac{1}{2}}\geq 0 \\ w = i\Delta rN_{i,j+1} & (W_z)_{i,j+\frac{1}{2}} < 0 \end{cases} \tag{A12} $$

$$ S_i = 2i\pi\Delta r^2 \tag{A13} $$

$$ (W_r)_{i+\frac{1}{2},j} = \frac{(W_r)_{i,j} + (W_r)_{i+1,j}}{2} \tag{A14} $$

$$ (W_z)_{i,j+\frac{1}{2}} = \frac{(W_z)_{i,j} + (W_z)_{i,j+1}}{2} \tag{A15} $$

where i and j are the sequence number of node along r and z directions, respectively.

(2) to obtain high-order flux ${F^H}_{i+\frac{1}{2},j}$, ${G^H}_{i,j+\frac{1}{2}}$

$$ \begin{aligned} {F^H}_{i+\frac{1}{2},j} = (2i+1)\pi\Delta z\Delta r\Delta t\Bigg[&\frac{533}{840}\left(f_{i+1,j}+f_{i,j}\right) - \frac{139}{840}\left(f_{i+2,j}+f_{i-1,j}\right) \\ &+\frac{29}{840}\left(f_{i+3,j}+f_{i-2,j}\right) - \frac{1}{280}\left(f_{i+4,j}+f_{i-3,j}\right)\Bigg] \end{aligned} \tag{A16} $$

$$ \begin{aligned} {G^H}_{i,j+\frac{1}{2}} = S_i\Delta t\Bigg[&\frac{533}{840}\left(g_{i,j+1}+g_{i,j}\right) - \frac{139}{840}\left(g_{i,j+2}+g_{i,j-1}\right) + \frac{29}{840}\left(g_{i,j+3}+g_{i,j-2}\right) \\ &-\frac{1}{280}\left(g_{i,j+4}+g_{i,j-3}\right)\Bigg] \end{aligned} \tag{A17} $$

$$ f_{i,j} = i\Delta rN_{i,j}(W_r)_{i,j} \tag{A18} $$

$$ g_{i,j} = i\Delta rN_{i,j}(W_z)_{i,j} \tag{A19} $$

(3) to define antidiffusion flux

$$ A_{i+\frac{1}{2},j} \equiv {F^H}_{i+\frac{1}{2},j} - {F^L}_{i+\frac{1}{2},j} \tag{A20} $$

$$ A_{i,j+\frac{1}{2}} \equiv {G^H}_{i,j+\frac{1}{2}} - {G^L}_{i,j+\frac{1}{2}} \tag{A21} $$

(4) to obtain the temporary solution

$$ w^{td}_{i,j} = w^n_{i,j} - \Delta V^{-1}_{i,j}\left[F^L_{i+\frac{1}{2},j} - F^L_{i-\frac{1}{2},j} + G^L_{i,j+\frac{1}{2}} - G^L_{i,j-\frac{1}{2}}\right] \tag{A22} $$

$$ w_{i,j} = i\Delta rN_{i,j} \tag{A23} $$

$$\Delta V_{i,j} = \Delta z S_i \tag{A24}$$

(5) to restrict the antidiffusion flux

If $A_{i+\frac{1}{2},j}\left(w_{i+1,j}^{td} - w_{i,j}^{td}\right) < 0\&\left\{A_{i+\frac{1}{2},j}\left(w_{i+2,j}^{td} - w_{i+1,j}^{td}\right) < 0\|A_{i+\frac{1}{2},j}\left(w_{i,j}^{td} - w_{i-1,j}^{td}\right) < 0\right\}$

$$A_{i+\frac{1}{2},j} = 0 \tag{A25}$$

If $A_{i,j+\frac{1}{2}}\left(w_{i,j+1}^{td} - w_{i,j}^{td}\right) < 0\&\left\{A_{i,j+\frac{1}{2}}\left(w_{i,j+2}^{td} - w_{i,j+1}^{td}\right) < 0\|A_{i,j+\frac{1}{2}}\left(w_{i,j}^{td} - w_{i,j-1}^{td}\right) < 0\right\}$

$$A_{i,j+\frac{1}{2}} = 0 \tag{A26}$$

$$A_{i+\frac{1}{2},j}^{C} = C_{i+\frac{1}{2},j}A_{i+\frac{1}{2},j} \quad 0 \le C_{i+\frac{1}{2},j} \le 1 \tag{A27}$$

$$A_{i,j+\frac{1}{2}}^{C} = C_{i,j+\frac{1}{2}}A_{i,j+\frac{1}{2}} \quad 0 \le C_{i,j+\frac{1}{2}} \le 1 \tag{A28}$$

$$C_{i+\frac{1}{2},j} = \begin{cases} \min\left(R_{i+1,j}^{+}, R_{i,j}^{-}\right) & A_{i+\frac{1}{2},j} \ge 0 \\ \min\left(R_{i,j}^{+}, R_{i+1,j}^{-}\right) & A_{i+\frac{1}{2},j} < 0 \end{cases} \tag{A29}$$

$$C_{i,j+\frac{1}{2}} = \begin{cases} \min\left(R_{i,j+1}^{+}, R_{i,j}^{-}\right) & A_{i,j+\frac{1}{2}} \ge 0 \\ \min\left(R_{i,j}^{+}, R_{i,j+1}^{-}\right) & A_{i,j+\frac{1}{2}} < 0 \end{cases} \tag{A30}$$

$$w_{i,j}^{a} = \max\left(w_{i,j}^{n}, w_{i,j}^{td}\right) \tag{A31}$$

$$w_{i,j}^{\max} = \max\left(w_{i-1,j}^{a}, w_{i,j}^{a}, w_{i+1,j}^{a}, w_{i,j-1}^{a}, w_{i,j+1}^{a}\right) \tag{A32}$$

$$w_{i,j}^{b} = \min\left(w_{i,j}^{n}, w_{i,j}^{td}\right) \tag{A33}$$

$$w_{i,j}^{\min} = \min\left(w_{i-1,j}^{b}, w_{i,j}^{b}, w_{i+1,j}^{b}, w_{i,j-1}^{b}, w_{i,j+1}^{b}\right) \tag{A34}$$

$$P_{i,j}^{+} = \max\left(0, A_{i-\frac{1}{2},j}\right) - \min\left(0, A_{i+\frac{1}{2},j}\right) + \max\left(0, A_{i,j-\frac{1}{2}}\right) - \min\left(0, A_{i,j+\frac{1}{2}}\right) \tag{A35}$$

$$Q_{i,j}^{+} = \left(w_{i,j}^{\max} - w_{i,j}^{td}\right)\Delta V_{i,j} \tag{A36}$$

$$R_{i,j}^{+} = \begin{cases} \min\left(1, Q_{i,j}^{+}/P_{i,j}^{+}\right) & P_{i,j}^{+} > 0 \\ 0 & P_{i,j}^{+} = 0 \end{cases} \tag{A37}$$

$$P_{i,j}^{-} = \max\left(0, A_{i+\frac{1}{2},j}\right) - \min\left(0, A_{i-\frac{1}{2},j}\right) + \max\left(0, A_{i,j+\frac{1}{2}}\right) - \min\left(0, A_{i,j-\frac{1}{2}}\right) \tag{A38}$$

$$Q_{i,j}^{-} = \left(w_{i,j}^{td} - w_{i,j}^{\min}\right)\Delta V_{i,j} \tag{A39}$$

$$R_{i,j}^{-} = \begin{cases} \min\left(1, Q_{i,j}^{-}/P_{i,j}^{-}\right) & P_{i,j}^{-} > 0 \\ 0 & P_{i,j}^{-} = 0 \end{cases} \tag{A40}$$

(6) to solve the charge concentration

$$w_{i,j}^{n+1} = w_{i,j}^{td} - \Delta V_{i,j}^{-1}\left[A_{i+\frac{1}{2},j}^{C} - A_{i-\frac{1}{2},j}^{C} + A_{i,j+\frac{1}{2}}^{C} - A_{i,j-\frac{1}{2}}^{C}\right] \tag{A41}$$

where n indicates $n\Delta t$ and n+1 indicates $(n+1)\Delta t$.

Author details

Cheng Pan*, Ju Tang and Fuping Zeng

*Address all correspondence to: pancheng2036@gmail.com

School of Electrical Engineering, Wuhan University, China

References

[1] Morshuis PHF. Degradation of solid dielectrics due to internal partial discharge: Some thoughts on progress made and where to go now. IEEE Transactions on Dielectrics and Electrical Insulation. 2005;**12**(5):905-913

[2] Van Brunt RJ. Stochastic properties of partial-discharge phenomena. IEEE Transactions on Electrical Insulation. 1991;**26**(5):902-948

[3] Kreuger FH. Partial Discharge Detection in High-Voltage Equipment. Oxford: Butterworth-Heinemann; 1989

[4] Gutfleisch F, Niemeyer L. Measurement and simulation of PD in epoxy voids. IEEE Transactions on Dielectrics and Electrical Insulation. 1995;**2**(5):729-743

[5] Das P, Chakravorti S. Simulation of PD patterns due to a narrow void in different E-field distribution. Journal of Electrostatics. 2010;**68**(3):218-226

[6] Wu K, Suzuoki Y, Mizutani T, Xie H. A novel physical model for partial discharge in narrow channels. IEEE Transactions on Dielectrics and Electrical Insulation. 1999;**6**(2):181-190

[7] Illias HA, Chen G, Lewin PL. The influence of spherical cavity surface charge distribution on the sequence of partial discharge events. Journal of Physics D: Applied Physics. 2011;**44**(24):21

[8] Meek JM, Craggs JD. Electrical Breakdown of Gases. New York: Wiley; 1978

[9] Niemeyer L. A generalized approach to partial discharge modeling. IEEE Transactions on Dielectrics and Electrical Insulation. 1995;**2**(4):510-528

[10] Okamoto T, Kato T, Yokomizu Y, Suzuoki Y. PD characteristics as a stochastic process and its integral equation under sinusoidal voltage. IEEE Transactions on Dielectrics and Electrical Insulation. 2001;**8**(1):82-90

[11] Hikita M, Yamada K, Nakamura A, Mizutani T. Measurements of partial discharges by computer and analysis of partial discharge distribution by the Monte Carlo method. IEEE Transactions on Electrical Insulation. 1990;**25**(3):453-468

[12] Achillides Z, Georghiou GE, Kyriakides E. Partial discharges and associated transients: The induced charge concept versus capacitive modeling. IEEE Transactions on Dielectrics and Electrical Insulation. 2008;**15**(6):1507-1516

[13] Forssen C, Edin H. Partial discharges in a cavity at variable applied frequency part 2: Measurements and modeling. IEEE Transactions on Dielectrics and Electrical Insulation. 2008;**15**(6):1610-1616

[14] Hazlee AI. Measurement and simulation of partial discharges within a spherical cavity in a solid dielectric material [thesis]. Southampton: University of Southampton; 2011

[15] Kupershtokh AL, Karpov DI, Medvedev DA, Stamatelatos CP, Charalambakos VP, Pyrgioti EC, Agoris DP. Stochastic models of partial discharge activity in solid and liquid dielectrics. IET Science, Measurement and Technology. 2007;**1**(6):303-311

[16] Illias HA, Tunio MA, Bakar AHA, Mokhlis H, Chen G. Partial discharge behaviours within a void-dielectric system under square waveform applied voltage stress. IET Science, Measurement and Technology. 2014;**8**(2):81-88

[17] Illias HA, Tunio MA, Bakar AHA, Mokhlis H, Chen G. Partial discharge phenomena within an artificial void in cable insulation geometry: Experimental validation and simulation. IEEE Transactions on Dielectrics and Electrical Insulation. 2016;**23**(1):451-459

[18] Wu K, Suzuoki Y, Dissado LA. The contribution of discharge area variation to partial discharge patterns in disc-voids. Journal of Physics D: Applied Physics. 2004;**37**:1815-1823

[19] Novak JP, Bartnikas R. Effect of dielectric surfaces on the nature of partial discharges. IEEE Transactions on Dielectrics and Electrical Insulation. 2002;**7**(1):146-151

[20] Bartnikas R. Some observations concerning the influence of dielectric surfaces upon the PD behavior. IEEE Transactions on Dielectrics and Electrical Insulation. 2008;**15**(6):1488-1493

[21] Nikonov V, Bartnikas R, Wertheimer MR. The influence of dielectric surface charge distribution upon the partial discharge behavior in short air gaps. IEEE Transactions on Plasma Science. 2001;**29**(6):866-874

[22] McAllister IW. Electric field theory and the fallacy of void capacitance. IEEE Transactions on Electrical Insulation. 1991;**26**(3):458-459

[23] Pedersen A, Crichton GC, McAllister IW. The theory and measurement of partial discharge transients. IEEE Transactions on Electrical Insulation. 1991;**26**(3):487-497

[24] Pedersen A, Crichton GC, McAllister IW. The functional relation between partial discharges and induced charge. IEEE Transactions on Dielectrics and Electrical Insulation. 1995;**2**(4):535-543

[25] Gibalov VI, Pietsch GJ. The development of dielectric barrier discharges in gas gaps and on surfaces. Journal of Physics D Applied Physics. 2000;**33**:2618-2636

[26] Li J, Dhali SK. Simulation of microdischarges in a dielectric-barrier discharge. Journal of Applied Physics. 1997;**82**(9):4205-4210

[27] Sjoberg M, Serdyuk YV, Gubanski SM, Leijon MAS. Experimental study and numerical modelling of a dielectric barrier discharge in hybrid air–dielectric insulation. Journal of Electrostatics. 2003;**59**:87-113

[28] Georghiou GE, Morrow R, Metaxas AC. The effect of photoemission on the streamer development and propagation in short uniform gaps. Journal of Physics D: Applied Physics. 2001;**34**(2):200-208

[29] Segur P, Bourdon A, Marode E, Bessieres D, Paillol JH. The use of an improved Eddington approximation to facilitate the calculation of photoionization in streamer discharges. Plasma Source Science and Technology. 2006;**15**(4):648-660

[30] Morrow R, Lowke JJ. Streamer propagation in air. Journal of Applied Physics. 1997;**30**(4): 614

[31] Georghiou GE, Morrow R, Metaxas AC. A two-dimensional, finite-element, flux-corrected transport algorithm for the solution of gas discharge problems. Journal of Applied Physics. 2000;**33**(19):2453

[32] Ducasse O, Papageorghiou L, Eichwald O. Critical analysis on two-dimensional point-to-plane streamer simulations using the finite element and finite volume methods. IEEE Transactions on Plasma Science. 2007;**35**(5):1287-1300

[33] Zalesaka ST. Fully multidimensional flux-corrected transport algorithms for fluids. Journal of Computational Physics. 1979;**31**(3):335-362

[34] Sato N. Discharge current induced by the motion of charged-particles. Journal of Physics D Applied Physics. 2000;**13**(1):L3

[35] Nikonov V, Bartnikas R, Wertheimer MR. Surface charge and photoionization effects in short air gaps undergoing discharges at atmospheric pressure. Journal of Physics D: Applied Physics. 2001;**34**(19):2979-2986

[36] Pan C, Wu K, Du Y, Tang J, Tao XT, Luo Y. Simulation of cavity PD sequences at DC voltage by considering surface charge decay. Journal of Physics D: Applied Physics. 2017; **50**(20):205202

[37] Illias HA, Tunio MA, Mokhlis H. Determination of partial discharge time lag in void using physical model approach. IEEE Transactions on Dielectrics and Electrical Insulation. 2015; **22**(1):463-471

[38] Hozumi N, Michiue H, Nagae H. Time-lag measurement of void discharges for the clarification of the factor for partial discharge pattern. Proceedings of the Conference on Electrical Insulation and Dielectric Phenomena. 2000

[39] Allen NL, Ghaffar A. The variation with temperature of positive streamer properties in air. Journal of Physics D: Applied Physics. 1995;**28**(2):338-343

[40] Pan C, Wu K, Du Y, Tang J. Comparison of Sato's equation and Pedersen's theory to obtain gas discharge current. IEEE Transactions on Dielectrics and Electrical Insulation. 2016; **23**(3):1690-1698

Chapter 7

Numerical Approach to Dynamical Structure Factor of Dusty Plasmas

Aamir Shahzad, Muhammad Asif Shakoori,
Mao-Gang HE and Yan Feng

Additional information is available at the end of the chapter

http://dx.doi.org/10.5772/intechopen.78334

Abstract

The dynamical structure factor [$S(k,\omega)$] gives the information about static and dynamic properties of complex dusty plasma (CDPs). We have used the equilibrium molecular dynamic (EMD) simulations for the investigation of $S(k,\omega)$ of strongly coupled CDPs (SCCDPs). In this work, we have computed all possible values of dynamical density with increasing and decreasing sequences of plasma frequency (ω_p) and wave number (k) over a wide range of different combinations of the plasma parameters (κ, Γ). Our new simulation results show that the fluctuation of $S(k,\omega)$ increases with increasing Γ and it decreases with an increase of κ and N. Moreover, investigation shows that the amplitude of $S(k,\omega)$ increases by increasing screening (κ) and wave number (k), and it decreases with increasing Γ. Our EMD simulation shows that dynamical density of SCCDPs is slightly dependent on N; however, it is nearly independent of other parameters. The presented results obtained through EMD approach are in reasonable agreement with earlier known results based on different numerical methods and plasma states. It is demonstrated that the presented model is the best tool for estimating the density fluctuation in the SCCDPs over a suitable range of parameters.

Keywords: plasma parameter, current correlation function, dynamical structure factor, strongly coupled complex (dusty) plasmas, equilibrium molecular dynamics, complex dusty plasma

1. Introduction

Dynamical structure factor $S(k,\omega)$ is very actively investigated through theoretical, experimental and computer simulation for simple liquids as well as complex systems (dusty plasmas). The

$S(k,\omega)$ of the dusty plasmas is very significant to understand the dynamic behavior of dust particles in the complex plasmas. The subject of dusty plasma containing micron-size charged condensed particles has recently been actively investigated in the fields of science and technology. In addition, the investigation of dynamical behaviors is also studied in the areas of physics and chemistry of plasmas, ionized gases, and the space environment, environmental sciences, semiconductor plasma processing industries, nuclear energy generation and materials research. Dust in the atmosphere and in the entire universe exists in different shapes and sizes. Mostly dust particles are observed in solid form and sometimes also in liquid and gaseous forms. Current correlations and wave spectra's in the dusty plasma are generated due to dynamic motion of charged dust particles. The dust particles increase remarkable and unique fundamental physical property of ionized gases and dense plasma. Further dust particle increases future application of dusty plasma in industrial fields including nuclear fusion energy, material modification, and synthesis, environmental remediation, nano, aerospace and medical technologies.

1.1. History of dusty plasma

Current research on the dusty (complex) plasma becomes an interesting field in sciences and technologies. The term "plasma" was first used by the Langmuir in 1924. The most thrilling events in the field of dusty plasmas were occurred in 1980 for planet Saturn mission. Mendis in 1997 observed the bright comet in distant ancestor, which was the excellent cosmic laboratory for the investigation of dusty plasma and their dynamical and physical consequences. The other appearances of dusty plasmas were zodiacal light, the origin nebula, the noctilucent clouds, etc. At the laboratory level, dusty plasma is available in terrestrial laboratory at the remote past. Lyman Spitzer along with Hannes Alfven was recognized that dust in the universe was not merely a hindrance to a visual opinion but it was an essential component of the universe. A dust particle image taken a shape of spokes that rotating around Saturn ring and last surveys designated that these spokes were fine dust material. In 2005, Cassini spacecraft was made a new and improved observation of spokes with a feature that would provide an improved considerate of their source. In 1992, the European spacecraft Ulysses flew by the planet Jupiter and detected the dust particles and measured their masses and impact speed. Again in 1995, NASA spacecraft, Galileo perceived the origin of dust streams around Jupiter. The current enormous interest in complex plasmas started in the mid of 1980s, and started by laboratory investigation of dynamical structure factor of dusty plasmas. Current situation (2000–2017) of dusty plasma is stable in laboratory conditions and it is very significant in industries, science, and technology, medical science, and energy sectors, etc. Different characteristics of dust particles are investigating via theoretical, computer simulation and experimental techniques [1, 2].

1.2. Characteristics and types of dusty plasma

Dusty plasma is the plasma that contains an addition of dust charged particles along with electrons, ions, and neutral particles. The dynamical properties of plasma become more complex when we insert dust particle, so it is also known as complex plasma [3]. Dust particles are much larger in charge as compared to electrons, ions, and neutral particles and its size vary from hundreds of millimeter to 10 of nanometers and having a mass approximately 3×10^{-11} kg. The dynamic behaviors are easily observed experimentally by CCD camera. Dusty plasmas exist in space as well as in the laboratory and these dust particles are negatively charged but sometimes positive charge as well. It is charged through photoionization, electron bombardment, etc. the amount of charge on dust particles depends upon shape and size of dust particles. It has mostly spherical shape but sometimes also having rod type shape and irregular [4]. Dusty plasmas are classified on the basis of density, temperature, potential and thermal energies. The Coulomb coupling parameter describes the classification of complex dusty plasmas. The Coulomb coupling that is defined as "the ratio of average potential energy to the average thermal energy" and mathematically it can be expressed as: $\Gamma = \frac{\text{P.E}}{\text{K.E}} = \frac{Q^2}{4\pi\varepsilon_0 k_B}\frac{1}{T}$ [5], where k_B is the Boltzmann constant, T is system temperature and ε_0 is permittivity of free space.

1.2.1. Weakly coupled dusty (complex) plasmas (WCCDPs)

When dust charged particles have average thermal energy due to neighboring particles much larger than the average potential energy, then that plasma is known as weakly coupled complex dusty plasma (WCDPs). The WCCDPs have a high temperature but low density and value of Coulomb coupling parameter less than 1 (mathematically $\Gamma < 1$). In case of WCCDPs, any structural form does not exist. The background of WCCDPs is considered as ionized gases.

1.2.2. Strongly coupled dusty (complex) plasma (SCCDPs)

The SCCDPs is speedily emerging filed from last three decades. It is very significant to astrophysical plasma and quickly progressing in laboratory experiments. In this type of dusty plasma average thermal energy of charged particles due to neighboring particles are much smaller than average potential energy and mathematically $\Gamma \geq 1$. In case of SCCDPs, it has high density and low temperature and can be specified in structural farm [6–8].

1.3. Dusty plasma in atmosphere and laboratory

There are many systems in the atmosphere where dust particles are established. Spaces between the stars are filled with a large amount of dust and gases. Dust particles in the interstellar region are metallic i.e. graphite, magnetite, and amorphous of carbon, dielectric material i.e. silicates and ices, etc. Comets, planetary space, planetary ring, and earth atmosphere are the region of our solar system. Gossamer ring, halo ring, and main rig are the three systems of Jupiter's ring. In the Saturn rings systems are mostly ices and its size vary from meter to micron. A Uranian rings system has major rings such as 6, 5, 4, 3, α, β, η, γ. In Neptune's ring systems appear in curios twisted materials and structure is dirty ice and composition such as iron, nickel, sulfur, earth atmosphere dusty ice, etc. [9]. A simple device for producing dusty plasmas is a dusty plasma device which is a single-ended *Q*-machine modified to allow the dispersal of dust grains. Dusty plasmas are produced by suspending micron-sized dust particles in a stratum of a dc neon glow discharge. Dusty plasma has been for the first time confined in cylindrical symmetric radio frequency plasma (RF) system also in the semiconductor industry.

1.4. Role of dust particles and applications

Dust particles have charge and chemically active species, it is formed and growth in dusty plasma devices. Sometimes in the form of a mixture of gases such as SiH_4, silane, oxygen, O_2, and Ar, etc.

Secondly dust particles are formed in devices when atoms and molecules are spurted from walls and electrodes into the plasma by electron and ion bombardment. Moreover, the growth of dust particles in the plasma is coagulation, nucleation and surface growth. Thermal fluctuations and Coulomb interaction play the significant role for determination structure of SCCDPs. When the values of Γ (>1) increase then system organized from nonideal gases phase to ordered condensed phase. Dust particles are suspended in the gaseous plasma phase with few electrons temperature and charge up to 10^4 ordered. Interestingly, the dust clouds in a dusty plasma formed into the structural form even at room temperature. Dust particle has large mass as compared to ion and an electron which gives the results slowly downtime scale and it can easily observed the macroscopic structure and its dynamical behavior directly study in space and real time [10]. Dusty plasma used for nanocrystalline silicon particles grow in the silane plasmas used to increase efficiency and lifetime of the silicon solar cells. It is used for thin film coating applied in plasma-enhanced chemical vapor deposition (PECVD) for the improvement of material surface properties. Carbon-based nanostructure growth in the hydrocarbon plasmas or fluorocarbons used to produce thin carbon films. It is used to improve material properties such as chemical inertness, high hardness and wear resistance. Self-lubrication coating and wear resistance using different compound as a dust particle (MoS_2). Ar/CH_4 plasma used for making the nanocrystalline diamond films fabricated which as exceptional properties such as chemical inertness, high hardness, and extreme smoothness which used to improve the performance of cutting tools. Diamond whiskers fabricated by the etching in RF plasmas for the enhancement of electron field emission. The reactive ion etching (RIE) process are used to a precise efficiently sharpen micro-tips of diamond [11]. Complex (dusty) plasmas (CDPs) have various advantages in a different industries, technologies, and energy sector due to the existence of dust particles. CDPs are stable under the laboratory condition. The CDPs can also be used for the diagnostic purpose because dust particles are trapped at the room temperature and keep their desire dynamical state for hours. Dusty plasma frequency in the range several hertz and easily observed through CCD cameras. It is produced in the gas discharge tube with natural gas pressure range that varies from 1 to 100 pa, which is subject to moderate damping [8]. Moreover, magnetized dustyplasma device used to produce a number of the verity of magnetic fields configuration with the help of four independent superconducting coils [12]. Magnetized glow discharge dusty plasma device, RF plasma device, ISS experiment and DC glow discharge devices used for different applications in industries and diagnostic purpose of dusty plasma. Dust particles are found in a tokamak (fusion plasma) and dusty plasmas depositions techniques devices [11, 13].

1.5. Dynamical structure factor

The dynamical structure factor $S(k,\omega)$] gives the information about static and dynamic properties of the fluid in simple and complex systems. In hydrodynamic condition, the $S(k,\omega)$ provides experimental calculable quantities such as the thermal diffusivity, adiabatic sound velocity and the ratio of specific heats [8]. These properties of the fluids are measured through light scattering, x-ray and inelastic neutron experiments on a substance such as dense plasmas, liquids, and glasses. Sound waves are generated through $S(k,\omega)$ in strongly coupled CDPs (SCCDPs) and density is more strongly damped at liquid phase [14, 15]. The SCCDP is many body dynamical systems that show different collective excitations and their properties investigated through numerical simulation and theoretical approaches [16]. In condensed matter physics, the pair correlation function usually cannot be determined directly. Rather, the structure factor is determined by scattering of x-rays or neutrons. In dusty plasmas, we are able to measure the pair correlation function directly and to calculate the structure factor in order to compare with condensed matter experiments. The $S(k,\omega)$ is just the Fourier transform of the pair correlation function. Fluctuation of the dynamical density of dusty plasma generates current correlation spectra such as longitudinal and transverse currents [17]. Dynamic ion structure factor of warm dense matter and dense plasma consist of the complete information of ions in strongly interacting systems and also influenced by the electrons property. It is closely associated with density fluctuation, thus determines transport properties and many relaxations such as electrons ion temperature equilibrium and stopping power and also the equation of state. It is also used for diagnostics of the extreme states of matter like a warm dense matter of x-ray Thomson scattering [18]. Dynamical scattering function is given through times correlation function, fluctuated density of liquid argon and light scattering function [19].

2. Numerical model and simulation techniques

In this section, we have implemented molecular dynamic (MD) simulation code with Ewald summation for forces and energies which makes it possible to account the long-range Coulomb interparticle interactions. We trace the motion of single charge species and integrated through leapfrog method and assume that the presence of neutralizing homogenous background. In this plasma environment, random fluctuating forces and friction forces are acting on a charged particle in addition to which forces initiating from the interaction of charged particle. Length of simulation cubic box is defined as $\left(\frac{4\pi N}{3}\right)^{1/3}$ and particles have a random spatial configuration for the beginning of simulation [15, 20]. Fluctuation of microscopic density is observed for different plasma parameters approaching near the equilibrium state [21]. The presented study includes the solution of the equation of motion of a system and particle interacts with each other through Yukawa potential. Provided that an accurate potential can be established for the system of attention under study and equilibrium MD (EMD) can be used irrespective of the phase condition and thermodynamic of the system involved. Yukawa potential is most commonly used potential (screened Coulomb) for SCCDPs including many physical systems such as physics of chemical and polymer, medicine and biology systems, astrophysics, environmental, etc. Major advantage for using this potential is that it reduces the calculation time compared other potentials [22]. The interaction potential energy of a charged particle in Yukawa liquid is given

$$\phi(\mathbf{r}) = \frac{Q^2}{4\pi\varepsilon_o}\frac{e^{\frac{-r}{\lambda_D}}}{|\mathbf{r}|} \quad (1)$$

Here Q is the charge on dust particle, r is the distance between interacting particles, λ_D is Debye screening length that accounts for the screening of interaction of other plasma species and ε_o is permittivity of free space. The scaling (dimensionless) parameters, which fully characterized the system, one is known as Coulomb coupling parameter [23],

$$\Gamma = \frac{Q^2}{4\pi\varepsilon_o a_{ws}} \frac{1}{k_B T} \tag{2}$$

where, a_{ws} is the "Wigner-Seitz" radius and it is defined as $\pi n^{\frac{-1}{2}}$ with n is the equilibrium dust number density, k_B is the Boltzmann constant and T is absolute system temperature. It is noted that the Γ is measured as the ratio of average potential energy to average kinetic energy per particle. Second scaling parameter is Debye screening parameter and it is given as $\kappa = a_{ws}/\lambda_D$.

The EMD simulations are performed for a particle number that is chosen between 500 and 1000 particles in a microcanonical ensemble using periodic boundary conditions and minimum image convention of the dust particles. It is to be mentioned that the number density (n) is defined as $n = N/V$, here N is the number of particles and V is the volume of simulation box and it is calculated as $V = 4\pi N/3$. On our case, most of simulations are performed for $N = 500$ and it is observed that the mentioned number of particles is suitable for EMD computations with statistical uncertainty limits. In presented case, the simulation time step is $dt = 0.0001$ and total simulation time limit was 425,000 step units. The EMD simulations are run between 4.25 10^6 ($1/\omega_p$) to 3.25 10^6 ($1/\omega_p$) time units for each combination of (Γ, κ) in the series of recording dynamical structure factor (DSF), $S(k,\omega)$, of SCCDPs. It can be seen that the first patch of $S(k,\omega)$ results is obtained after the time limit of (38000) step unit. For our case, 13 patches of $S(k,\omega)$ results are obtained and results show that each patch has nearly behavior of $S(k,\omega)$ showing the accuracy of numerical algorithm. In this work, the dynamical structure factor computations of SCCDPs are reported for a wide range of Coulomb coupling parameters $\Gamma \equiv (1, 200)$ and the Debye screening strength $\kappa \equiv (1.4, 4)$.

2.1. Model of DSF [$S(k,\omega)$]

The number density of single species also known as mass density and dimensionless quantity in molecular dynamic units can be written as

$$\rho(r,t) = \sum_j \delta(r - r_j(t)) \tag{3}$$

here r is the point at time t. from the practical point of view, it is known as local density. The average occupancy in the small volume of r space and it calculates over the short interval of time. The definition of density is satisfied matter conserved requirement.

$$f\rho(r,t)dr = N_m \tag{4}$$

The space and time-dependent density correlation are explain from the van Hove correlation function which is define as $G(r, t) = G_s(r, t) + G_s(r, t)$ and further detail is given in Ref. [23]. Fourier transform of density fluctuation and new expression is given as

$$\rho(k,t) = f\rho(r,t)\, e^{-ik.r}\, dr = \sum_j e^{-ik.r(t)} \tag{5}$$

where k is the wave number and becomes equal to $k = 2\pi/L$ and L is the length of simulation box. The intermediate function can be defined as

$$F(k,t) = \langle \rho(k,t)\rho(-k,t) \rangle \tag{6}$$

The relation between static structure factors with the dynamic structure factor is

$$S(k,\omega) = \frac{1}{2\pi}\int_{-\infty}^{\infty} F(k,t)\, e^{i\omega t}\, dt \tag{7}$$

This satisfies the sum rule as

$$\int_{-\infty}^{\infty} S(k,\omega) d\omega = S(k) \tag{8}$$

The dynamical structure function is related to longitudinal current correlation function, which is expressed as [22].

$$S(k,\omega) = \frac{1}{\omega^2} C_L(k,\omega) \tag{9}$$

3. Results and discussion

The EMD simulation has been used for the calculation of $S(k,\omega)$ to understand the dynamic phenomenon of particles for 3D SCCDPs. We have analyzed our simulation results of $S(k,\omega)$ in term of frequency, amplitude and fluctuation rate with respect to these parameters (κ, Γ, and k).

This section shows an overview of our results obtained through EMD simulations for dynamical structure factor $S(k,\omega)$ function of SCCDPs at $\kappa = 1, 2, 3$ and 5 with $N = 500$. **Figures 1–4** shown our results for $S(k,\omega)$ over a wide suitable range of plasma states (Γ, κ) at four values of wave vectors $\boldsymbol{k}$ = (0, 1, 2 and 3). A sequence of dynamical structure factor in increasing of wave vector ($\boldsymbol{k}$) is computed to determine the suitable equilibrium values of $S(k,\omega)$. We have performed 16 EMD simulation with $N = 500$ for each screening parameters at different four values of $\boldsymbol{k}$. There are 64 simulations are carried out for different combination of plasma parameters in order to observe complete behaviors of DSF $S(k,\omega)$ at higher varying frequency (ω_p) as compared to earlier simulation results [16]. It is observed that the presented results obtained through EMD simulations for varying parameters have suitable signal-to-noise ratio of the DSF. Our results are satisfactory good agreement with earlier EMD estimations and show that the presented EMD results at higher ω_p and earlier EMD simulations have comparable performance with small system size, both yielding the close values of the DSF $S(k,\omega)$. Moreover, the system temperature ($1/\Gamma$), strength of Debye screening (κ), system run time (total time), system size (N), and wave vector ($\boldsymbol{k}$) are changed to observe how efficiently the presented EMD method computes the DSF $S(k,\omega)$ of SCCDPs.

In addition, in each panels of **Figures 1–4**, we have shown the behaviors of $S(k, \omega)$ at four values of $\boldsymbol{k}$. All these data are excellent due to the good statistics and allow the analysis of the structure–function within a broad dynamical range. It is examined that the peaks of $S(k,\omega)$ are decay at lower Γ values for different four values of κ and $\boldsymbol{k}$. It can be noted that the peaks survive up to higher k values at intermediate to higher Γ values and shift toward the higher frequency (ω_p). It is further observed that the wave spectra of $S(k,\omega)$ shifts toward sinusoidal waves form, at the intermediate values of $\Gamma = 50$ and wave spectra shifts toward square type waves forms at higher values of (Γ = 100,200). Moreover, the panels (a) to (d) of each figures represent the results of $S(k,\omega)$ in nonideal gases state to liquid and crystalline order state

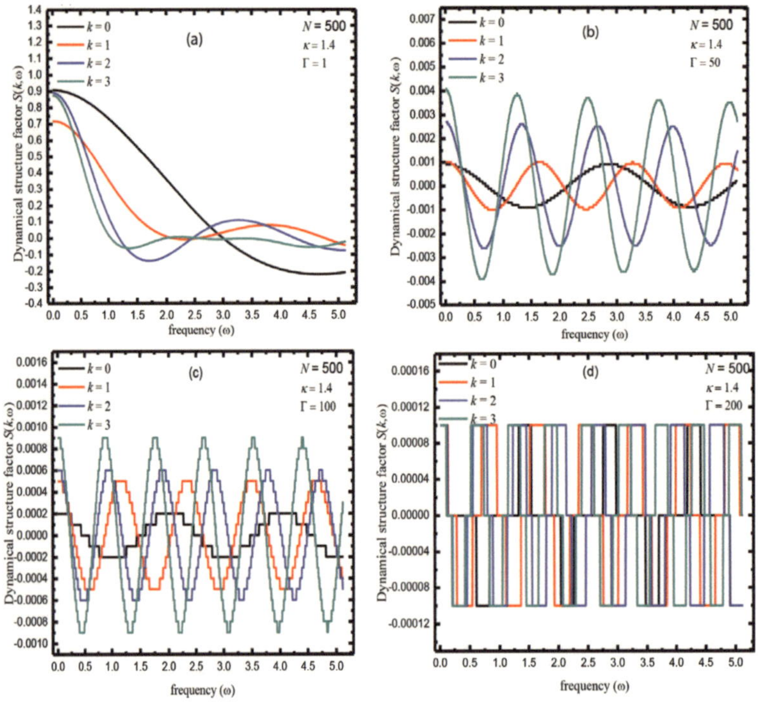

Figure 1. Variation of dynamical structure factor $S(k,\omega)$ as a function of plasma frequency (ω) of strongly coupled complex plasma at $\kappa = 1.4$, $N = 500$ and waves number $k = 0, 1, 2,$ and 3 for (a) $\Gamma = 1$, (b) $\Gamma = 50$, (c) $\Gamma = 100$, (d) $\Gamma = 200$.

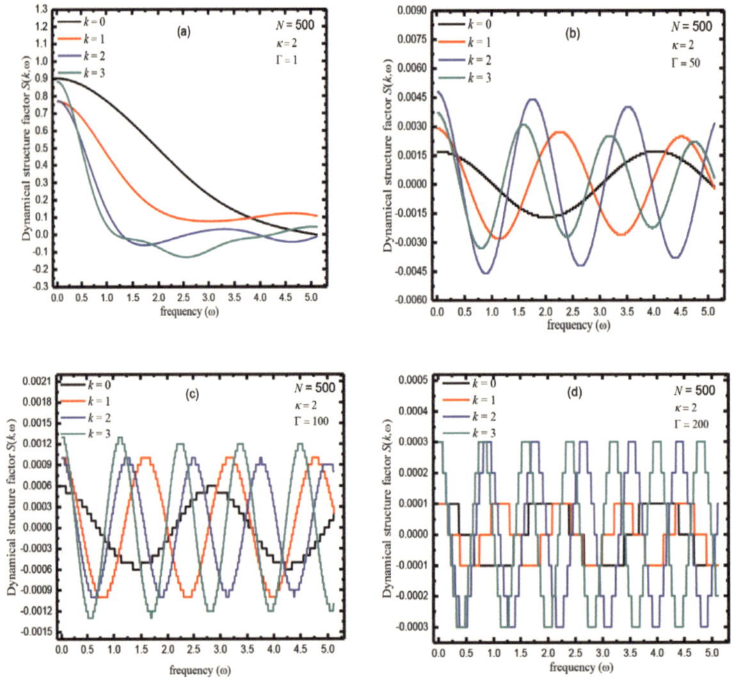

Figure 2. Variation of dynamical structure factor $S(k,\omega)$ as a function of plasma frequency (ω) of strongly coupled complex plasma at $\kappa = 2$, $N = 500$ and waves number $k = 0, 1, 2,$ and 3 for (a) $\Gamma = 1$, (b) $\Gamma = 50$, (c) $\Gamma = 100$, (d) $\Gamma = 200$.

http://dx.doi.org/10.5772/intechopen.78334

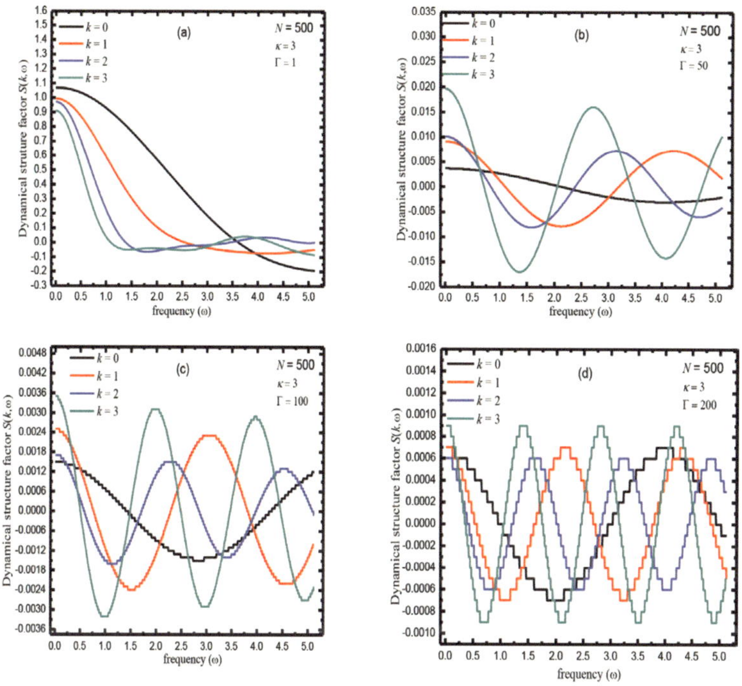

Figure 3. Variation of dynamical structure factor $S(k,\omega)$ as a function of plasma frequency (ω) of strongly coupled complex plasma at $\kappa = 3$, $N = 500$ and waves number $k = 0$, 1, 2, and 3 for (a) $\Gamma = 1$, (b) $\Gamma = 50$, (c) $\Gamma = 100$, (d) $\Gamma = 200$.

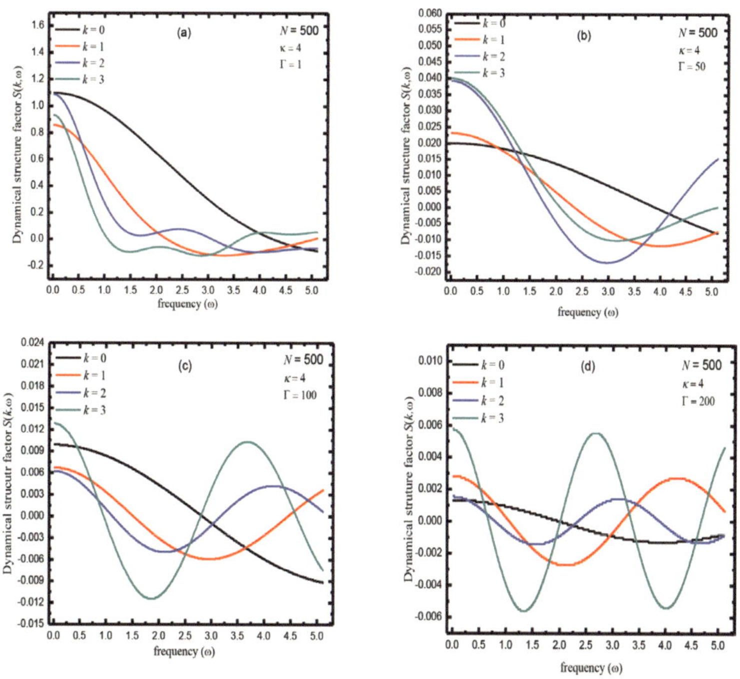

Figure 4. Variation of dynamical structure factor $S(k,\omega)$ as a function of plasma frequency (ω) of strongly coupled complex plasma at $\kappa = 4$, $N = 500$ and waves number $k = 0$, 1, 2, and 3 for (a) $\Gamma = 1$, (b) $\Gamma = 50$, (c) $\Gamma = 100$, (d) $\Gamma = 200$.

corresponding to Γ (1, 200). In case of crystalline order state, the amplitude of vibrating particle decreases significantly and it converts completely in square waveform. It is interesting to note that the fundamental behavior of dust particles is different and it shows decaying trends in nonideal gases state, sinusoidal form in liquid state and square wave form in crystalline state at fixed screening value. Figures show that the results of $S(k,\omega)$ depend on the plasma parameters (κ, Γ), as expected. Furthermore, it is investigated that in the ideal form of dusty plasma with high temperatures (low coupling values) the dust particles are exponentially decaying from high to low amplitude. One of justification is the dust particles transferred their energy to the surrounding particles that at high plasma temperature values. Furthermore, in panels (b) represent the EMD results of $S(k,\omega)$ in the liquefied state of dusty plasma. In this regime, dust particles comparatively less transfer their energy to the system. At this regime amplitude is maximum for simulation box size (L = max) and frequency of oscillating of dust particles is low and in the sinusoidal waveform. The frequency and amplitude of oscillating dust particles increase with increasing wave numbers k = 2, 3 and exhibit in the periodic wave form. Panels (c) show the results of $S(k,\omega)$ of dusty plasma in the nearly crystalline states. We have analyzed that in this regime the particles are tightly bound. In this case, the dust particles have small amplitude as compared to the gaseous and liquefy forms of dusty plasma.

It is observed from each panel of **Figures 1–4**, the dynamic of dust particles increases with increasing wave number. It is to be noted that the values of amplitude of the $S(k,\omega)$ are 3 = 0.7406, 0.6662, 0.8004 and 0.7902 respectively, for four wave numbers (k = 0, 1, 2, and 3) at the same values of κ, Γ and N. It is observed that the dynamical structure factor of SCCDPs depends on plasma parameters (κ, Γ). The frequency mode of $S(k,\omega)$ is high at low κ for SCCDPs. It is observed that κ is equally affecting on the dynamic of dust particles either the dusty plasma in any phase (gaseous, liquid and crystalline).

4. Summary

We have employed EMD simulations for the investigations of $S(k,\omega)$ over a wide range of Coulomb and Debye screening parameters (κ, Γ). It has been shown that the presented EMD technique and previous numerical methods have equivalent performance for the wide range of plasma state points, both yielding satisfactory data for plasma $S(k,\omega)$. New simulations provide more consistent and inclusive results for the plasma $S(k,\omega)$ over a complete range of Γ (1, 200) and κ (1.4, 4) than the previously known numerical results. Our investigations show that the dynamic of dust particles are exponentially decayed, sinusoidal form and crystalline form, respectively, in the gaseous, liquefy regime and crystalline states of SCCDPs. Moreover, the dynamical spectra of dusty plasma do not observe at very high values of Γ and low values of κ. Moreover, our results indicate that the dynamical structure factor in SCCDPs depends on κ, Γ and k. It has been shown that the presented simulation has comparable performance with the earlier simulation of $S(k,\omega)$ over the wide domain of plasma states. For future work, it is suggested that presented EMD technique based Ewald summation can be used to investigate and explore dynamical structure factor behaviors in other simple liquid, dipolar and ionic materials.

http://dx.doi.org/10.5772/intechopen.78334

Acknowledgements

The authors are grateful to the National Advanced Computing, National Center of Physics (NCP), Pakistan, and National High Performance Computing Center of X'ian Jiaotong. The authors thank the University, P.R. China, for allocating computer time to test and run our MD code.

Abbreviation and symbols

EMD	equilibrium molecular dynamics
MD	molecular dynamics
SCCDPs	strongly coupled complex dusty plasmas
WCCDPs	weakly coupled complex dusty plasmas
CDPs	complex (dusty) plasma
DSF	dynamical structure factor
PECVD	plasma-enhanced chemical vapor deposition
$S(k,\omega)$	dynamical structure factor
k	wave number
N	number of particles
ω_p	plasma frequency
κ	Debye screening strength
Γ	Coulomb coupling

Author details

Aamir Shahzad[1,2]*, Muhammad Asif Shakoori[1], Mao-Gang HE[2] and Yan Feng[3]

*Address all correspondence to: aamirshahzad_8@hotmail.com

1 Molecular Modeling and Simulation Laboratory, Department of Physics, Government College University Faisalabad (GCUF), Faisalabad, Pakistan

2 Key Laboratory of Thermo-Fluid Science and Engineering, Ministry of Education (MOE), Xi'an Jiaotong University, P. R. China

3 Center for Soft Condensed Matter Physics and Interdisciplinary Research, College of Physics, Optoelectronics and Energy, Soochow University, Suzhou, China

References

[1] Shukla PK, Mendis DA, Desai T. Conference Proceeding. Advances in Dusty Plasmas. Singapore: World Scientific; 1997. QB791.153671996

[2] Gregori G. Molecular dynamics simulations of the equilibrium dynamics of non-ideal plasmas. (Doctoral dissertation). University of Oxford;2012

[3] Merlino RL. Dusty plasmas and applications in space and industry. Iowa City USA. Plasma Physics Applied. 2006;**81**:73-110. ISBN: 81-7895-230-0

[4] Shahzad A, Shakoori MA, He M-G, Bashir S. Sound waves in complex (Dusty) plasmas. In: Computational and Experimental Studies of Acoustic Waves. Rijeka, Croatia: InTech; 2018. DOI: doi.org/10.5772/intechopen.71203

[5] Shahzad A, H M-G, Shakoori MA. Thermal transport and non-newtonian behaviors of 3D complex liquids using molecular simulations. In: Applied Sciences and Technology (IBCAST), 2017 14th International Bhurban Conference on (pp. 472-474). USA: IEEE; 2017 January. DOI: 10.1109/IBCAST.2017.7868096

[6] Bellan PM. Fundamentals of Plasma Physics. UK: Cambridge University Press; 2008. ISBN: 9780521528009

[7] Shahzad A, He MG. Thermal conductivity of three-dimensional Yukawa liquids (dusty plasmas). Contributions to Plasma Physics. 2012;**52**(8):667. DOI: 10.1002/ctpp.201200002

[8] Killian T, Pattard T, Pohl T, Rost J. Ultracold neutral plasmas. Physics Reports. 2007;**449**: 77. DOI: 10.1016/j.physrep.2007.04.007

[9] Shukla PK, Mamun AA. Introduction to Dusty Plasma Physics. New York: CRC Press; 2015. ISBN 0 7503 0653

[10] Bonitz M, Henning C, Block D. Complex plasmas: A laboratory for strong correlations. Reports on Progress in Physics. 2010;**73**(6):066501. DOI: 10.1088/0034-4885/73/6/066501

[11] Schoen M, Vogelsang R, Hoheisel C. Computation and analysis of the dynamic structure factor S(k, ω) for small wave vectors: A molecular dynamics study for a Lennard-Jones liquid country. Molecular Physics. 1986; **57**(3):445-471. DOI: doi.org/10.1080/00268978600100351

[12] Thomas E, Konopka U, Artis D, Lynch B. The magnetized dusty plasma experiment. Cambridge University Press; 2015. DOI:http://doi.org/10.1017/s0022377815000148

[13] Alley WE, Alder BJ, Yip S. The neutron scattering function for hard spheres. Physical Review A. 1983;**27**(6):3174. https://doi.org/10.1103/PhysRevA.27.3174

[14] Magyar P, Hartmann P, Kalman GJ, Golden KI, Donkó Z. Factorization of 3-point static structure functions in 3D Yukawa liquids. Contributions to Plasma Physics. 2016;**56**(9): 816-829. DOI: 10.1002/ctpp.201500062

[15] Arkhipov YV, Askaruly A, Davletov AE, Dubovtsev DY, Donkó Z, Hartmann P, Korolov I, Conde L, Tkachenko IM. Direct determination of dynamic properties of coulomb and

Yukawa classical one-component plasmas. Physical Review Letters. 2017;**119**(4). DOI: 045001, https://doi.org/10.1103/PhysRevLett.119.045001

[16] Donkó Z, Hartmann P, Kalman GJ, Golden KI. Simulations of strongly coupled charged particle systems: Static and dynamical properties of classical bilayers. Journal of Physics A: Mathematical and General. 2003;**36**(22):5877. DOI: https://doi.org/10.1088/0305-4470/36/22/307

[17] Vorberger J, Donko Z, Tkachenko IM, Gericke DO. Dynamic ion structure factor of warm dense matter. Physical Review Letters. 2012;**109**(22):225001. DOI: https://doi.org/10.1103/PhysRevLett.109.225001

[18] Enciso E, Almarza NG, del Prado V, Bermejo FJ, Zapata EL, Ujaldón M. Molecular-dynamics simulation on simple fluids: Departure from linearized hydrodynamic behavior of the dynamical structure factor. Physical Review E. 1994;**50**(2):1336. DOI: https://doi.org/10.1103/PhysRevE.50.133

[19] Donkó Z, Hartmann P, Kalman GJ. Molecular dynamics simulations of strongly coupled plasmas: Localization and microscopic dynamics. Physics of Plasmas. 2003;**10**(5):1563-1568 https://doi.org/10.1063/1.1560612

[20] Piel A, Nosenko V, Goree J. Experiments and molecular-dynamics simulation of elastic waves in a plasma crystal radiated from a small dipole source. Physical Review Letters. 2002;**89**(8):085004. DOI: https://doi.org/10.1103/PhysRevLett.89.085004

[21] Shahzad A, He M-G. Numerical experiment of thermal conductivity in two-dimensional Yukawa liquids. Physics of Plasmas. 2015;**22**(12):123707 https://doi.org/10.1063/1.4938275

[22] Rapaport DC, Blumberg RL, McKay SR, Christian W. The art of molecular dynamics simulation. Computers in Physics. 1996;**10**(5):456-456 https://doi.org/10.1063/1.4822471

[23] Shahzad A, Aslam A, He M-G. Equilibrium molecular dynamics simulation of shear viscosity of two-dimensional complex (dusty) plasmas. Radiation Effects and Defects in Solids;**169**(11):931-941 https://doi.org/10.1080/10420150.2014.968852

Section 2

Modern Applications of Plasma

Chapter 8

A Test Facility to Investigate Sheath Effects during Ion Cyclotron Resonance Heating

Kristel Crombe, Rodolphe D' Inca, Eric Faudot, Helmut Faugel, Ana Kostic, Mariia Usoltceva, Jean-Marie Noterdaeme, Anton Nikiforov, Helmut Fuenfgelder, Stephane Heuraux, Jonathan Jacquot, Fabrice Louche, Roman Ochoukov, Ilya Shesterikov and Dirk Van Eester

Additional information is available at the end of the chapter

http://dx.doi.org/10.5772/intechopen.76730

Abstract

Nuclear fusion is a promising candidate to supply energy for future generations. At the high temperatures needed for the nuclei to fuse, ions and electrons are no longer bound into atoms. Magnetic fields confine the resulting plasma. One of the heating methods is the ion cyclotron resonant absorption of waves emitted by an external Ion Cyclotron Radio Frequency (ICRF) antenna. The efficiency of ICRF heating is strongly affected by rectified RF electric fields at antenna and other in-vessel components (so-called 'sheath effects'). The chapter presents an overview of ICRF principles. Attention is given to characterising the detrimental sheath effects through experiments on a dedicated test facility (IShTAR: Ion cyclotron Sheath Test ARrangement). IShTAR has a linear magnetic configuration and is equipped with an independent helicon plasma source. The configuration and capabilities of the test-bed and its diagnostics are described, as well as an analysis of the plasmas.

Keywords: plasma, helicon, radio frequency, sheaths, diagnostics

1. Introduction

IShTAR (Ion cyclotron Sheath Test ARrangement) is a linear magnetised plasma test facility for RF sheaths studies at the Max-Planck-Institut fuer Plasmaphysik in Garching (IPP-Garching), Germany. The test facility consists of a cylindrical vacuum vessel with a diameter of 1 m and length of 1.1 m. The plasma is created by an external cylindrical plasma source equipped with a helical antenna that has been designed to excite the m = 1 helicon mode. In contrast to a tokamak, a test stand provides more liberty to impose the parameters and gives better access for the instrumentation and antennas. The project also supports the development of diagnostic methods for characterising RF sheaths and validating and improving the theoretical predictions.

The present specific application is to analyse the formation of RF sheaths when waves in the Ion Cyclotron Range of Frequencies (ICRF) are injected in the plasma. In tokamaks, they have been successfully used to heat the plasmas to the nuclear fusion relevant temperatures of around 10 keV [1], but experimental and theoretical studies have shown that several spurious edge plasma interactions can prevent an optimal coupling if effects are forgotten or misunderstood. Particularly, the RF sheaths are created on the antenna surface and on limiters by the large $E_\parallel$ (parallel to the background magnetic field) component of the Slow Wave (SW) from the antenna [2, 3]. Under this field, the electrons with a lighter mass are more prone to be expelled from the plasma than the heavier ions: this creates a net positive DC voltage, called rectified sheath potential which conserves the charge ambipolarity. The additional DC potential raises the energy of the ions knocking the wall, the sputtering increases, hot spots are generated and more impurities are released into the plasma. At the same time the RF power losses become more pronounced [4–6]. Even though extensive studies have been carried out in the last years and the understanding has improved [7], a definitive solution is still pending because a systematic investigation cannot be done efficiently in the main fusion experiments. There are several reasons: the experimental time allocated to the topic is limited; the access, location and operations of diagnostics dedicated to RF sheath properties is constrained by other instruments, by the machine operation parameters and by the plasma itself. The application of corrective measures or the test of new antenna designs is constrained by the limited opening time of the tokamak and by the interfaces with the access port and the surrounding wall. Therefore, a dedicated test-bed offering more working time has been developed and assembled, in which it is possible to launch the waves using different antenna designs. More time can be devoted to the analysis of RF sheaths effects. New diagnostics to measure plasma parameters and electromagnetic fields in front of the ICRF antenna can be tested and the results can be compared with existing theoretical models of the sheath rectification effect. Different types of solutions emerging from modelling can be tested. In addition, other interesting phenomena related to antenna-plasma coupling can be analysed: sheath-plasma waves and resonances, effect of fast ions, plasma breakdown and wall conditioning by the ICRF antenna. In the first part, the constraints, which guided the initial choice and the operating parameters, are discussed. It is followed by a description of the resulting design and architecture of the facility. In the last part, first operations and measurements are illustrated, to conclude with the next steps and extension plans for the diagnostic instrumentation.

2. Requirements and constraints

In this section the requirements, which were used to guide the design of the test-bed, are discussed. IShTAR is designed with the main purpose of investigating the RF sheath effects [8]. It is thus important to relate its functional parameters to the characteristics of a tokamak in the vicinity of the ICRF antenna.

2.1. Functional requirements

The main requirement for a test-bed dedicated to the analysis of ICRF antenna/plasma coupling, independent of the type of topic studied, is the geometry of the configuration, which has to be as similar as possible to the tokamak edge. This requires a vacuum vessel with a curved wall, a support to mount the ICRF antenna and a port for the RF feeding lines. A magnetised plasma is present and located only few centimetres away from the antenna (typically 15–20 cm). To simplify the configuration the magnetic lines are straight. Two main design factors have an important impact for the sheath studies: the plasma parameters and the antenna frequency. The former constrains the design of the plasma source and the second the choice of generator and matching system for the ICRF antenna. These factors determine the type of waves that are coupled and whether they are able to propagate, or not, in the plasma. The main parameters for the plasma are: the gas type, magnetic field, density and temperature. In a homogeneous, magnetised plasma, the wave equation reads:

$$\bar{k} \times (\bar{k} \times \bar{E}) + k_0^2 \bar{K} \cdot \bar{E} = 0, \tag{1}$$

where $\bar{k}$ is the wave vector, $\bar{E}$ the electric field associated with the wave, ω the angular frequency of the wave in vacuum, $k_0 = \omega / c$ the vacuum wavelength with c the speed of light and $\bar{K}$ is the plasma dielectric tensor. A detailed study of the wave propagation in IShTAR can be found in [9]. The main purpose of the test-bed is to provide conditions where the dispersion relation of the fast wave (used for heating) and of the slow wave (which produces the RF sheath) are similar to the tokamak case. Cold plasma calculations have been done for the reference case of a tokamak. It is plotted in **Figure 1b**. The operational parameters (magnetic field, ICRF frequency) of the test-bed have been varied to get similar wave behaviour as for the tokamak case, where slow and fast waves are well distinguished (**Figure 1a**).

2.2. Operational requirements

The main operational constraint for the operation and maintenance of the machine is the flexibility and accessibility: this is what makes the difference between the tokamak and the test-bed. We will see in the design section that the requirements lead to a machine with an important level of complexity. It is necessary that it does not impede an easy access to the machine or impose long time of preparation before the experiments. Therefore we have the following requirements on the operations and on the maintenance:

1. Start-up sequence duration shorter than an hour: from initial sleep phase at atmospheric pressure to first plasma in vacuum with full instrumentation. This has an impact on the

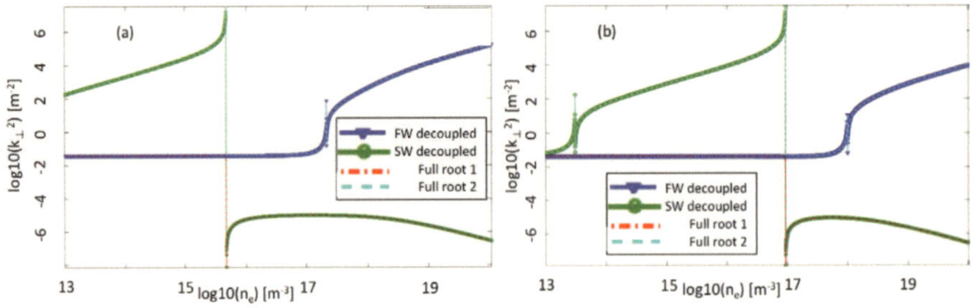

Figure 1. $k^2_{\perp}$ values for full and decoupled solutions as a function of the plasma density. (a) The reference case of *ISh*TAR for typical operating conditions: $f = 5$ MHz, $Bt = 0.1$T, $Te = 10$ eV and plasma composition: 100% Ar. (b) A tokamak scenario with: $f = 50$ MHz, $Bt = 2$T, $Te = 10$ eV, plasma composition: 95% D, 5% H. The notation "log10" denotes sign (F) log10 (|F|) when |F| > 1 and 0 when |F| < 1. It is introduced to enable capturing the very different length scales of the modes the cold plasma supports while allowing to identify regions of wave propagation and evanescence by a mere glance of the sign of log10 ($k\perp^2$). In the low density region, both roots approach the vacuum limit, $k\perp^2 = k0^2 - k//^2$, as expected, while at higher densities representative for the core one root is propagative (fast wave FW) and the other is deeply evanescent (slow wave SW).

speed of the vacuum pumps and on the control system, which has to be fully automatized. The purpose is to be able to do short iterations between experiments and analysis.

2. Short vessel opening and closing durations: it should be possible to open the vessel, install new components (like an antenna) and close the vessel in a couple of days; here again, it helps to have short iteration times between the installation of new solutions, their test and the analysis of their impact.

3. The "plug and play" instrumentation platform: the test-bed enables the analysis of many different phenomena requiring different types of diagnostics, either of in-house origin or provided by external teams. The control and acquisition system has to be universal enough to reduce the time between the installation of the instrument and the exploitation of its data in the centralised discharge database.

2.3. Cost control

The last constraint comes from the costs. To keep the costs under control, not all components were designed from scratch. Several have been refurbished from previous experiments. Among the most important ones is the main vacuum vessel with its 8 kA magnetic coils, the smaller coils and the ICRF generator, which comes from the WEGA stellarator. The downside is that their range of operation or geometry can limit the operational parameter space. This overview of the requirements clarifies the guidelines for the design of the facility.

3. Design and setup

The main design choice concerns the creation of the plasma in front of the ICRF antenna: either the ICRF itself can breakdown the gas or an external source can be used. Both solutions have been retained for the final design.

The ICRF antenna itself can create plasma. The electrons are heated by the parallel component (parallel to the B-field) of the RF electric field. The advantage would be that no other means than the ICRF components are needed to create the plasma. But there are three disadvantages: first, the plan is to test different types of antenna and it will be difficult to reproduce the same plasma with different geometries. Second, the physics of plasma breakdown with ICRF antennas is still not completely understood. And third, the parallel electric field used for the breakdown is the same that creates the RF sheath: it may be difficult to disentangle both phenomena [10].

An external plasma source solves this problem by separating the plasma creation from the plasma heating. The concept of a helicon discharge [11–13] has been retained: it is compatible with the magnetic field and it creates high densities in large volumes. However, the solution has a price: the physics of helicon sources is still an area of investigation; it requires large amount of power to ionise the plasma volume required; it adds a new wave inside the plasma, which can interact—depending on the frequency—with the ICRF waves. Therefore, a backup solution exists with the use of a more classical inductive coil.

The test-bed has therefore two large components: the main vessel where the ICRF antenna is installed and connected to the power transmission lines; and the plasma source, connected with an open port to the main vessel, which generates the plasma that will flow in front of the antenna. These systems are connected to power generators, gas feeding lines, DC current modules and real-time controllers to monitor the operations and the safety of the test-bed. An overview of the facility is depicted in **Figure 2**; the characteristics of different components are explained in the next subsections.

3.1. Vacuum vessel

The vacuum vessel has a cylindrical geometry. The length is 1126 mm and the diameter (D) is 1000 mm. There are five ports in the wall: two horizontal ports (D = 225 mm) on one side which

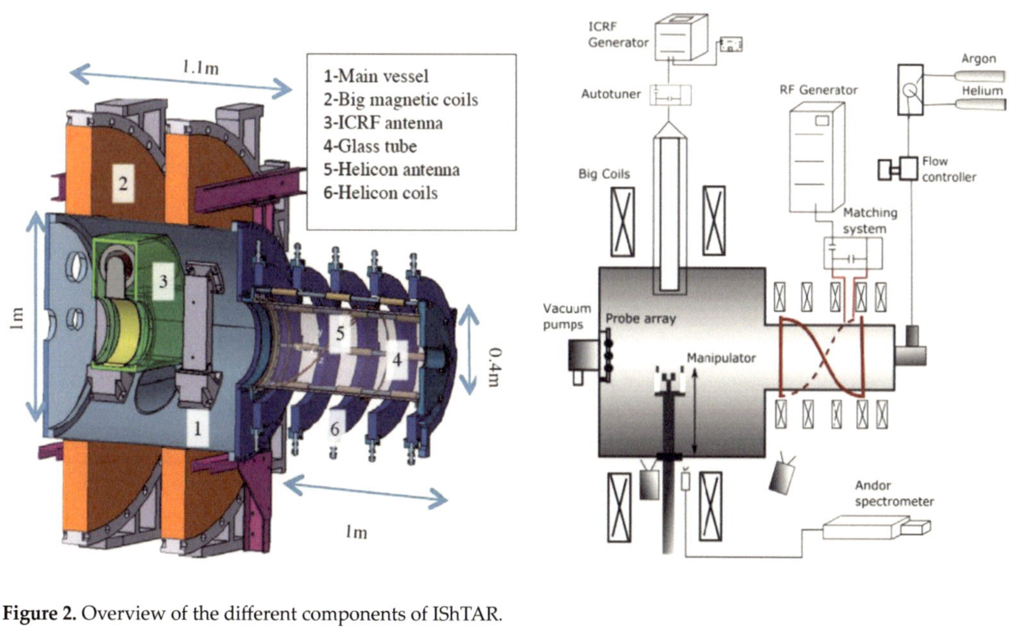

Figure 2. Overview of the different components of IShTAR.

can be used to connect the RF transmission line for the ICRF antenna and pressure gauges and valves, and, on the opposite side, one large horizontal port (D = 225 mm) and two smaller ports (D = 100 mm) angled at approximately 30 degrees with respect to the horizontal line. All ports have the same axial position, at 391 mm from the back end of the cylinder. On both ends of the vessel, there are two flanges with the following glass windows: at the back end, three large windows (D = 160 mm) and two small panels (D = 105 mm) at one end; at the front end, two large ports (D = 400 mm) and two small windows (D = 105 mm). The plasma source is connected to one of the front large ports. In the first months of operation it was not centred with respect to the axis of the main vessel, which resulted in instable operating conditions. Therefore later on the connecting flange was changed and now the helical source and main vessel are aligned. The vacuum system is connected to the back flange and consists of a pre-vacuum pump to reach a pressure of 10^{-2} mbar and a turbo molecular pump to create a high vacuum at 10^{-6} mbar. These pumps have a large flow rate making; the minimal pressure can be reached in 30 min time.

3.2. Large magnetic field coils

The main vessel is equipped with two magnetic coils in a Helmholtz configuration. The distance between the coils is about 815 mm. Some of the characteristics are given in **Table 1**. The outer/inner diameter is 2200/1200 mm and the thickness varies between 120 and 180 mm. The coils are connected to the central high current supply of the institute, which uses two modules connected either in parallel or in series using regular electrical voltage at 10.5 kV at 50 Hz. The magnetic (B) field in the centre of the vacuum vessel is expected to be 0.1 T for a current of 2.4 kA in the main coils. Higher currents and fields are possible. The current generator can sustain 4 kA for 10s and 12kA for 1 s, corresponding with B-fields up to 0.275 and 0.4 T respectively.

3.3. External plasma source

The plasma source is based on the concept of ionisation by helicon waves, which offer the possibility for large volumes to reach high densities (10^{18} m^{-3} and above) with uniform radial profiles

	Large coils	Small coils
Inductance L	0.4 mH	1.74 mH
Resistance R	0.86 mΩ	1.54 mΩ
Maximal current	47 kA	10 kA
Maximal voltage	3 kV	12 V
Time constant τ	0.5 s	n.a.
Pulsed current	4 kA	1 kA
Nominal current	1.5 kA	n.a.
Nominal voltage	600 or 340 kV	n.a.
Pulse length	10 s	10 s
Duty cycle	5.6%	n.a.

Table 1. Characteristics of the large and small coils.

[10–12]. The price for this objective is the high level of injected power (several kW), which can lead to dramatic damages if the flow of the power in the source is not controlled. To reduce the risks the operations will be carried out first in inductive mode and, by progressively increasing the available power on the generator, we will try to reach the mode transition to the helicon mode. The source includes a glass tube as vacuum vessel, four magnetic coils (the helicon wave is a magnetised wave), a helical antenna with its own power supply and a gas injector.

The most risk prone component is the glass tube, which faces the following dangers:

i. Arcing: the risk is to hit and break the glass tube. If only inductive discharges would be generated, it would be possible to install a shield to protect the tube (with the advantage to remove spurious capacitive effects in the plasma); however, with helicon plasmas, the electrostatic component is necessary as well. Therefore, before increasing the power, we need to investigate the distribution of RF fields in the plasma source and check that we are able to have a good absorption in the plasma (decreasing the voltage on the antenna). Even with that, an arc detection system may be necessary to prevent in a very short time the development of an arc.

ii. Heat fluxes: the glass tube is made of borosilicate glass Duran with a safe maximum temperature of 150 and 500°C if it is submitted to homogeneous fluxes on its surface. Given its specific heat capacity Cp = 750 J/(kg °C), a mass of 17 kg and a discharge duration of 8 s, it can sustain a flux of 250 W at 150°C.

iii. Sputtering: this is the unknown and will require a better analysis. Previous experiments show an increase of the sputtering during helicon discharges, but variable from machine to machine.

The characteristics of the helicon magnetic coils are displayed in **Table 1**. These small coils are fed with a power supply DC10, with current between 0 and 1 kA, and voltage of 10 V. The maximum field inside the source is 0.1 T. In standard operation, this field is superimposed on the field generated by the large coils. The structure with the helicon in its central position is presented in **Figure 3** for a current of 1kA in the large coils and 0.45 kA in the small coils. The resulting B-field in the centre of the vessel is around 40 mT. By adapting the ratio in the field in small and large coils the plasma performance can be influenced. The optimisation of the performance is the subject of on-going research.

The test-bed flexibility makes it possible to test different types of antennas. The presently mounted antenna is a Shoji III half turn [14]. In other experiments, it proved to have a better coupling, especially with the mode m = 1. It has a length L_a = 1 m and diameter d_a = 0.6 m. The dispersion relation of the helicon wave sets a relation between the minimal density for which the wave can propagate and the magnetic field [14]:

$$kk_z = \frac{e\,\mu_0\,n_e\,\omega}{B_0}, \tag{2}$$

k is the total wave number, k_z is the longitudinal wave number, e is the electron charge, μ_0 is the vacuum permeability, n_e is the electron density, ω is the pulsation of the generator and B_0 is the static magnetic field.

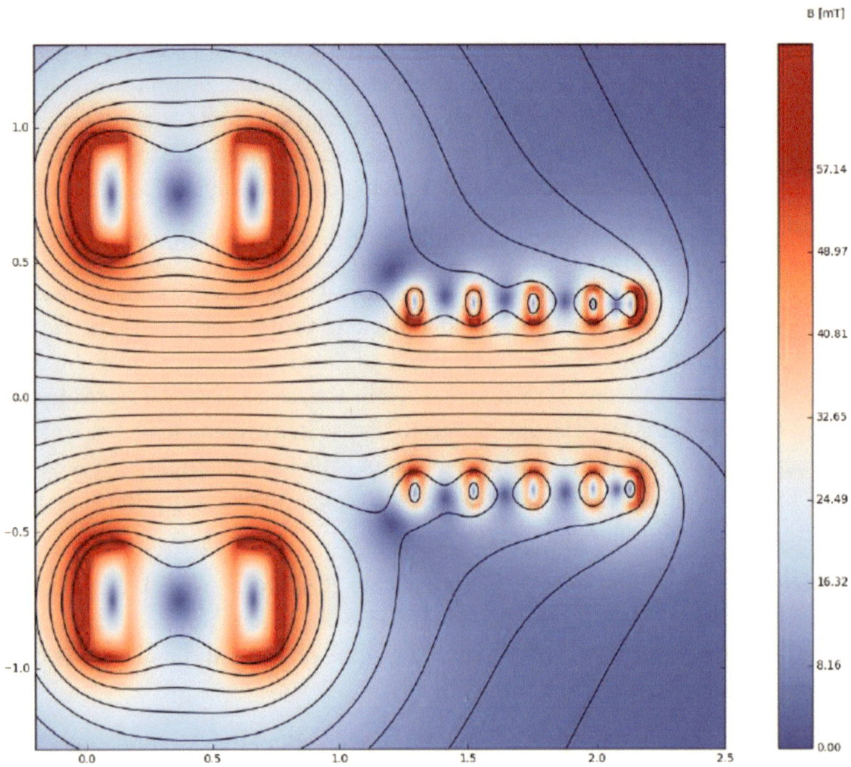

Figure 3. Example of a typical magnetic field topology for IShTAR (in mT). It is generated with currents of 1 kA in the large coils and 0.45 kA in the small coils.

k_z is determined by the geometry of the antenna:

$$k_z = (2\kappa + 1)\frac{\pi}{L_A}, \tag{3}$$

where κ is the longitudinal mode number and L_A is the length of the antenna.

The radial wavenumber k_r is determined from the calculation of the electrical field in a cylindrical geometry (in a simple case with constant density). This relation is represented for the mode m = 0 in **Figure 4**. For a maximum field B = 0.1 T in the plasma source, a density $n_i \approx 7\ 10^{16}\ m^{-3}$ is necessary for the helicon wave to propagate. The low frequency has been chosen based on simulations with the electromagnetic code MicroWave Studio (MWS) [15] (with a simple dielectric) to minimise the electric fields and the risk of arcing.

The plasma source is equipped with a gas valve at the back of the tube with a flow rate range of 5–5000 sccm. Three gases can routinely be used: argon, helium and hydrogen with three different feeding lines.

3.4. Control system

The control system automates the experiments, enables a fast start and remote operations, and it is possible to monitor the status of the different parameters (pressure, temperature,

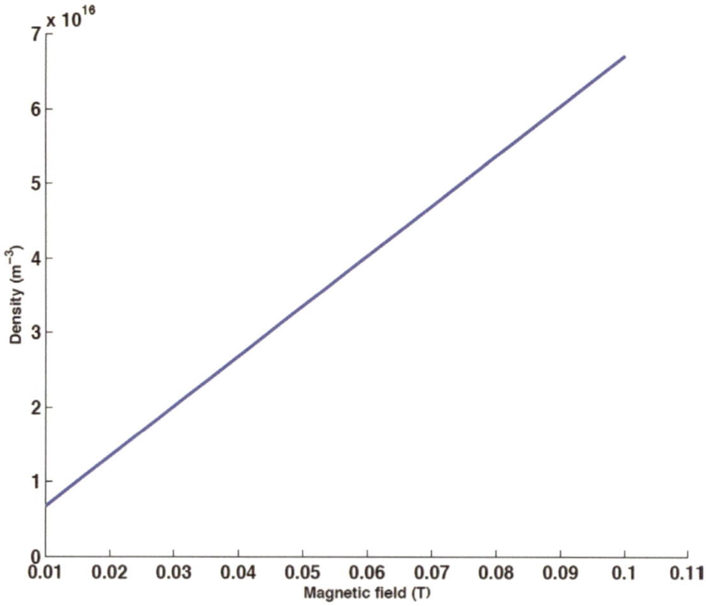

Figure 4. Minimum density required for the helicon mode $m = 0$ to propagate as a function of the magnetic field for the chosen antenna parameters $f = 11.7$ MHz, length = 1 m, radius $r_a = 0.3$ m.

current) of the test-bed. The control is based on a Simatic system for the hardware and WinCC applications for the software with a time step of 10 ms; it controls the automatic opening and closing of the valves and pumps during the vacuum build-up and re-pressurisation. It also enables the operator to define ramps in the coil current. It ensures the safety of operation by keeping the access doors locked below a predefined vacuum level in the vessel, checking the flow of coolant in the different components.

3.5. ICRF system

A mock-up of an operational ICRF antenna can be installed on the wall of the vessel and connected on one of the side ports to the transmission line, which is equipped with a ceramic vacuum window. Two RF power sources are presently available for the ICRF system:

i. One ASDEX Upgrade RF generator (30–120 MHz) with a power up to 2 MW at 30–80 MHz and 1 MW at 80–120 MHz.

ii. A broadband amplifier with an output power of 1 kW in the frequency range 100 kHz to 100 MHz.

The matching of the ASDEX Upgrade generator is insured by a system of two stub tuners. The broadband amplifier requires the installation of an additional capacitor-base matching network more suited to low levels of power.

3.6. ICRF antenna

An ICRF antenna was designed at the Laboratory for Plasma Physics in Brussels, Belgium (LPP-ERM/KMS) and installed in IShTAR. The MicroWave Studio (MWS) [15] model of the

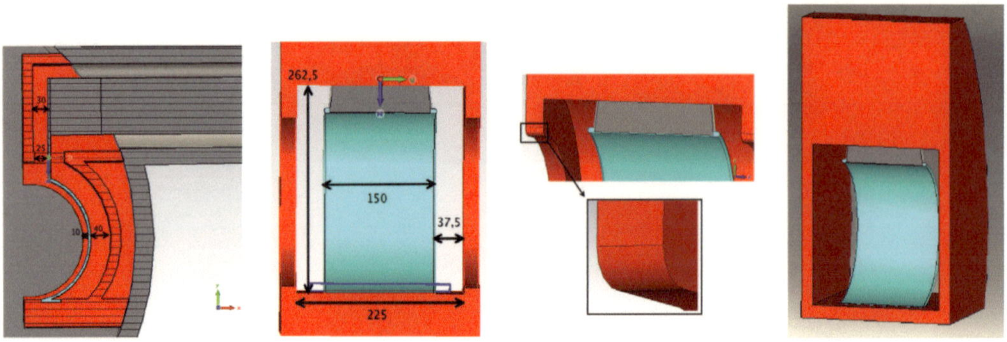

(a) (b) (c) (d)

Figure 5. MicWs model of the ICRF antenna and its feeding line; (a) lateral cut view of the initial geometry; (b) front view; (c) zoom of the angular part of the limiter; (d) simplified geometry; All dimensions are given in mm.

retained geometry (with initial dimensions in mm) of the IShTAR antenna is visible in **Figure 5**. The strap is fed from the top feed through by a coaxial transmission line connected to a vertical plate inserted into the strap. Unlike usual ICRF antennas the IShTAR antenna is not designed to couple as much power as possible to the plasma, but rather to generate the typical electric fields structures expected in a tokamak. The first simulations with MWS clearly demonstrated some issues due to the presence of sharp angles in the lateral limiters. Even by further smoothing these angles (see **Figure 5c**) the electric fields maps were dominated by peaked radiations at the level of the limiters, which could also be partly due to numerical issues (meshing). We therefore decided to replace these profiled limiters by rectangular plates: see **Figure 5d**, resulting in a more regular electric field distribution. This change also considerably simplifies the building of the antenna box [16]. The antenna was installed in IShTAR and is visible in **Figure 6b**.

3.7. Diagnostics

In the present configuration, each diagnostic dedicated to physics applications has its own acquisition system. A centralised acquisition system is being built with a NI Chassis and a set of digitizer boards to offer a common time reference, shared data storage and network access. The instrumentation is still under development. At present the two main tools to diagnose plasma parameters and electric fields are probes and spectroscopy. In future other diagnostics will be added, for example an interferometer for density measurements has already been designed, in order to bench mark the data obtained by the Langmuir probes. In addition, more RF compensated probes will be added in the helical plasma source to study better the helicon physics and the effect of the magnetic topology.

The probes are depicted in **Figure 6**. Different Langmuir and B-dot probes have been installed. **Figure 6a** shows an inside view of the back flange, with an array of three probes in the middle, to the right a planar non-compensated Langmuir probe with a diameter of 2 cm can be seen. A manipulator arm (**Figure 6b**) can carry up to four probes simultaneously, which allows a partly 2D scan of the plasma profile on a shot-to-shot basis. In **Figure 6b** also the ICRF

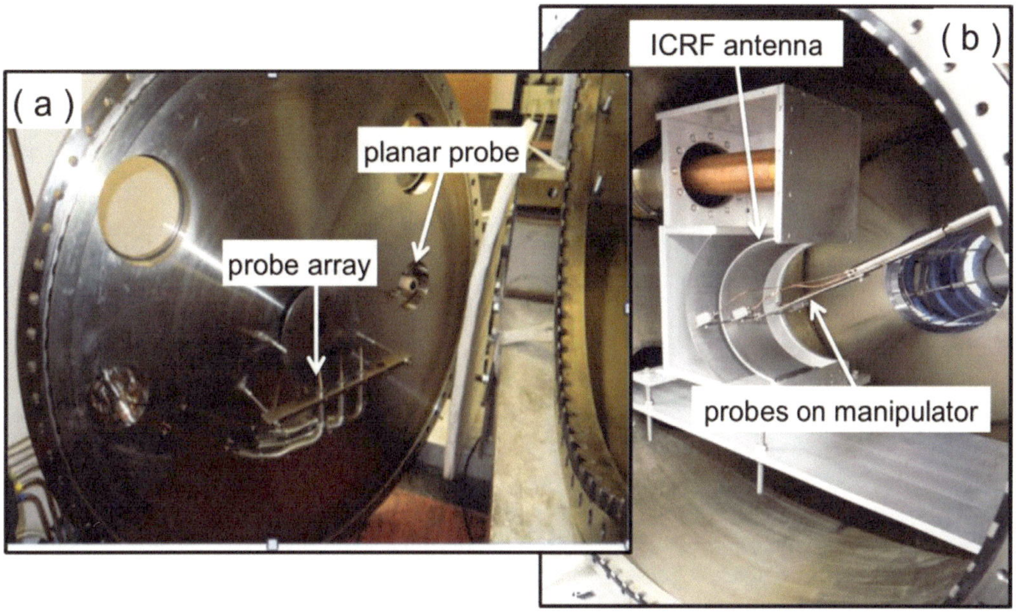

Figure 6. Inside view of the main vessel with the different probes: (a) on the back flange and (b) on a movable manipulator arm. The helical plasma source is visible on the right.

antenna can be seen, it has a single antenna strap, which is curved to follow the plasma shape. In this picture the antenna box was not yet closed and the connection to the feeding line on the top is visible. For the sheath physics, an important parameter to investigate is the electric (E) field created at the antenna and plasma facing components. The ions accelerated in the sheath potential can damage the components by creating a local overheating (hot spots) and or sputtering, as it is has been observed during ICRF experiments in different fusion machines.

After a theoretical study, two different spectroscopic methods have been retained and are presently being developed in parallel.

Passive optical spectroscopy is selected as the first approach to directly measure electric fields in the vicinity of an ICRF antenna, without disturbing the plasma environment [17–20]. This technique enables studying the perturbation of the electronic structure of an atom caused by an external electric field, the Stark effect. These perturbations are detectable as a shift of the central wavelength of a spectral line, and the occurrence of forbidden components of the fine structure in a spectral line profile. We have focused our research on the Stark effect on 4^3D - 2^3P transitions in helium. The method requires a high-resolution spectrometer with high dynamic range detector capable to resolve the allowed and forbidden lines of the transition under the study. To this end a 0.75 m Andor spectrometer equipped with an Andor ICCD camera has been installed. In the test work the He n = 4^3D - 2^3P transition has been selected for the E-field measurements, since the signal-to-noise ratio of the recorded spectra was acceptable and the line Stark broadening is strongly affected by the E-field. Once the time-averaged spectra are recorded, the spectral line profiles are compared to the simulated ones and the electric field amplitude is extracted with the method of least squares. To simulate the spectra perturbed by an electric field, in the presence of a background magnetic field, the Explicit Zeeman Stark Spectral Simulator (EZSSS) [17] was used. This code generates the discrete spectrum by solving the Schrödinger equation in electric dipole approximation, with external electric and magnetic fields as perturbations. In the second step, by convoluting the discrete spectra with Gaussian and/or Lorentzian profiles to mimic the broadening mechanisms, the continuous spectrum is obtained.

Figure 7 depicts the modelled triplet line profile corresponding to the 4^3D - 2^3P He-I transition with no electric field externally imposed on the system. The discrete spectra, calculated with the EZSSS code is shown as a set of lines in the mirror image of the intensity scale. The continuous spectra, presented in the positive part of the intensity axes is convoluted with a Gaussian distribution corresponding to the Doppler broadening with a temperature of the radiator of 0.7 eV. The distinct feature of these spectra is the occurrence of the second spectral line red-shifted from the main component, corresponding to the fine structure of the He triplet transition. This separation of the components allows estimating the E-field in the plasma.

An alternative, but more complicated approach, is to use Doppler-free saturation spectroscopy (DFSS) to eliminate disturbing effects such as the Doppler broadening and to highlight the E-field influence [17, 21]. The basic principle is to create a cross section in the plasma of overlapping pump and probe beams from the same laser source tuned to specific absorption

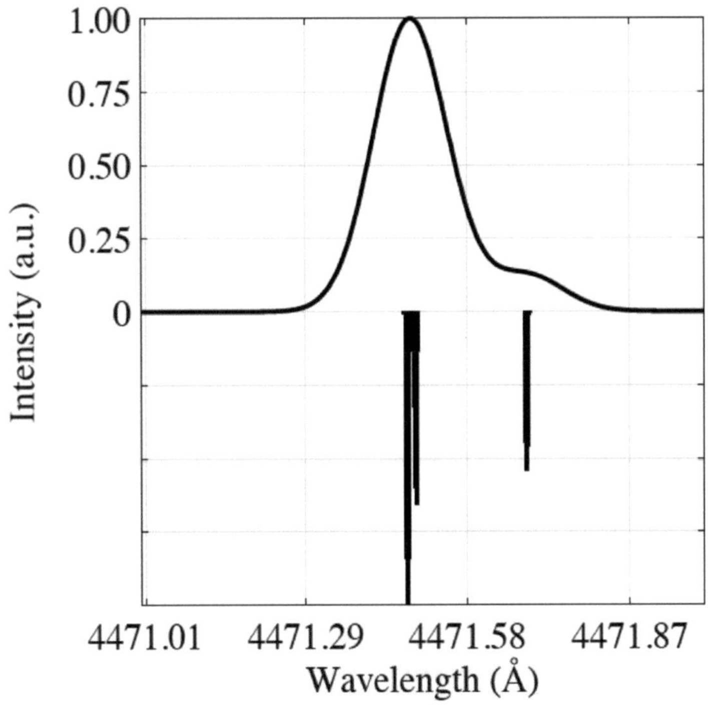

Figure 7. Triplet He-I 4^3D-2^3 P line profile simulated with the EZSSS code with no external electric field (reference scenario) shows discrete and continuous results, with the main component at wavelength $\lambda = 4471.49$ Å and the second component at 4471.6 Å.

http://dx.doi.org/10.5772/intechopen.76730

transition of He or H atoms. The pump beam depletes the ground state; the probe beam passes the plasma with reduced absorption. By depleting the ground state, the fine structure of the spectral line should become more clearly visible, in the form of local dips in the Doppler broadened absorption line. This allows measuring line profile with eliminated effect of the Doppler broadening and precisely estimate E-field with high sensitivity.

4. Operation and results as preparation for the sheath studies

4.1. Sequence of a typical discharge

A typical discharge sequence is presented in **Figure 8a**, as it is seen by the different diagnostics; one camera view is shown in **Figure 8b**. The vessel is prefilled at the operating pressure before the start of the sequence. The RF power is matched for plasma with magnetic field. Therefore the coils are usually activated before the antenna. However, the reversed sequence, as used on the discharge in **Figure 8a** shows how the magnetic field affects the plasma and the measurements of the Langmuir probe. We notice that the ion saturation current starts to ramp up only when the field is activated. On the cameras, the visible light shows that the plasma is first confined in the plasma source and then, when the coils are powered, a plasma tube develops in the main vessel and is displaced towards the centre, further away from the Langmuir probe, following the field lines represented in **Figure 3**. The minimum amount of power required to ignite the plasma was evaluated at 50 W (with and without magnetic field). During the phase without the main field, arcing is noticed in the main vessel, which is then wiped out with the start-up of the coils. No arcing has been noticed on the helicon antenna

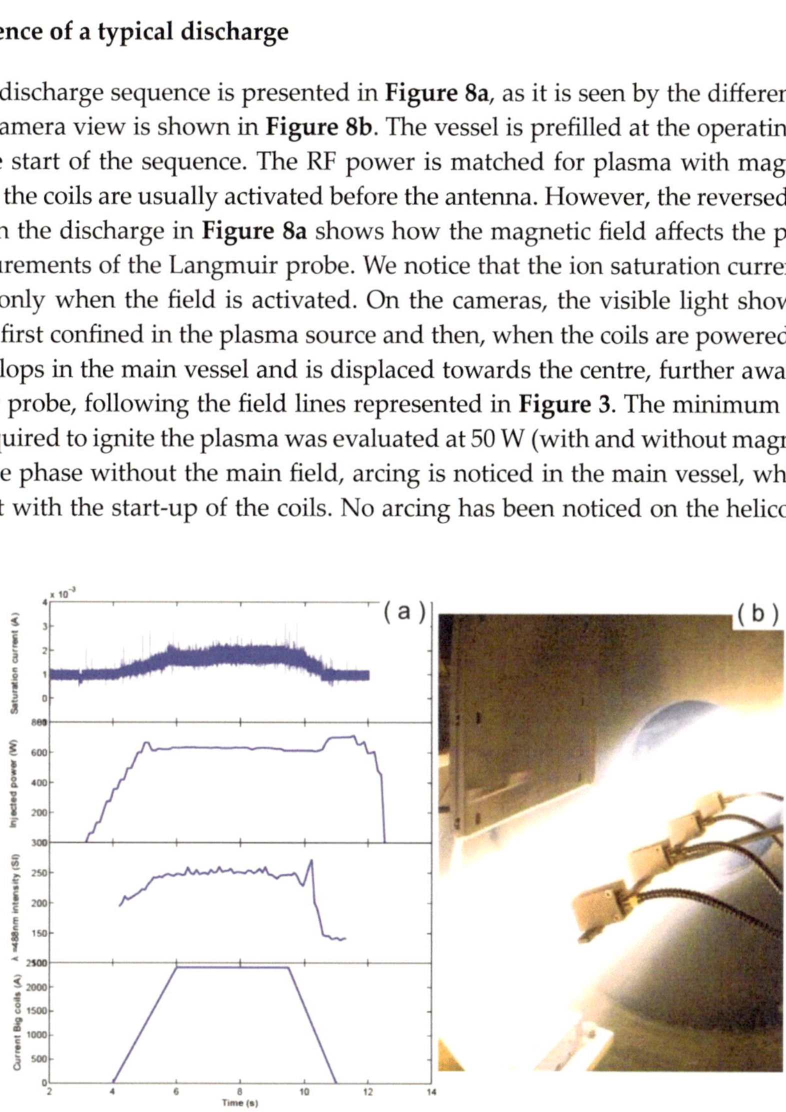

Figure 8. (a) Evolution of the main parameters for an injected helicon power of 700 W. (b) Plasma inside the main vessel in front of the ICRF antenna.

with the camera. However, post-experiment inspection revealed that the copper layer has been damaged on some parts of the line.

4.2. Characterisation and optimisation of the plasma

The operating parameters (injected RF power, neutral gas pressure and magnetic field) were scanned to observe the effect on the plasma density and temperature measured in the centre of the main vessel. The purpose is to select the set of parameters that optimises the density. The results of the parametric studies are presented in **Figure 9** [22–24]. At a constant magnetic field, the density increases almost linearly with the power. A small drop around 1.8 kW is observed, which may be due to an internal resonance. The nature of the resonance is not yet clear and needs further investigation. The optimisation of the IShTAR plasma operations is subject of on-going research. However no step in the scan is seen, which would correspond to a change of regime from inductive to helicon mode. This is probably due to a too low level of power with respect to the volume to ionise. The electron temperature does not feature any evolution inside a band between 4 and 7 eV, except for a peak at the same assumed resonance at 1.8 kW. This would confirm that the increased injected power is used to generate denser plasmas and not to heat. At constant injected power, two different trends are observed when the magnetic field is ramped up. At low pressure, the density reaches a plateau at 0.06 T and saturates before starting to drop at 0.14 T. At higher pressure, the density increases almost linearly with the magnetic field. The temperature presents a slight trend towards an increase, but there is not much difference between the high- and low-pressure cases. The cause for the saturation of the density is not yet clear and will require comparison with theoretical models, which are presently not yet available. The radial profiles of the plasma density and of the

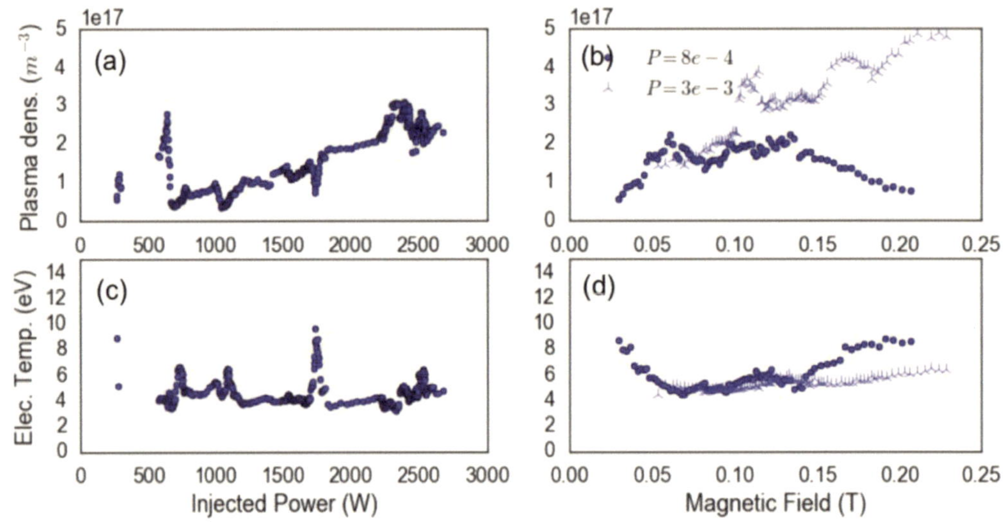

Figure 9. (a) and (c) Scan of argon plasma parameters as a function of the injected power at constant magnetic field (0.1 T) and neutral gas pressure P (10^4 mbar). (b) and (d) Plasma parameters as a function of the magnetic field at constant injected power (2.5 kW) for two levels of neutral gas pressure P (8×10^{-4} and 3×10^{-3} mbar).

RF fields were measured with the manipulator over several discharges and are presented in **Figure 10** for two different magnetic fields at constant injected power of 2.5 kW and a neutral pressure of 10^{-4} mbar. The magnetic field strengths of 0.12 and 0.06 T, indicated in the figure, refer to the centre of the main vessel. The limit of 0.1 T, which was mentioned before, is for the helical plasma source. It is caused by the 1kA current limit in the small coils. The large coils around the main vessel can generate fields up to 0.4 T. The density profile has a Gaussian shape with a mean width of 10 cm, half the length of the plasma source geometrical radius. The profile does not change with the magnetic field, which corresponds to the "saturation" effect described above. Optimisation of the profile shape and peak density is still possible by adapting the magnetic field strength and topology. It seems to be connected to the helicon wave dispersion relation. A detailed study is on-going in order to create the best possible conditions at the location of the ICRF antenna to be relevant for the intended sheath studies.

In the last set of experiments high-resolution passive optical emission spectroscopy was used to estimate an E-field in typical conditions of the IShTAR device operation [18]. To detect the expected Stark-effect-induced changes on the spectral line profile two experimental scenarios have been performed to obtain the results presented here. The reference data were recorded without the voltage applied to the electrode placed directly in the Helicon source, while the second set was obtained for the set of experiments with the electrode biased to a positive DC-voltage of $U_{el} = 1$ kV. The obtained line profiles corresponding to the 4^3D - 2^3P transition in He-I were recorded over the whole duration of a discharge, with an exposure time of 0.5 s, in a kinetic series of 24 scans per discharge. The processed data shows a reproducible shift of the He-I 4^3D - 2^3P line (**Figure 11**) when the external electric field is present in the thermal sheath in front of the DC-biased electrode. For a rough estimate of the electric field expected in a thermal sheath in front of the electrode we can assume that it originates from the difference in potentials between the electrode and the plasma, over the distance corresponding to the sheath thickness. Compared to the potential applied on the electrode, $U_{el} = 1$ kV, the plasma potential is $U_{pl} = 0$ kV. The sheath thickness corresponding to the density of helium plasma in IShTAR of $n_e = 10^{16}$ m^{-3}, and the electron temperature of $T_e = 5$ eV is proportional to the Debye length of $\lambda_D = 5\times10^4$ m. Therefore the electric field in the vicinity of the electrode biased to $U_{el} = 1$ kV, can be estimated to be the applied voltage drop over the sheath thickness. To complete the

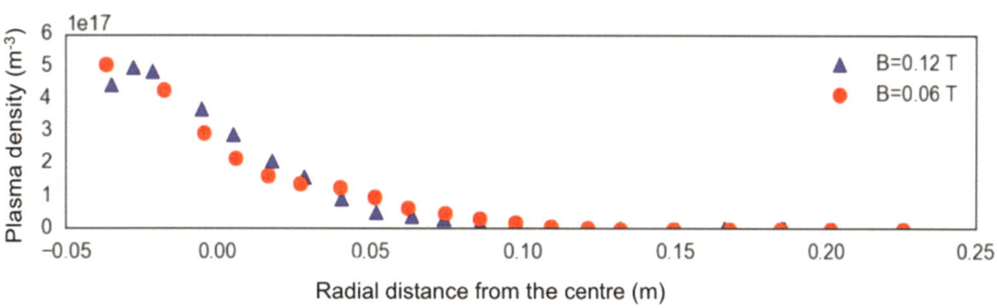

Figure 10. Radial profile measurements of the plasma density. The magnetic field strengths refer to centre of the main vessel.

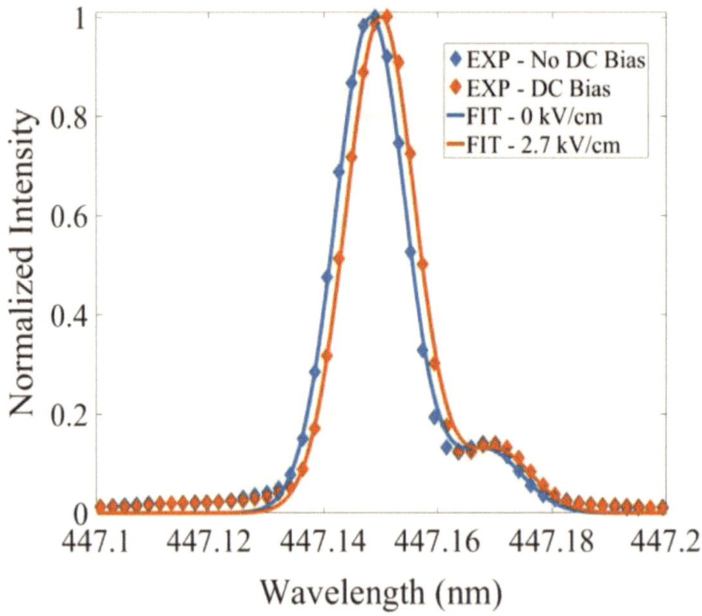

Figure 11. Experimental 4^3D-2^3 P He I line profile for E_{DC} = 0 kV/cm and E_{DC} = 2.7 kV/cm (black and red diamonds, respectively) fitted with the simulated theoretical spectra calculated with the EZSSS code (solid lines).

study, the He-I 4^3D - 2^3P transition is simulated with the EZSSS code for several values of the DC electric field, and the best match between the measured and simulated lines corresponds to a simulated spectral line exposed to the external electrical field of E = 2.7 kV/cm, as depicted in **Figure 3** with the solid lines. The estimated electrical field agrees with theoretical prediction and showing possibility of passive emission spectroscopy with high resolution to be used for E-field measurements [16, 17].

5. Conclusions and future plans

A dedicated test facility has been constructed for studying RF sheath effects [24]. For this purpose plasma conditions representative of the edge of a tokamak are needed. A helical antenna creates plasmas with the required parameters. Helium and argon operation are used routinely, also hydrogen is foreseen, as well as gas mixtures. Plasma densities of the order of 10^{17}–10^{18} m^{-3} and temperatures around 5–10 eV have been obtained. The main limiting factor is the generator power (3 kW) of the helical antenna. Performance optimisation is possible by adapting the magnetic field strength and topology; the detailed study of the helical plasma source will be the subject of future research.

A simple ICRF antenna is installed and operational, it is coupled to a broadband generator with frequencies in the range of 100 kHz to 100 MHz and a maximal power of 1 kW. If needed

http://dx.doi.org/10.5772/intechopen.76730

in the future a much more powerful generator can be used, when it will be coupled to the ASDEX RF system, with an available power up to the MW level.

Attention is also given to developing diagnostics for characterising the plasma parameters and electric fields, especially in the vicinity of the ICRF antenna, since they are the key ingredients for the sheath theories and modelling codes. Different probes and spectroscopic methods are used.

With the probes the behaviour of the plasma electron density and the three magnetic field components of the injected RF fields are measured under different operating conditions. Modelling is on-going using the COMSOL multi-physics environment [25, 26]. The measurements will also serve more advanced sheath and edge simulation codes in the future.

For measuring electric fields in the plasma caused by the RF antenna sheaths two approaches are followed. Passive optical emission spectroscopy monitors Stark effects on spectral lines with a high-resolution spectrometer, provided that the local electric fields are strong enough to overcome the broadening of the lines. Doppler-free saturation spectroscopy is more powerful; a laser beam depletes the ground state, eliminates the line broadening effects and makes smaller electric fields visible. However, the more complicated set-up, with a careful alignment of laser beams, makes the measurements much more challenging. After a first test on a glow discharge plasma, the design of the optical path and the installation of the laser at IShTAR have started.

Acknowledgements

The authors want to thank the technical staff at IPP - Garching, LPP-ERM/KMS and Ghent University, in particular F. Fischer, G. Siegl, M. Berte and J. Peelman.

Conflict of interest

There are no conflicts of interest.

Notes/Thanks/Other declarations

The work received support from the Research Foundation Flanders (G0B3115N). This work has been carried out within the framework of the EUROfusion Consortium and has received funding from the Euratom research and training programme 2014–2018 under grant agreement No 633053. The views and opinions expressed herein do not necessarily reflect those of the European Commission.

Author details

Kristel Crombe[1,2*], Rodolphe D' Inca[3], Eric Faudot[4], Helmut Faugel[3], Ana Kostic[3], Mariia Usoltceva[4], Jean-Marie Noterdaeme[3], Anton Nikiforov[1,3], Helmut Fuenfgelder[2], Stephane Heuraux[1], Jonathan Jacquot[3], Fabrice Louche[3], Roman Ochoukov[1,3,4], Ilya Shesterikov[2] and Dirk Van Eester[1,3]

*Address all correspondence to: kristel.crombe@ugent.be

1 Department of Applied Physics, Ghent University, Ghent, Belgium

2 Laboratory for Plasma Physics, ERM-KMS, Brussels, Belgium

3 Max-Planck-Institut fuer Plasmaphysik, Garching, Germany

4 Université de Lorraine, France

References

[1] Chen FF. High Density Plasma Sources. 1st ed. USA: William Andrew; 1995

[2] Noterdaeme J.-M, Van Oost G. The interaction between waves in the ion cyclotron range of frequencies and the plasma boundary. Plasma Physics and Controlled Fusion. 1993;**35**:1481 and references therein

[3] Bobkov V et al. ICRF operation with improved antennas in ASDEX upgrade with W wall. Nuclear Fusion. 2013;**53**(9):093018

[4] Jacquet P et al. Ion cyclotron resonance frequency heating in JET during initial operations with the ITER-like wall. Physics of Plasmas. 2014;**21**:061510

[5] Klepper CC et al. RF sheath-enhanced beryllium sources at JET's ICRH antennas. Journal of Nuclear Materials. 2013;**7**(438):S594-S598

[6] Lerche E et al. Impact of minority concentration on fundamental (H)D ICRF heating performance in JET-ILW. Nuclear Fusion. 2014;**54**(7):073006

[7] Colas L et al. Self-consistent radio-frequency wave propagation and peripheral direct current plasma biasing: Simplified three-dimensional non-linear treatment in the "wide sheath" asymptotic regime. Physics of Plasmas. 2012;**19**:092505

[8] Crombé K et al. Studies of RF sheaths and diagnostics on IShTAR. AIP Conference. Proc. 1689, Proceedings of the 21st Topical Conference on Radiofrequecy Power in Plasmas, UCLA Conference Center at Lake Arrowhead, California 2015. 030006

[9] Crombé K et al. Coupled and decoupled solutions of the cold plasma dispersion relation. Journal of Plasma Physics. 2016;**82**:905820203

[10] Tripsky M et al. Discharge initiation by ICRF antenna in IShTAR. EPJ Web of Conferences. 2017;**157**:03056

[11] Raizer YP, Kisin VI, Allen JE. Gas Discharge Physics. Vol. 1. Berlin: Springer-Verlag; 1991

[12] Lieberman MA, Lichtenberg AJ. Principles of plasma discharges and materials processing. MRS Bulletin. 1994;**30**:899

[13] Miao TT et al. The study of helicon plasma source. Review of Scientific Instruments. 2010;**81**(2):02B105

[14] Chen FF, Boswell RW. Plasma Science. IEEE Transactions. 1997;**25**:1245

[15] CST STUDIO SUITE®, CST AG, Germany, www.cst.com

[16] Louche F et al. Designing the IShTAR antenna: Physics and engineering aspects. AIP Conference Proc. 1689. In: Proceedings of the 21st Topical Conference on Radiofrequency Power in Plasmas. UCLA Conference Center at Lake Arrowhead, California. 2015

[17] Martin EH. Electric Field Measurements of the Capacitively Coupled Magnetized RF Sheath Utilizing Passive Optical Emission Spectroscopy [PhD thesis]. North Carolina State University; 2014

[18] Kostic A et al. Feasibility study of passive optical emission spectroscopy for the electric field measurements in IShTAR. EPJ Web of Conferences. 2017;**157**:03025

[19] Crombé K et al. Helium operation of IShTAR in preparation of E-field measurements. ECA. 2017;**41F**:5-144

[20] Lee C, Graves D, Lieberman M, Hess D. Journal of the Electrochemical Society. 1994;**141**:1546

[21] Wieman C, Hansch T. Doppler-free polarization spectroscopy. Physical Review Letters. 1976;**17**:1170

[22] D'Inca R, Jacquot J, et al. Development and first operations of the IShTAR test facility. In: 42nd EPS Conference on Plasma Physics (EPS), Vol. 39E; Lisbon, Portugal: ECA. 2015. P1.154

[23] D'Inca R, Kostic A, et al. Characterization of the RF plasma on the IShTAR testbed. In: 43rd EPS Conference on Plasma Physics, Vol. 40A; 4-8 July 2016; Leuven, Belgium: ECA. 2016. O5.129

[24] Crombé K et al. Characterization of the RF plasma on the IShTAR testbed. In: 22nd IAEA FEC Conference; 17-22 October 2016; Kyoto, Japan

[25] Usoltceva M et al. IShTAR: A helicon plasma source to characterize the interactions between ICRF and plasma. In: COMSOL Conference; 12-14 October 2016; München, Germany

[26] Usoltceva M et al. IShTAR ICRF antenna field characterization in vacuum and plasma by using probe diagnostic. EPJ Web of Conferences. 2017;**157**:03058

Chapter 9

Plasma-Assisted Combustion

Siyin Zhou, Haiqing Wang, Wansheng Nie and
Xueke Che

Additional information is available at the end of the chapter

http://dx.doi.org/10.5772/intechopen.80959

Abstract

As a promising technology, plasma-assisted combustion (PAC) has attracted many researchers to explore the effect of PAC on improving the combustion in propulsion devices, such as scramjet, detonation engines, internal engines, and so on. In this chapter, we aim to exhibit the influence of quasi-DC discharge plasma on the operating performance of scramjet combustor and find the internal mechanisms, which may contribute to the development of PAC technology in supersonic combustion. For case one, a plasma filament is generated upstream of fuel jet through quasi-DC discharge in a scramjet combustor; for case two, the plasma is formed across the backward facing step of a flame holding cavity to improve the flame stabilization of the cavity in the scramjet combustor. The two cases are investigated in detail through three-dimensional numerical simulation based on the dominant thermal blocking mechanism. Important parameters including temperature distribution, separation zone, water production, stagnation pressure loss, combustion efficiency, cavity drag, mass exchange rate, and cavity oscillating characteristics are obtained and analyzed. It shows that the quasi-DC discharge plasma does benefit for the improvement of the combustion in a scramjet combustor.

Keywords: quasi-DC discharge, plasma-assisted combustion, supersonic combustion, scramjet, transverse fuel jet, cavity, numerical simulation

1. Introduction

As one of the qualified candidates for hypersonic propulsion system, scramjet combustor has attracted a large amount of attention all over the world. As is now well known, realizing high efficiency and steady combustion in a supersonic combustor always remains as a critical issue for scramjet engines. Numerous researches have indicated that the additions of cavity in combustor and the transverse fuel injection upstream of the cavity can promote the stabilization of combustion and flame [1–3]. Whereas, it is difficult yet for the fuel to reach properly mixing within the supersonic flow by the mechanical methods [4, 5]. Moreover, it is a great challenge for matching the transverse fuel jet up with the cavity under off-design condition [4]. Furthermore, certain stagnation pressure loss will appear using the approaches. Taken all account, some new methods are imminently needed for keeping stable and highly efficient combustion in combustor with the least penalties adding to the flowfield.

It is rather promising to adopt a discharge plasma for supersonic flow and/or combustion control in aerospace field [6–8]. It has been widely considered that plasma-assisted combustion is one of the most promising approaches for enhancing ignition and combustion in the environment of scramjet combustion [4, 6, 9]. As a further promising method, there are three advantages, that is: rapid response, less inertia, and flexibility [10]. Former studies clearly show that the quasi-DC discharge plasma can availably modify supersonic flow in a controllable manner among the discharge plasma mentioned above [8, 11], whereas the DBD plasma is commonly used in low-speed flow environment [12, 13]. If we can combine the quasi-DC discharge plasma with cavity and transverse fuel jet together, some new phenomena will surely appear, which may help in the improvement of combustor performance.

There in three primary mechanisms of the plasma effect on flow and combustion can be summarized: (1) momentum transfer, (2) fast local ohmic heating of the medium, and (3) active particles [6, 7, 14]. Ignoring the magnetic field, the mechanism of quasi-DC discharge plasma affecting a supersonic flow is mainly its fast local heating rather than the electrostatic force (i.e., the momentum transfer mechanism) [10, 15, 16].

This chapter aims at investigating the changes of the fuel jet, cavity, and whole scramjet combustor led by the quasi-DC discharge plasma based on the dominant thermal blocking mechanism. Here, a short cavity downstream of a fuel jet orifice is considered in order to simulate the combustor flowfield more realistic. The plasma is set as a controllable heat source. The k-ω shear stress transport (SST) model together with finite rate chemical reaction model is used simulating the turbulent flow and combustion. The flow structures, equivalent ratio, product distribution, stagnation pressure, and combustion efficiency in the combustor are all obtained and analyzed using the 3D numerical simulation which can acquire some data that are difficult to measure in experiments.

2. Case one—effect of plasma on fuel jet

2.1. Numerical methods

2.1.1. Plasma model and its configuration

By applying a high voltage across the anode and cathode, a bright plasma filament appears above the two electrodes. With the impacting of incoming flow, the plasma occurs not only between the two electrodes but also several centimeters downstream on the wall [10]. In consideration of the oscillation character of its electrical parameters and mean temperature in

http://dx.doi.org/10.5772/intechopen.80959

the discharge path, this type of discharge plasma is called "quasi-DC discharge plasma." However, it is a type of low temperature arc plasma in fact. The main properties of quasi-DC discharge plasma are described in [10–12, 15].

As depicted in **Figure 1**, the domain of quasi-DC discharge plasma filament is simplified as a cuboid region presented in red dashed lines for simulation based on its appearance. In **Figure 1**, the symmetric plotted line indicates that the cathode and anode are symmetrical for the central line of combustor wall and so does the plasma filament.

Because the quasi-DC discharge plasma releases a large amount of heat concentrated in its discharge path, the discharge path goes into very hot. This high temperature domain (i.e., the discharge path) obstructs the supersonic inflow due to a thermal blocking occurred. Hereby, the quasi-DC discharge plasma behaves as a virtual blockage in the high speed flow of scramjet combustor, which is called as "the dominant thermal blocking mechanism" [17]. On the foundation of dominant thermal blocking mechanism, the individual plasma filament can be simulated as a volumetric heat source [10, 12].

Generally speaking, the plasma input power and the effective heat power that transfers into circumambient gas are quite different. Besides, the percentage of loss commonly varies with the power supply and environment, so it is not a suitable way to use the plasma input power as the heat source value of quasi-DC discharge plasma in simulation. In order to straightforwardly describe the plasma heat strength, the average temperature of the plasma zone is acquired by using the numerical simulation here. This way is feasible when the average temperature of the plasma zone is within a reasonable range [10, 16, 18, 19]. Hence, based on the thermal blocking mechanism, a certain temperature which denotes the actuating strength (i.e., the input power) is specified for the plasma filament domain when the actuator works.

The electrodes are flush mounted and do not have influence on the main flow themselves. The length and the section dimensions of individual plasma filament heat source are 20 mm length and 3 × 3 mm, respectively. The plasma filament locates 40 mm upstream of the fuel orifice center. Besides, it is generated near the wall in the center of the combustor, which means the symmetric plane of plasma filament is within the symmetric *xy*-plane of combustor. The quasi-DC discharge plasma working under a pulsed mode shows better performance than a continuous mode [19], so a plasma control frequency F_c = 8 kHz with duty cycle ratio D = 2/5 is chosen here. And T_{pl} = 2500 K is specified as an optimal actuating strength.

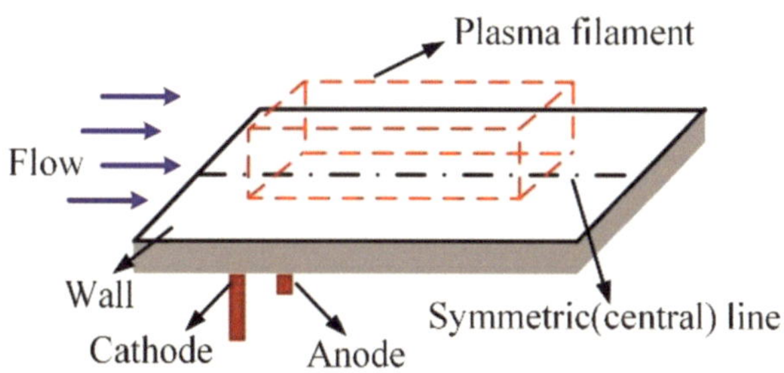

Figure 1. Computed configuration of quasi-DC discharge plasma in case one.

2.1.2. Governing equations, physical models, and numerical schemes

The unsteady Reynolds averaged 3D Navier-Stokes equations (URANS) with the k-ω SST two equation turbulence model [20] and four species (H_2, O_2, N_2, and H_2O) conservation equations are used as the governing equations of flow and combustion. The thermal conductivity of every species remains as constant and the mixture thermal conductivity is calculated using ideal gas mixing law. Based on the perfect gas assumption, it can be seen that the mixing gas satisfies the local thermodynamics equilibrium hypothesis and follows the state equation:

$$P = R_0 T \sum_{i=1}^{N_s} \frac{\rho_i}{W_i} \tag{1}$$

here, R_0 = 8.314 J/ (mol·K) is the gas constant. The specific heat capacity c_{pi} of species i is derived from a piecewise polynomial as follows:

$$c_{pi} = a_{1,i} + a_{2,i}T + a_{3,i}T^2 + a_{4,i}T^3 + a_{5,i}T^4 \tag{2}$$

where the coefficients can be found in [21]. The k-ω SST two equation turbulence model is used here because of its well behave in separation flow and free shear flow. Based on the Boussinesq assumption, viscosity coefficient $\mu = \mu_l + \mu_t$, where μ_l is the laminar viscosity coefficient and μ_t is the turbulence viscosity coefficient. μ_l can be received from the Sutherland law, whereas μ_t is received from the k-ω SST model.

$$\mu_l = \mu_{\mathrm{ref}} \left(\frac{T}{T_{\mathrm{ref}}} \right)^{2/3} \frac{T_{\mathrm{ref}} + S}{T + S} \tag{3}$$

where μ_{ref} is the reference viscosity coefficient under corresponding reference temperature T_{ref} and S is the Sutherland constant which can be got from **Table 1**.

Since this study focuses on the qualitative effect of plasma on the fuel jet and the scramjet combustor, the finite rate chemical model with the single step H_2/O_2 combustion kinetics model is applied. The reaction rate constant is derived from the Arrhenius formula. Hence, the computational time can be saved much and the well combustion flowfield can be obtained, too. The relevant parameters of H_2/O_2 one-step chemical model are shown in **Table 2**.

In order to capture the shock waves and other complex fluid structures better, the convective fluxes are evaluated using the advection upstream splitting method (AUSM) [22] with the

	T_{ref} (K)	S (K)	μ_{ref} (kg/m·s)
H_2	273.11	96.67	8.411×10^{-6}
O_2	273.11	138.9	1.919×10^{-5}
H_2O	416.67	861.11	1.703×10^{-5}
N_2	273.11	106.67	1.663×10^{-5}

Table 1. Parameters in Sutherland law.

http://dx.doi.org/10.5772/intechopen.80959

Equation	Pre-exponential factor	Temperature index	Activity energy [J/kg·mol]
$H_2 + 0.5O_2 \leftrightarrow H_2O$	9.87E + 8	0	3.1E + 7

Table 2. H_2/O_2 one-step chemical model.

second-order upwind approach at same time. The viscous fluxes are evaluated by using the second-order central differential scheme. Because the transport process of multispecies and the reaction both exist in the flowfield, a modified LU-SGS implicit method [23] is adopted for temporal integration.

2.1.3. *Computational zone and boundary conditions*

Figure 2 shows the entire computational domain which is the half of scramjet combustor together with a short cavity and a fuel jet orifice. The inlet of combustor is located at x = 0 mm, which is 33.064 mm high and 44 mm wide. The upper wall angle is set as 1° to prevent thermal block. For avoiding the interaction between multifuel jets, there is only one single fuel orifice arranged at the symmetric plane of combustor, which locates at 60 mm downstream of the inlet. The distance between the fuel orifice center and the leading edge of downstream cavity is 10 mm, which can strengthen the resisting back pressure capability. And a short cavity is adopted here with rear edge angle 45°, length 56 mm, width 44 mm, and depth 8 mm, respectively. Instead using a circular fuel orifice, a 1.772 mm × 1.772 mm square cross section orifice is adopted to acquire high-quality mesh, which has the same cross section area as a 2 mm diameter circular orifice.

The grids number is largely reduced by setting the symmetry face (xy-plane) of combustor as a symmetry boundary condition. In order to improve the quality of computational grids, the whole domain is divided into six parts to make all grids in structured type, except for the rear part of the cavity with unstructured grids. All grids aspect ratio and equisize skew are within 8 and 0.54, respectively. Grids are refined in key zones, such as the plasma domain, fuel orifice, wall, and shear layer. Besides, grids in zones with relatively large pressure gradient and the boundary layer are refined using adaptive mesh refining method based on the initial simulation

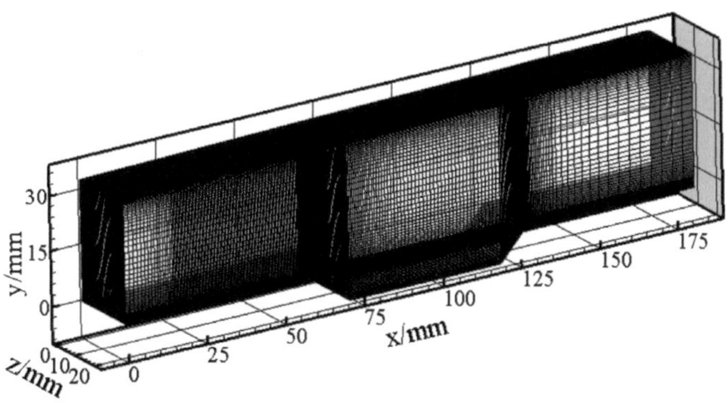

Figure 2. Computational domain and grids of the combustor.

results. After comparing the results of different grids size, a mesh scheme which consists of 955,166 cells in the whole computational zone is chosen, and it can be proved when the grids size increases, the flowfield remains almost unchanged all the same.

The entire computational domain is divided into 32 subdomains, and all assignments are completed by parallel computation on HP senior workstation, which takes about 320 h to obtain a convergence result. The inflow conditions in computation are as follows: $Ma = 2.2$, static pressure $P = 101$ kPa, static temperature $T = 823$ K, and boundary layer thickness $\delta = 2.0$ mm. In order to identify product water, dry air is set as the inflow (the mass fraction of oxygen and nitrogen are $Y_{O2} = 23.3\%$, $Y_{N2} = 76.7\%$, respectively). Here, hydrogen is used as fuel which is perpendicularly injected into the main flow at sonic speed. The parameters of fuel injection at the outlet of fuel orifice are given as: static pressure $P_{jst} = 334$ kPa, stagnation pressure $P_{j0} = 0.6$ MPa, and stagnation temperature $T_{j0} = 290$ K. At the exit boundary, the supersonic extrapolation condition is used. Besides, symmetry condition is used in central *xy*-plane ($z = 0$) for decreasing the calculation cost. There is no slip and adiabatic conditions, which are specified on the upper, bottom, and lateral walls of combustor including the cavity walls.

2.1.4. Simulation validation

The ability of our numerical methods for simulating the multispecies reaction flow field of the scramjet combustor with a cavity has been validated and can be found in [14, 18].

2.2. Results and discussion

Three representative times are selected during one plasma control cycle for comparison after computation converged. The end of the actuator duty time ($t = 2/5T_c$, where T_c is the period of one plasma control cycle), a certain time of the actuator free time ($t = 4/5T_c$) and the end of a whole plasma control cycle ($t = T_c$) are named A, B, and C, respectively.

2.2.1. Temperature and wall pressure distribution of the jet flowfield

The temperature distribution on the symmetric *xy*-plane of combustor is shown in **Figure 3**. It can be seen that the relatively high temperature zone appears mainly in the cavity when there is no plasma actuator arranged. But it moves downstream with the actuator working. The local temperature of the near wall downstream and the rear part of the cavity increases distinctly, which means the combustion centrality zone moves downstream. Besides, the lower temperature zone which denotes the fuel jet flow (i.e., the deep blue zone in **Figure 3**) prominently shrinks as the actuator works. However, the area of the lower temperature zone extended gradually at time B and C compared with time A. The unsteady phenomena should be due to the high heat nature of quasi-DC discharge plasma, which transfers its heat to the fuel jet resulted in changing the combustion zone downstream.

The distribution of wall pressure near the fuel orifice on the symmetric plane of the combustor is shown in **Figure 4**, which is normalized by the value of the static pressure of inflow at the inlet. Compared with no plasma case, the positions that pressure starts to rise at different times

http://dx.doi.org/10.5772/intechopen.80959

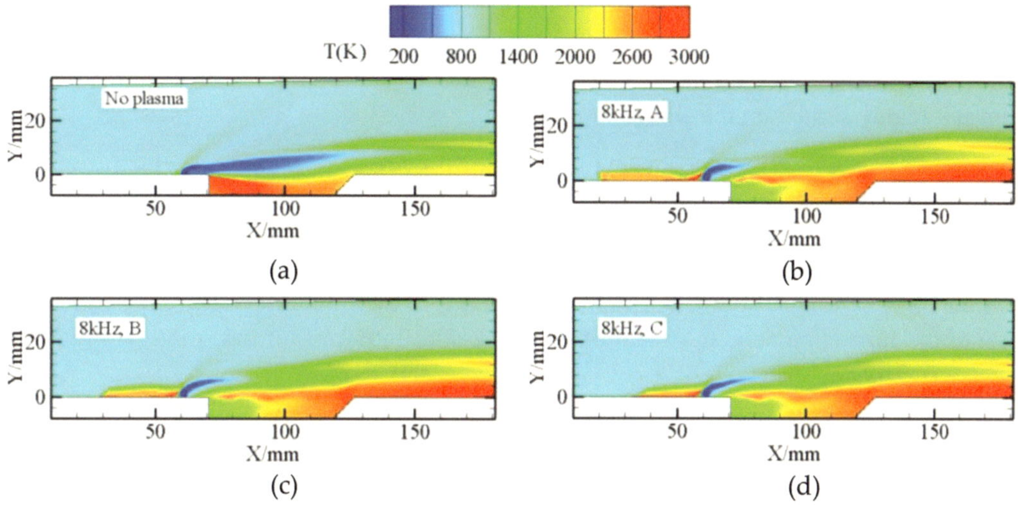

Figure 3. Contours of temperature on the symmetric *xy*-plane of combustor. (a) Plasma off. (b) Plasma on, time A. (c) Plasma on, time B. (d) Plasma on, time C.

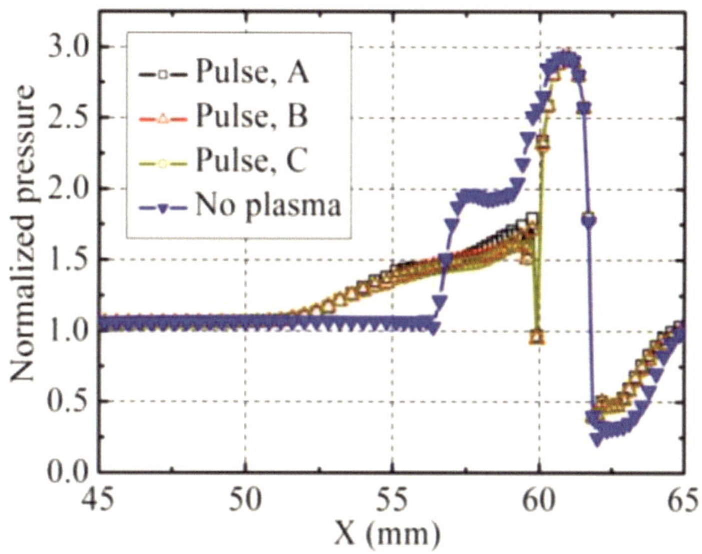

Figure 4. Wall pressure distribution on the symmetric plane.

move upstream from 56.6 to 52 mm due to the effect of plasma, which indicates the separation shock wave induced by the fuel jet moves upstream. And the first pressure peak decreases from 2.0 to 1.7, because of the weakening of the fuel jet induced shock on the symmetric plane. The results above are similar to the previous studies on nonreaction flow combustor [19]. More details indicate that the first pressure peak of time A is a bit higher than time B and C, but time B equals time C, which is due to the plasma control cycle too short for flow response and the duty cycle ratio also comparatively large in the flow condition here. Hence, the influence of plasma on the shock will be observed a little latter due to an inertia influence. Nevertheless, the effect of plasma on the shock wave for three typical times is highly similar on the whole.

Because the separation zone upstream of the fuel orifice is primarily controlled by the separation shock wave, the size of this separation zone can be regulated by changing the location of the separation shock wave. On the one side, this zone behaves as an main ignition zone which can provide a high temperature and low flow speed environment in the scramjet combustor. On the other side, it will bring in certain pressure loss to the combustor. Hence, we can make use of the separation zone upstream of the fuel orifice by means of using the plasma with proper control parameters.

2.2.2. *Characteristics of fuel mixing and combustion*

For the sake of determining the fuel mixing quality, the five cross-sectional planes distributed in equivalent ratio along the flow direction are given, as shown in **Figure 5**. The fuel jet arises with the actuator working, resulting in fuel decreasing near the wall including the bottom wall of the cavity. Whereas, the distribution shape of fuel in the cross-sectional planes varies from a distribution narrow and long profile into a circular profile in the upper space, which shows that the process of fuel spreading into the main supersonic flow is enhanced resulted from the plasma. Furthermore, the fuel distributions at time A are nearly the same as at time C, which is also correlated to the inertial effect mentioned above.

In **Figure 6**, the distribution of product water is shown. In order to distinguish the extent of product water between time A and C easily, the iso-surfaces of both are plotted by combining the half parts of them together as shown in **Figure 6(b)**. Similar to the changes in **Figure 5**, more water is generated in the upper space due to plasma. Contrasted with the case without plasma actuator, the iso-surface of water expands much in its center and shrinks near the combustor wall especially downstream of cavity. These above should be attributed to the change of fuel jet spread, as given in **Figure 5**. Compared time A with C, the only distinction is a little more water formed at time A, which indicates that the combustion becomes weaker in the free time of a plasma control cycle.

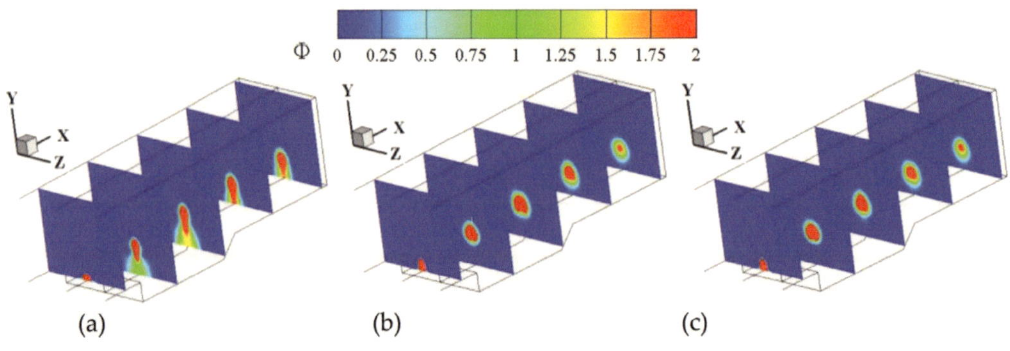

Figure 5. Distribution of equivalent ratio in cross-sectional planes. (a) Plasma off. (b) Plasma on, time A. (c) Plasma on, time C.

http://dx.doi.org/10.5772/intechopen.80959

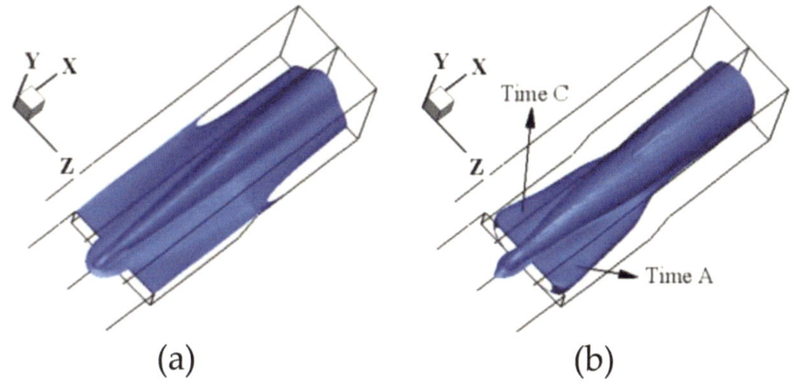

Figure 6. Iso-surfaces of product water, $Y_{H2O} = 0.015$. (a) Plasma off. (b) Plasma on.

It can be seen that the fuel entrainment into the cavity is decreased and the combustion in cavity becomes weaker due to the plasma, as given in **Figures 5** and **6**.Whereas, the two figures also show that more fuel penetrates into the supersonic air flow, so the mass exchange between the inside cavity and its outside is enhanced due to the plasma. Hence, the plasma improves the fuel mixing above the cavity prominently, which can also be realized as the calculation results of combustion efficiency in Section 2.2.4.

2.2.3. *Combustor stagnation pressure*

The stagnation pressure recovery coefficient can indicate the pressure loss in a combustor, so it is an important index. When the stagnation pressure recovery coefficient goes up, the capability of combustor outflow will be enhanced. So, it is defined as the ratio of combustor outlet stagnation pressure to inlet stagnation pressure [24]. In fact, calculating the mass flow weighted mean stagnation pressure can gain the stagnation pressure in a cross section.

$$\eta_{p_0} = \frac{\overline{p}_{0\_outlet}}{\overline{p}_{0\_inlet}} \tag{4}$$

where η_{p0} and $\overline{p}_0$ are the stagnation pressure recovery coefficient and the combustor stagnation pressure, respectively.

$$\overline{p}_0 = \frac{\int p_0 \overline{\rho} u \mathrm{d}y}{\int \overline{\rho} \mathrm{d}y} \tag{5}$$

where u and $\overline{\rho}$ are the flow velocity across a certain plane and the density of selected plane, respectively. Based on above, we get the stagnation pressure loss coefficient:

$$\eta_{p_0, \text{loss}} = 1.0 - \frac{\overline{p}_{0\_outlet}}{\overline{p}_{0\_inlet}} \tag{6}$$

The calculated data about the stagnation pressure are given in **Table 3**. In order to analyze the tendency of stagnation pressure loss coefficient further, the average stagnation pressure at the outlet is replaced by the average stagnation pressure at different positions, as shown in **Figure 7**. At the actuator working time or even at the free time, the stagnation pressure loss can both be increased due to the plasma, as given in **Table 3**. In (7), the relative change rate of η_{p0} is defined to realize the variation degree of the stagnation pressure recovery coefficient.

$$\varepsilon_i = \left| \frac{\eta_i - \eta_b}{\eta_b} \right| \times 100\% \tag{7}$$

where η_i is the stagnation pressure recovery coefficient of a certain case and η_b is the base case.

In **Figure 7**, the difference of stagnation pressure loss between plasma cases is shown where no plasma case enlarges along the flow direction (x-direction). At time A, B, and C, the relative change rate of stagnation pressure recovery coefficient at outlet point are 3.7, 3.1, and 3.4%, respectively, which indicates that the stagnation pressure loss varies little during a whole plasma control cycle and the loss is relatively little. The reasons resulted in these changes are

Index of performance	Case			
	Plasma off	Plasma on, time A	Plasma on, time B	Plasma on, time C
P_{0_inlet} /Pa	1,098,093			
P_{0_outlet} /Pa	916,532	882,270	888,143	885,268
η_{p0}	0.83466	0.80346	0.80880	0.80619
$\eta_{p0,loss}$	0.16534	0.19654	0.19120	0.19381

Table 3. The inlet mean stagnation pressure P_{0_inlet}, outlet mean stagnation pressure P_{0_outlet}, stagnation pressure recovery coefficient $\eta_{p0\ z}$, and stagnation pressure loss coefficient in the combustor.

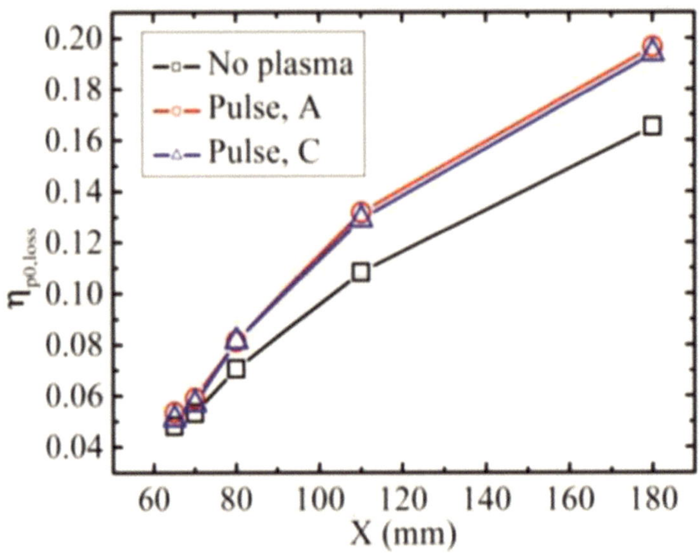

Figure 7. Distribution of the stagnation pressure loss coefficient.

in three aspects: (1) the strength of separation shock upstream of the jet is changed due to plasma as mentioned above, which will affect the stagnation pressure distribution; (2) the quasi-DC discharge plasma behaves as a virtual blockage in supersonic flow resulted in new shock waves or compression waves forming. Then, the pressure loss is increased; (3) the heat release from combustion can result in the stagnation pressure decreasing. Hence, the increase of stagnation pressure loss in the combustor is correlated to all the changes in flowfield which is due to the comprehensive effect mentioned above. To ensure the maximal combustion efficiency while keeping the stagnation pressure loss as little as possible, the optimal design is important. So the calculation of combustion efficiency is essential to choose the proper plasma actuator parameters.

2.2.4. Combustion efficiency downstream of the fuel orifice

The combustion efficiency η_c is commonly denoted by the amount of combustion product. Thus, the amount of water is employed [24]:

$$\eta_c = \frac{(\dot{m}_{H_2O,X} - \dot{m}_{H_2O,I})/W_{H_2O}}{\dot{m}_{H_2}/W_{H_2}} \tag{8}$$

where $\dot{m}_{H_2O,I}$ is the mass flow rate of water at inlet cross section and $\dot{m}_{H_2O,x}$ is x-direction cross section, respectively. $\dot{m}_{H_2}$ and W_i are the mass flow rate of hydrogen and the mole mass of specie i, respectively. And dry air is assumed at the inlet in the simulation $\dot{m}_{H_2O,I} = 0$.

Figure 8 plots the distribution of combustion efficiency in x-direction. At both time A and C, the rise rates of combustion efficiency go into larger along x-direction. At the outlet, η_c reaches 0.81341, 0.76008, and 0.60278 for time A, C, and no plasma case, respectively. Namely that η_c at time A and C are 1.35 and 1.26 times of no plasma case. As a result, the quasi-DC plasma does obviously improve the combustion in combustor on the whole level, as shown in **Figure 8**, even if the most water forms in the upper space rather than in the cavity. But the η_c at time A and time C are almost same. From **Figure 6**, the increase of combustion efficiency can also be realized, which indicates that the mixing performance of the fuel jet in scramjet combustor is improved due to the quasi-DC discharge plasma.

In order to define the cost to effectiveness of quasi-DC plasma, it is calculated in value $E_f = 0.00811$ for the ratio of deposited plasma energy to the increased combustion heat release. Accordingly, for improving the combustion of scramjet combustor, the quasi-DC plasma shows good capability as costing a little.

2.3. Conclusions

The main results in this section are as follows: (1) The distribution of relatively high temperature zone moves downstream prominently due to the heat release from the quasi-DC discharge plasma. The separation shock wave induced by the fuel jet is partly weakened and moves upstream due to the plasma, which can regulate the size of recirculation zone upstream of the fuel orifice. (2) The fuel jet moves upward integrally resulted from the plasma heating effect. The fuel spread wider along the spanwise and penetrates into the leading flow deeper,

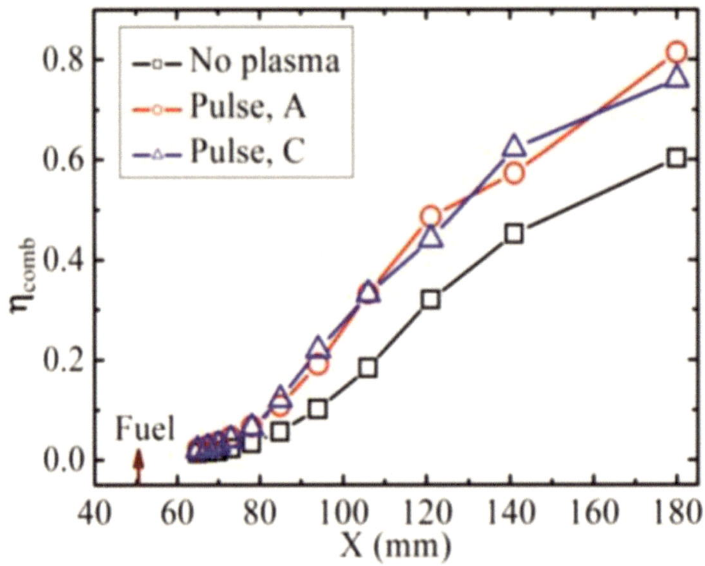

Figure 8. Distribution of combustion efficiency downstream of fuel orifice.

resulting in the cross-section shape of the fuel jet varying from a narrow and long profile to a circular profile. Because of the variation of fuel mixing, in the upper space, more water forms while less appears near the wall compared with the case without plasma. (3) The stagnation pressure loss of combustor increases a little as actuator works, but the combustion efficiency in the combustor rises obviously. These above can be summarized as the comprehensive effects of flow structure changes caused by the plasma, including waves induced and heat transfer. Since it is negligible for the relative change of stagnation pressure recovery coefficient in the actuator working cases, and the ratio of deposited plasma energy to the increased combustion heat release is very little, it can be obtained that the quasi-DC discharge plasma can make more benefits than penalties for the scramjet combustor, when proper adopting control parameters of the plasma actuator.

3. Case two—effect of plasma on cavity

3.1. Plasma model

On account of the direction of the quasi-DC discharge path for the inflow, the discharge modes include two types: longitudinal mode and transversal mode [12]. In case one, the quasi-dc discharge operates under transversal mode, while in case two, a longitudinal mode is adopted for the configuration. As shown in **Figure 9**, there are five anodes set upstream of the backward wall of cavity, while in the bottom wall of cavity the five cathodes are set. The five pairs of identical electrodes are arranged parallel and symmetrically. Besides, the flow will not be disturbed directly resulted from all the electrodes flush mounted in the wall. A filament plasma forms when applied a high voltage (generally 150–1200 V) between anode and cathode, which

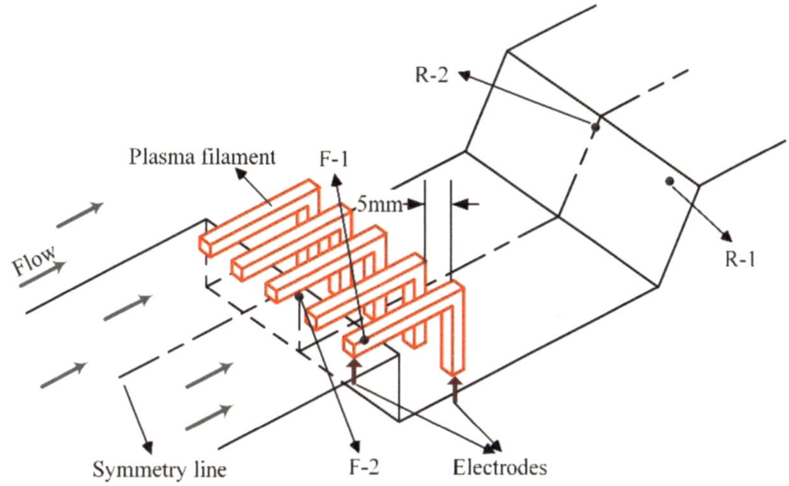

Figure 9. The schematic of plasma filaments in the cavity.

acts as strongly oscillating and bright and looks like an inverted "L" crossing the backward wall of cavity. Since the major heat energy of the quasi-DC discharge plasma focuses on the bright filament domain [10, 25], the filament plasma domain can be established in such a simplified shape as given in **Figure 9**.

Just like the way given in case one, every quasi-DC plasma filament is dealt as a volumetric heat source. To represent the plasma heat strength reasonably, the mean temperature of the plasma domain T_{pl} is specified in this simulation too. Considering the real size of the plasma filament, all the heat source is modeled with a section of dimensions 3 × 3 mm. the distance between the backward wall and left side of plasma filament upstream of cavity is 6 mm as depicted in **Figure 9**. While the distance between the backward wall and the right side of plasma filament downstream of cavity is 25.5 mm. Based on the previous research, $T_{pl} = 3000$ K is specified as the actuating strength, and five actuators work together in a pulsed mode with plasma actuation frequency $F_c = 5$ kHz and duty cycle $D = 1/5$. In this case, all configurations (e.g., combustor, cavity) and simulation conditions are identical to those used in case one, except for the plasma. Therefore, the numerical methods, including physical models, numerical schemes, computational zone, etc., could be found in Section 2.1.

3.2. Results and discussion

Three representative times are chosen from one plasma cycle, which consists of the later actuator free duration and the actuator working duration, for comparison after computations converged. In the simulation, the duration of one plasma cycle is $T_c = 200$ μs. The end of an actuator working duration $t = 1/5T_c$, $t = 3/5T_c$, and the end of a cycle $t = T_c$ are named A, B, and C, respectively.

3.2.1. Typical parameters distribution of cavity flowfield

The Mach number distribution of local cavity flowfield on the symmetrical *xy* plane is shown in **Figure 10**. A distinct shear layer forms upon the cavity mouth and develops toward the

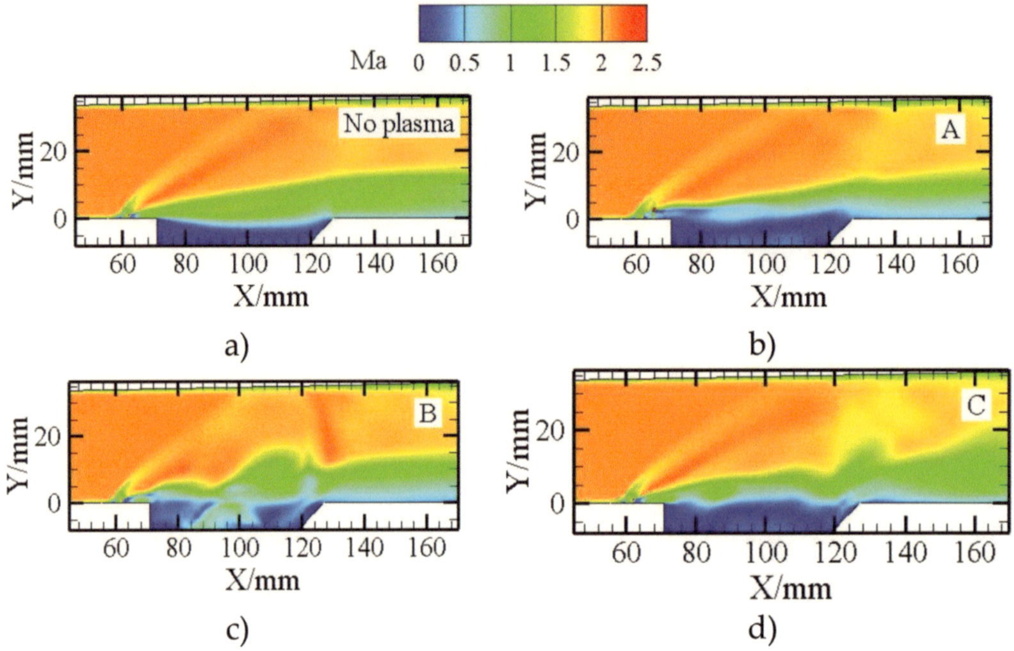

Figure 10. Mach number distribution: (a) no plasma; (b) with plasma, time A; (c) with plasma, time B; (d) with plasma, time C.

main flow, which is well known as the typical combustion flowfield of cavity. Contrasted with the flowfield of the no plasma case, the cavity shear layer fluctuates abruptly, particularly in the *y* direction. And it develops unsteadily resulted in the division of cavity recirculation zone at time B, as depicted in **Figure 10c**. Resulted from the plasma observed the Mach number distribution from time A to C, the main flow velocity downstream of the cavity decreases to a certain degree. On the whole, the oscillation phenomenon of cavity shear layer strengthens firstly and then weakens in pace with the plasma cycle.

Owing to the periodic heat release from the plasma filaments and the thermal blocking functions, the plasma filaments behave as five knives cut the cavity shear layer. Hence, the mass transportation process will be disturbed by this "cutting" effect, and the moving direction of original shear layer has to be changed also. And then, the turbulence intensity and vorticity magnitude both are increased, so the fuel and air existing around the original shear layer can exchange through the cavity mouth more easily. Furthermore, the combustion enhancement downstream or over the cavity causes the decrease of local flow velocity, which results in the rise of local static pressure and blocks the incoming flow.

Figure 11 presents the distributions of wall pressure near the rear edge of cavity on $z = 0$ and $z = 16$ mm plane. The pressure value on *y*-axis is normalized by the inflow static pressure. Being the same as the Mach number in **Figure 10**, the plasma influence on the wall pressure is strong and unsteady. On $z = 0$ mm plane, the pressure peaks of both time A and C move upstream compared with the no plasma case, which are shown in **Figure 11a**). At time A, its peak value is nearly equal to the case without plasma, and decreases more gently from its peak

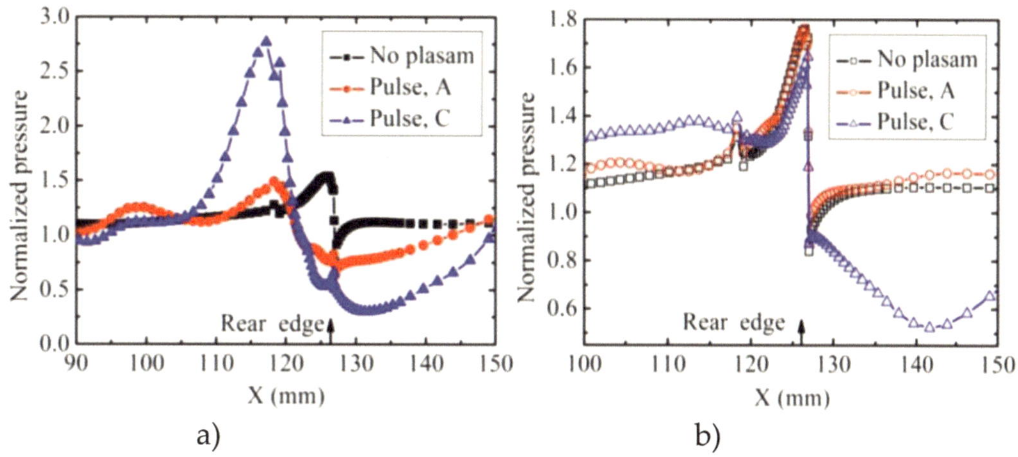

Figure 11. Wall pressure distribution near the rear edge of cavity. (a) z = 0 mm. (b) z = 16 mm.

to minimum value. But at time C, its pressure peak value is nearly 2.8 times as the no plasma case. Results about show that the original waves around the rear edge are altered which results from local combustion zone induced by the periodic disturbance of plasma and the variation of cavity shear layer.

As presented in **Figure 11b**, the wall pressure of cavity rises distinctly on z = 16 mm, which illuminates that the temperature goes up in the cavity. Besides, the pressure peak value at time A equals the value of the no plasma case and is lower than the value at time C. This suggests clearly that the plasma filaments can weaken the waves around the cavity rear edge during plasma actuators free period (means the actuators shutdown.), which is the same to the study of nonreaction flow [19]. Moreover, because of the effect of expansion waves (as it is known that the abrupt pressure reduction near a cavity rear edge is induced by expansion waves), the normalized pressure reduces to around0.8 and then drops in a relative gentle way until $x \approx 140.0$ mm at time C, which is due to the change of unsteady local waves and combustion.

Because the distributions of combustion products may be impacted by the change of cavity shear layer and waves, the mass fraction iso-surface of water (Y_{H2O} = 0.05) is shown in **Figure 12**. Firstly, around the fuel orifice for the no plasma case some water forms, but there is no water forming during the plasma actuator working duration. Secondly, compared with the no plasma case having a smooth iso-surface, the iso-surface looks wrinkled distinctly due to the influence of plasma filaments, especially in the shear layer zone. Several symmetry structures shaped like "branch pipes" can be found obviously because of plasma. The "branch pipes" generate from the front or middle part of the cavity, and then move downstream. And "branch pipes" also move upward in the process mentioned above. Hence, a "branch pipes" produced cycle is established from time A to time C. The wide extent of product water shortens in the z direction, while more water forms in the y direction at the same time.

The above phenomena can be attributed to the reasons as follows: (1) Plasma filaments existed upstream of the cavity front edge release a large amount of heat, so the local static temperature increases and then the movement of fuel jet is promoted toward upper space. Because of the

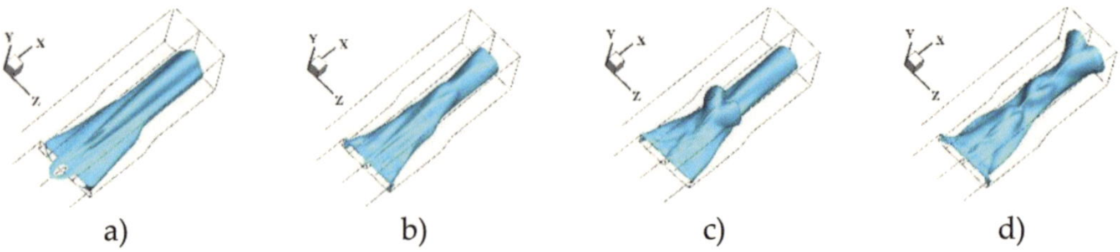

Figure 12. Iso-surface of product water, $Y_{H2O} = 0.05$: (a) no plasma; (b) with plasma, time A; (c) with plasma, time B; (d) with plasma, time C.

larger velocity far from the bottom wall (i.e., flow moves faster in the middle height of the combustor), the original product water will surely move downstream quickly. Therefore, the little water is found around the orifice when plasma actuators work, and the symmetry structures are formed in higher space downstream. It can be verified from the "branch pipes" presented in **Figure 12b–d**. (2) As stated above, the periodic fluctuation of cavity shear layer is related to the plasma filaments "cutting" effect on it, and then the mass transportation process is blocked which leads to the obvious change of combustion over the cavity. Hence, it affects the distribution of product water. Meanwhile, notice that the "branch pipes" curve structures match with the typical spanwise reverse vortex pairs in shape. Generally speaking, the spanwise reverse vortex pairs can extend the contact area between air and fuel by the methods of entraining air into its fuel core, which makes the mixing between fuel and air better and also the local combustion efficiency arisen. Whereas, the periodic disturbance from the plasma filaments impels the air and fuel transporting in the y direction. Hence, at certain periods, the efficiency of vortex pairs rises in several zones, which leads to the unsteady variation of product water, as shown in **Figure 12**. (3) The phenomena that more water forms in higher space and the less exists near the cavity mouth and combustor wall, which can be attributed to the heating effect on the whole cavity by the plasma filaments.

3.2.2. *Analysis of cavity drag and mass exchange rate*

The drag and mass exchange rate of cavity are two important cavity performance parameters [26]. The combustor's drag is mainly generated by the cavity as it has a nearly constant cross section. The cavity drag includes two types: pressure drag and friction drag. Pressure drag is defined as the difference value between the force on cavity front wall and rear wall. Usually, the friction drag is too small compared with pressure drag so it always can be ignored. So the pressure drag is regarded as the cavity drag here. The drag coefficient of flame holding cavity is defined as:

$$C_D = \frac{2F_D}{A\rho V^2} \tag{9}$$

where F_D is the cavity drag; A is the section area of combustor inlet; ρ is mass density; and V is velocity of inflow.

In **Table 4**, the calculation results of the cavity drag and its drag coefficient are listed. It can be observed that the drag coefficient is 0.060 at the no plasma case, which is smaller than time B

	Front wall	Rear wall	Drag	Drag coefficient
No plasma	−19.8	34.3	14.5	0.060
Plasma, A	−19.6	33.1	13.5	0.056
Plasma, B	−19.6	37.6	18.0	0.075
Plasma, C	−18.1	33.5	15.4	0.064

Table 4. Calculation results about the cavity drag, unit: N.

and C but larger than time A. In view of the change of cavity shear layer in **Figure 10**, it should be due to the moving upward of the shear layer as actuator works. As the shear layer moves upward, it deflects to the main flow and no more impacts on the cavity rear wall so the drag decreases. However, at time B and C, because of the variation of the shear layer fluctuation, stronger impact may appear the rear wall of cavity, and the combustion may be boosted in the cavity rear part, which can both lead to the increase of back pressure in the cavity. Considering the drag coefficient values at the three typical times, the time weighted average cavity drag coefficient is calculated to be 0.0668 which means the cavity drag usually increases by the plasma.

As we know, the cavity drag in reaction flow is closely related to the combustion heat release zone of a cavity. So only reducing the cavity drag may not bring benefit for the whole combustor. It can be seen that the further detail analysis about the combustion distribution in flowfield should be done to understand the effect of plasma on the cavity performance.

The mass exchange rate of cavity m' is another important parameter, which stands for the fluid mass that is inhaled into cavity from the main flow every second. The shedding vortex from cavity shear layer primary conducts the mass exchange process which can effectively inhale air and fuel into cavity and then takes them away. More shedding vortices and quicker movement can advance the flame holding ability of cavity with efficient mixing. The mass exchange rate can be analyzed by monitoring the mass flux that passes through the cavity mouth when the shear layer just covers the cavity. The mixing gas in the cavity is marked as "Y" in the unsteady flowfield at a time with stable result, and the dynamic process until all the gas "Y" leave the cavity can be observed and then the whole time involved "τ" in this process (i.e., residence time in cavity) has been recorded easily. Afterward, the mass exchange rate is $m' = m/\tau$, where m is the total mass of fluid in a cavity. Basically, the fuel exchange between internal and external of cavity is mostly affected by the interaction between the back wall and shear layer of cavity and the fuel distribution in cavity shear layer.

The mass exchange rate of cavity is given in **Table 5**. Because the species transportation is achieved through the vortices in the cavity shear layer, it is very significant for enhancing the diffusion ability of cavity shear layer. Owing to the break on the original stable structure of the shear layer resulted from the plasma filaments "cutting" effect, the mass exchange rates at plasma existing cases at times A, B, and C are 9.2, 197.2, and 107.8 times, respectively, than the "no plasma" case. In addition, the time-weighted average mass exchange rate is 8.217 g/s, which is 122.6 times than the "no plasma" case. And the variation of the mass exchange rate magnitude from time A to C is in related to fluctuation degree of cavity shear layer. As a result, the quasi-DC plasma does obviously promote the species transportation between external and internal cavity.

	No plasma	Pulse, A	Pulse, B	Pulse, C
m'	0.067	0.614	13.013	7.223

Table 5. Calculation results of cavity mass exchange rate, unit: g/s.

3.2.3. *Analysis of pressure oscillation in cavity*

In the research fields related to "sound cavity," "embedded magazine" and "flame holding cavity," the dynamic study on pressure oscillation in a cavity is always greatly valued around the world [27]. According to the cavity shear layer distribution, four monitor points are chosen to capture the dynamic pressure feature, as shown in **Figure 9**. F-1 is located on the front wall of cavity with $z = 16$ mm (refers to a lateral section of the combustor) and F-2 is located on $z = 0$ mm (refers to the symmetry section of the combustor), and they are 0.5 mm beneath the front edge. R-1 is located on the rear wall with $z = 16$ mm and R-2 is with $z = 0$ mm, and they are 0.71 mm away from the rear edge. To ascertain the intensity of pressure oscillations, the sound pressure level (SPL) is used to present the time averaged pressure fluctuation magnitude, with dB as its unit, which is defined as:

$$L_{sp} = 20 \log \frac{\overline{p'}}{p_{ref}} \tag{10}$$

where $\overline{p'}$ and P_{ref} are the RMS of dynamic pressure and the reference sound pressure, respectively, and here P_{ref} is 2×10^{-5} Pa.

The SPLs of F-1, F-2, R-1, and R-2 are 168, 193, 188, and 191 dB, respectively. However, at the no plasma case, their SPL values are 120, 118, 124, and 106 dB, respectively. It indicates that the SPL of all the monitor points go up sharply, and the relationship between point location and magnitude of SPL changes prominently. The points are near the mouth of cavity and the effect of plasma filaments on the cavity edges is strongest, so the SPL magnitude increases here. Moreover, through the analysis on cavity drag in **Table 4**, the conclusion can be obtained that the pressure disturbance on the monitor points is controlled by combustion, and the shear layer strongly affects the change of local combustion.

In **Figure 13**, the frequency spectrum characteristics of R-1 and R-2 by FFT are depicted. The several dominant frequencies of pressure oscillation are marked with red circles here, which show that the first dominant frequency is almost the double of the plasma actuation frequency 5 kHz. Whereas, it is barely equal to the plasma actuation frequency under nonreaction condition [19]. Furthermore, nearly every dominant frequency is an integer multiple of the plasma actuation frequency. The results above indicate that the pressure oscillation in the cavity is controlled by both local combustion flowfield and the plasma actuation frequency.

3.2.4. *Analysis of stagnation pressure loss and combustion efficiency*

The combustor stagnation pressure recovery coefficient of time A, B, C, and the no plasma case is 0.75, 0.81, 0.65, and 0.83, respectively. And the calculated time weighted average stagnation pressure recovery coefficient is 0.73. It suggests that the overall stagnation pressure loss is

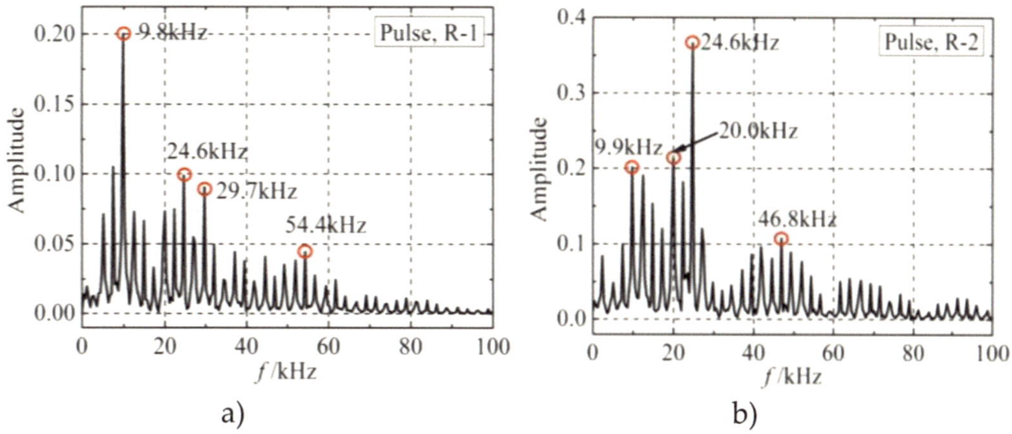

Figure 13. The normalized pressure frequency spectrum, with plasma. (a) R-1. (b) R-2.

raised due to the plasma filaments, especially for time A and C. The increase of the combustor stagnation pressure loss is caused by several factors, such as the enlargement of separation zones, the rise of back pressure induced by local combustion and the strengthening of shock waves, which is based on the analysis of cavity shear layer and wall pressure distribution.

Considering the interesting variation of product water distribution as shown in **Figure 12**, the distribution of combustion efficiency in the flow direction may also vary influenced by the plasma filaments. Hence, the combustion efficiency η_c is calculated. The distribution of combustion efficiency along x-direction is presented in **Figure 14**. Usually, the combustion efficiency goes on increasing from time A to time C due to the plasma. It can be seen the combustion efficiency of time A is little lower at most positions, but prominently higher at time B in the upstream of cavity rear wall and time C on the whole compared with the no plasma case. As involved above, the transportation ability of the spanwise reverse vortex pairs are improved by the plasma, so the local combustion efficiency during certain periods increases.

In a word, the combustion efficiency increases downstream of the front wall of cavity in most of the time due to the plasma. It becomes lower than the no plasma case at several positions for few moments, but the decrease is comparatively small. Furthermore, during one plasma cycle, the combustion efficiency gradually increases from beginning to end, which indicates that the plasma performs better during actuator free period for the pulsed actuation mode.

The ratio of power consumption of the plasma actuator to the increased combustion heat release of the combustor is calculated in value $E_f = 0.0016$ to depict the cost to effectiveness of quasi-DC plasma. Therefore, the assistance of plasma to the combustion process presents improvements in performance that outweighs the additional cost.

3.3. Conclusions

The main results are as follows: (1) Because of the "cutting" effect of quasi-DC plasma filaments, the cavity shear layer becomes fluctuating and the local combustion changes which also results

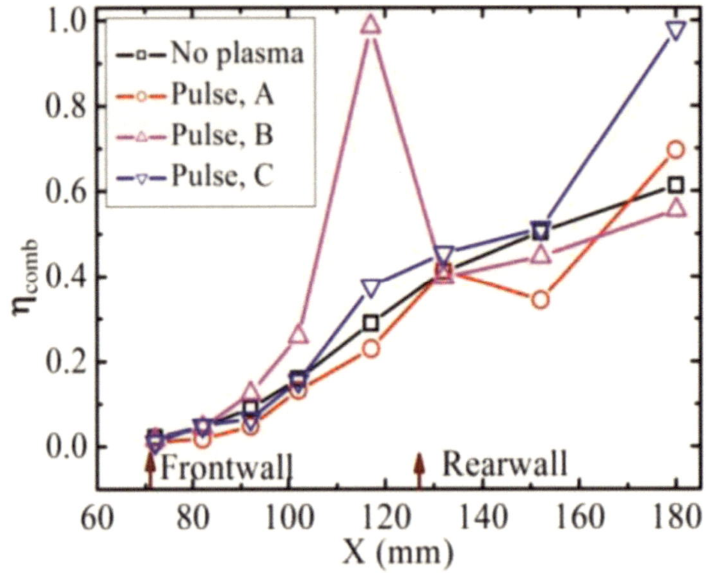

Figure 14. Combustion efficiency along *x*.

in the variation of wall pressure distribution near the cavity rear edge. On the mass fraction isosurface of product water several "branch pipes" curve structures form periodically, which can be put down to the influence of plasma on the local combustion. (2) The effect of plasma on the cavity drag is relatively perplexing and the drag coefficient value is observed to be unsteady. Nevertheless, the mass exchange rate goes up prominently due to plasma, and the magnitude change from time A to C is in agreement with the fluctuation degree of cavity shear layer. (3) The SPLs of four cavity monitor points rise, and the frequency spectrum of monitor points near the rear edge presents that the first dominant frequency is twice the plasma actuation frequency, under the effect of plasma. Hence, the pressure oscillation in the cavity is controlled by both the plasma and local combustion flowfield. (4) On the one side, the combustion efficiency is usually increased due to plasma, and on the other side, the certain pressure loss increases resulted from the changes of combustion and waves in the flowfield by the plasma. Because the ratio of deposited plasma energy to the increased combustion heat release is extraordinarily small, it is considered that if optimal actuation parameters of the actuator are chosen the quasi-DC plasma can bring more benefits than penalties for the cavity in scramjet combustor.

Acknowledgements

This work is sponsored by the National Natural Science Fund of China, No. 51876219, No. 91441123, and No. 51777214. The authors would like to acknowledge the help of Dr. W. Feng in numerical simulation.

Conflict of interest

The authors declared that they have no conflicts of interest to this work.

Author details

Siyin Zhou, Haiqing Wang, Wansheng Nie* and Xueke Che

*Address all correspondence to: nws1969@126.com; 286982061@qq.com

Space Engineering University, Beijing, China

References

[1] Mathur T, Gruber M, Jackson K, et al. Supersonic combustion experiments with a cavity-based fuel injector. Journal of Propulsion and Power. 2011;**17**:1305-1312

[2] Ben YA. Experimental investigation of mixing and ignition of transverse jets in supersonic crossflows [thesis]. Palo Alto: Stanford University; 2000

[3] Ding M. Research on the flame holding technology based on cavity in supersonic combustion [thesis]. Changsha: National University of Defense Technology; 2005

[4] Li G, Li H, Yang L, et al. Review of the plasma ignition and assisted combustion in Russia. Science & Technology Review. 2012;**30**:66-72

[5] Kim KM, Back SW, Han CY. Numerical study on supersonic combustion with cavity-based fuel injection. International Journal of Heat and Mass Transfer. 2004;**47**:271-286

[6] Starikovskiy A, Aleksandrov N. Plasma-assisted ignition and combustion. Progress in Energy and Combustion Science. 2013;**39**:61-110

[7] Starikovskaia SM. Plasma assisted ignition and combustion. Journal of Physics D: Applied Physics. 2006;**39**:265-299

[8] Do H, Im S, Cappelli MA, et al. Plasma assisted flame ignition of supersonic flows over a flat wall. Combustion and Flame. 2010;**157**:2298-2305

[9] Do H. Cappelli MA, Mungal G: Plasma assisted cavity flame ignition in supersonic flows. Combustion and Flame. 2010;**157**:1783-1794

[10] Leonov SB, Firsov AA, Yarantsev DA, et al. Plasma effect on shocks configuration in compression ramp. In: 17th AIAA International Space Planes and Hypersonic Systems and Technologies Conference; 11–14 April 2011. San Francisco: AIAA; 2011. pp. 2011-2362

[11] Leonov SB, Savelkin CV, Yarantsev DA, et al. Experiments on plasma-assisted combustion in M=2 hot test-bed PWT-50h. In: 46th AIAA Aerospace Sciences Meeting and Exhibit; 7–10 January 2008. Reno: AIAA; 2008. pp. 2008-1359

[12] Leonov SB, Yarantsev DA, Gromov VG, et al. Mechanisms of flow control by near-surface electrical discharge generation. In: 43rd AIAA Aerospace Sciences Meeting and Exhibit; 10–13 January 2005. Reno: AIAA; 2005. pp. 2005-2780

[13] Enloe CL, McLaughlin TE, VanDyken RD, et al. Mechanisms and responses of a single dielectric barrier plasma. In: 41st AIAA Aerospace Sciences Meeting and Exhibit; 6–9 January 2003. Reno: AIAA; 2003. pp. 2003-1021

[14] Zhou S, Nie W, Che X. Numerical modeling of quasi-dc plasma-assisted combustion for flame holding cavity. Combustion Science and Technology. 2016;**188**:1640-1654

[15] Leonov SB, Yarantsev DA, Gromov VG, et al. Mechanisms of flow near surface electrical discharge generation. Vacuum. 2006;**80**:1199-1205

[16] Cheng Y, Nie W, Li G. Numerical study of plasma aerodynamic actuation mechanism. Acta Physica Sinica. 2012;**61**:060509

[17] Wang J, Li Y, Cheng B, et al. The mechanism investigation on shock wave controlled by plasma aerodynamic actuation. Acta Physica Sinica. 2009;**58**:5513-5519

[18] Zhou S, Nie W, Che X. Numerical investigation of influence of quasi-dc discharge plasma on fuel jet in scramjet combustor. IEEE Transactions on Plasma Science. 2015;**43**:896-905

[19] Zhou S. Studies on plasma assisted combustion/flame stabilization in scramjet [thesis]. Beijing: Equipment Academy; 2014

[20] Menter FR. Two-equation eddy-viscosity turbulence models for engineering applications. AIAA Journal. 1995;**33**:1418-1425

[21] Esch DD, Siripong A, Pike RW. Thermodynamic properties in polynomial form for carbon, hydrogen, nitrogen and oxygen systems from 300 to15000°K. NASA RFL-TR-70-3 (NASA CR-111989), Nov. 1970

[22] Liou MS, Steffen CJ. A new flux splitting scheme. Journal of Computational Physics. 1993; **107**:23-39

[23] Yoon S, Jameson A. Lower-upper symmetric-gauss-Seidel method for the Euler and Navier-stokes equations. AIAA Journal. 1988;**26**:1025-1026

[24] Rajasekaran A, Babu V. Numerical simulation of three-dimensional reacting flow in a model supersonic combustor. Journal of Propulsion and Power. 2006;**22**:820-827

[25] Jiang J, Weng J. Cathode Electronics and Gas Discharge. 1st ed. Beijing: Tsinghua University Press; 1980

[26] Sun M. Studies on flow patterns and flameholding mechanisms of cavity flameholders in supersonic flows [thesis]. Changsha: National University of Defense Technology; 2008

[27] Yang D, Li J, Fan Z, et al. Investigation on aerodynamic noise characteristics of cavity flow at high subsonic speeds. Acta Aerodynamica Sinica. 2010;**28**:703-707

Chapter 10

Plasma Generation and Application in a Laser Ablation Pulsed Plasma Thruster

Jianjun Wu, Yu Zhang, Yuqiang Cheng,
Qiang Huang, Jian Li and Xiaobin Zhu

Additional information is available at the end of the chapter

http://dx.doi.org/10.5772/intechopen.77511

Abstract

The laser ablation plasma thruster is a novel electric propulsion thruster, which combined the laser ablation and electromagnetic acceleration. In order to investigate the plasma expansion and ionization in the laser ablation plasma thruster, which was difficult to obtain from experiments, the heat conduction model and fluid dynamics model were established. The heat conduction model was established to calculate the target ablation, taking into account temperature-dependent material properties, phase transition, dielectric transition and phase explosion. The fluid dynamics model was used to calculate the plasma properties, taking into account ionization, plasma absorption and shielding. The ablation plasma velocity, temperature and electron number density were predicted by using the numerical method. The calculated results showed that the peak values of ablation plasma velocity, temperature and electron number density fraction were distributed at the front of the plasma plume. Moreover, the discharge characteristics and thrust performance were tested with different charged energy, structural parameters and propellants. The thrust performance was proven to be improved by electromagnetic acceleration.

Keywords: laser ablation plasma thruster, heat conduction, plasma expansion, ionization, discharge, thrust performance

1. Introduction

Because of the advantages of low-cost, low-mass, and high specific impulse, electric propulsion thrusters (EPTs) for spacecraft orbit correction and interplanetary spacecraft acceleration have recently become the front subject and focal point in space propulsion fields [1, 2]. As a member of EPTs, pulsed plasma thruster (PPT) has a broad prospect on small satellites for its small, compact, and low mass [3–5]. PPT has been studied decades on its performance and lifetime, and has been successfully applied to a number of satellites [6–8]. However, the problem of low efficiency and ignition failure still restricts the development of PPT.

Early in the year 2000, Horisawa et al. proposed a laser-assisted plasma thruster (LS-PPT), in which a laser-induced plasma was induced through laser beam irradiation onto a solid target and accelerated by electrical means [9–16]. Compared with the conventional PPT, the LS-PPT combines the laser ablation with electromagnetic acceleration means, which can significantly enhance the thrust performance. However, the phenomenon of "late ablation" is still inevitable in the LS-PPT, which significantly reduces the thrust efficiency of the thruster. In order to overcome the shortage of "late ablation," a novel laser ablation plasma thruster (LA-PPT) is proposed [17]. The LA-PPT separates the laser ablation from electromagnetic acceleration through a ceramic tube. As shown in **Figure 1**, the LA-PPT consists of a pair of rectangular electrodes, a ceramic tube and an insulator. The propellant is placed inside the ceramic tube. Because of the unique structure of this thruster, almost all types of solid matter can be applied as the propellant, such as metals, polymers and so on. Because laser-ablation plasma can has a directed initial velocity of tens of kilometers per second, which will be further accelerated by electrical means, a significant specific impulse can be expected [9]. Hence the LA-PPT is a promising candidate for small satellites propulsion, and the physical mechanisms of the thruster should be further investigated. The working process of the LA-PPT can be divided into two stages: laser-induced ablation and plasma-induced discharge. The ablation plasma expansion and ionization in the ceramic tube is the combination of the two stages, and it is crucial to understand the working process of the LA-PPT. However, the ablation plasma expansion and ionization is difficult to be experimentally investigated, especially when it occurs in a ceramic tube. Therefore, we utilize numerical method to investigate the ablation plasma expansion and ionization in the ceramic tube.

By using the numerical method to simulate the ablation plasma expansion and ionization in the ceramic tube, the relevant physics of propellant ablation needs to be implemented. The heating process within a propellant material during the irradiation of the laser pulse can be calculated by taking into account temperature-dependent material properties, melting, phase transition, dielectric transition, phase explosion, and the reflection of the laser beam at the surface of the propellant [18]. In addition, the ablation plasma absorbs part of laser

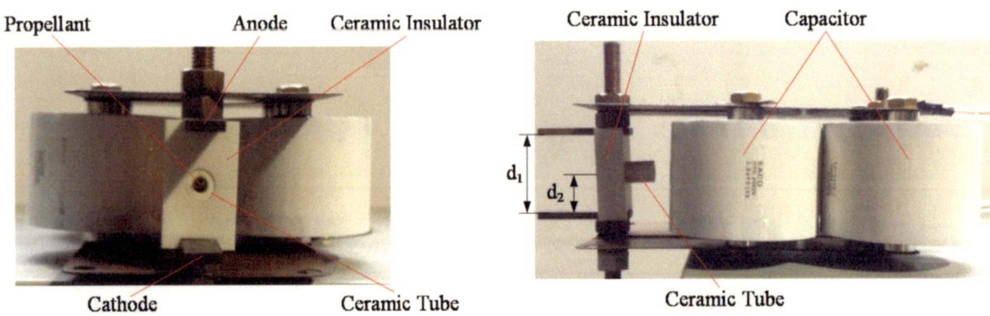

Figure 1. (a) Front and (b) right view of the laser ablation plasma thruster.

energy, and causes the density, temperature, pressure and components in the ablation plasma plume dramatically vary, which has an significant effect on the thrust performance. During the years, the expansion dynamics of pulsed laser generated plasma plume has been widely investigated through experimental and numerical methods [19–23]. Especially, several models have been proposed to describe the expansion of ablation plasma and the formation of plasma in the plume [21]. Aden et al. utilized hydrodynamic equations to describe the plasma expansion, which is called hydrodynamic model [24]. Afterwards, the hydrodynamic model had been widely applied to study various laser-solid interactions [25–28]. In order to calculate the laser energy loss caused by plasma absorption and shielding, the formation of plasma in the plume was also considered in several hydrodynamic models [22, 29–32]. Therefore, we establish a hydrodynamic model to investigate the ablation plasma expansion and ionization in the ceramic tube. The numerical model can give insight in the plasma dynamics and plasma behavior of LA-PPT, which is sometimes difficult to obtain from experiments.

In this paper, a novel laser ablation plasma thruster with a ceramic tube is developed and investigated. This thruster is expected to overcome the shortages of conventional laser-electric hybrid acceleration systems in the above paragraph. And the ionization rate, specific impulse and thrust efficiency of the thruster are expected to be increased. Specially, a numerical model for nanosecond laser ablation of aluminum propellant is presented, which contains the target ablation and plasma expansion in the ceramic tube. Heat conduction model is established to calculate the target ablation, taking into account temperature-dependent material properties, phase transition, dielectric transition and phase explosion. Meanwhile, Hydrodynamic model is established to calculate the plasma expansion, taking into account ionization, plasma absorption and shielding. Then both calculations of the target ablation and plasma expansion are coupled in each time step. Afterwards, the plasma properties (such as velocity, temperature and electron number density) in the ceramic tube are numerically investigated utilizing the model. Moreover, the thrust performance of the LA-PPT is also investigated by experimental methods.

2. Physical model

2.1. Heat conduction model

As shown in **Figure 2**, when the laser beam irradiates the target (e.g., aluminum), the temperature in the target rises, and melting occurs when the surface temperature reaches the melting temperature T_m. Then a part of molten materials begins to vaporize when the surface temperature reaches the boiling temperature T_b. With sufficient laser fluence, the target temperature may approach the critical temperature T_c, where dielectric transition occurs, and the dielectric layer is formed near the surface. Furthermore, the ablation plasma expands in the opposite direction of the incident laser beam, and absorbs part of laser energy before the incident laser beam reaches the target surface. The absorption of the laser energy in the plasma accelerates the plasma expansion, at the same time, the shielding of the laser energy by the plasma significantly affect the heat conduction of the target.

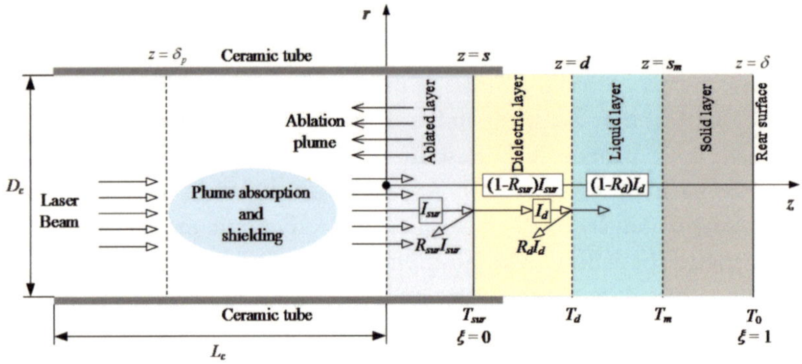

Figure 2. Schematic of the physical model and coordinates.

The initial length of the aluminum target is labeled as δ, as shown in **Figure 2**. The locations of the melting phase interface, the interface between dielectric layer and liquid phase, as well as the exposed ablation surface are labeled as s_m, d, and s, respectively. The length and the inner diameter of the ceramic tube are labeled as L_c and D_c.

The heat conduction equation in terms of the volumetric enthalpy H can be written as:

$$\frac{\partial H}{\partial t} = \frac{\partial}{\partial z}\left(K(T)\frac{\partial T}{\partial z}\right) + v_{sur}(t)\frac{\partial H}{\partial z} + S(t) \tag{1}$$

where $K(T)$ represents the temperature-dependent thermal conductivity. $v_{sur}(t)$ denotes the surface recession velocity, which can be calculated by Hertz-Knudsen [33, 34] and Clausius-Clapeyron equation [35]. The heat source term S is given by:

$$\begin{cases} S(t) = (1 - R_{sur})I_{sur}(t)\alpha(t)\exp\left(-\int_{s(t)}^{z}\alpha(t)dz\right), & s(t) \le z \le d(t) \\ S(t) = (1 - R_d)I_d(t)\alpha(t)\exp\left(-\int_{d(t)}^{z}\alpha(t)dz\right), & z > d(t) \end{cases} \tag{2}$$

where α is the absorption coefficient of the target, R_{sur} and I_{sur} are the reflectivity and laser intensity on the target surface, respectively. d, R_d and I_d are the location, reflectivity and laser intensity on the interface between dielectric layer and liquid, respectively. Herein, I_{sur} and I_d can be given by:

$$\begin{cases} I_{sur}(t) = I_0(t)\exp\left(-\int_{-\delta_p}^{s(t)}\beta(t)dz\right) \\ I_d(t) = (1 - R_{sur})I_{sur}(t)\exp\left(-\int_{s(t)}^{d(t)}\alpha(t)dz\right) \end{cases} \tag{3}$$

where $I_0(t)$ is the initial laser intensity, and β is the absorption coefficient of the plasma.

Considering the regression of the ablation surface, it is more convenient to transform the coordinate (z, t) into the fixed coordinate system (ξ, τ) by the following expressions:

$$\begin{cases} \tau = t \\ \xi = \dfrac{z - s(t)}{\delta - s(t)} \end{cases} \tag{4}$$

As shown in **Figure 2**, $\xi = 0$ (i.e. $z = s(t)$) indicates the ablation surface, and $\xi = 1$ (i.e. $z = \delta$) indicates the rear surface of the target. Moreover, the chain rule can be obtained as follows:

$$\begin{cases} \dfrac{\partial}{\partial z} = \dfrac{1}{\delta - s(t)} \dfrac{\partial}{\partial \xi} \\ \dfrac{\partial}{\partial t} = \dfrac{\partial}{\partial \tau} - \dfrac{\dfrac{\partial s(t)}{\partial t}(1-\xi)}{\delta - s(t)} \dfrac{\partial}{\partial \xi} \end{cases} \tag{5}$$

Combining with Eq. (5), the heat conduction Eq. (1) can be transformed into following form:

$$\frac{\partial H}{\partial \tau} = \frac{1}{(\delta - s)^2} \frac{\partial}{\partial \xi}\left(K \frac{\partial T}{\partial \xi} \right) + \frac{v_{sur}(2-\xi)}{\delta - s} \frac{\partial H}{\partial \xi} + S \tag{6}$$

The initial temperature of the target is considered to be equal to the ambient temperature. The boundary conditions on the rear and side surfaces are considered to be thermal insulation, and the boundary condition on the ablation surface is obtained by energy conservation. Therefore, the initial and boundary conditions are written as:

$$\begin{cases} T(r,z,t)\big|_{t=0} = T_0 \\ \left.\dfrac{\partial T}{\partial z}\right|_{z=\delta} = 0,\ K\left.\dfrac{\partial T}{\partial z}\right|_{z=s(t)} = L_v \rho v_{sur} \end{cases} \tag{7}$$

where T_0 is the initial temperature, L_v is the latent heat of vaporization, and ρ represents the density of the target.

2.2. Plasma expansion and ionization model

The ablation plasma generates with the target ablation, then expands in the opposite direction of the incident laser beam, as shown in **Figure 2**. Part of laser energy deposits in the plasma through inverse bremsstrahlung (IB) absorption, which causes the decrease of the laser intensity reaching on the target surface [31]. In other words, the calculated result of the plasma expansion has an significant effect on the calculation of the target ablation. Moreover, the ionization in the plasma also affects the properties of the plasma. Therefore, the plasma expansion should be calculated in detail, considering the ionization and plasma absorption.

Gas-dynamical equation is used to simulate the plasma expansion, and it can be written as follows:

$$\frac{\partial}{\partial t}\begin{bmatrix}\rho \\ \rho u \\ E\end{bmatrix}+\frac{\partial}{\partial x}\begin{bmatrix}\rho u \\ \rho u^2+p \\ (E+p)u\end{bmatrix}=\begin{bmatrix}0 \\ 0 \\ S_{plume}\end{bmatrix} \tag{8}$$

where the total energy per unit volume E is the sum of internal energy and kinetic energy.

$$E=\frac{3}{2}\rho\bar{R}T+\sum_{s\neq e}\frac{\rho_s R_s g_1^{(s)}\Theta_{el,1}^{(s)}\exp(-\Theta_{el,1}^{(s)}/T_e)}{\sum_{i=0}^{j^s} g_i^{(s)}\exp(-\Theta_{el,i}^{(s)}/T_e)}+\frac{1}{2}\rho u^2 \tag{9}$$

where ρ denotes density, $\bar{R}=R_0/\bar{M}$, $\bar{M}=\left[\sum(c_s/M_s)\right]^{-1}$, c_s and M_s are the mass fraction and molar mass of species s. $\Theta_{el,i}^{(s)}$ and $g_i^{(s)}$ are the characteristic temperature and degeneracy for species s at electronic energy level i.

The pressure p can be obtained from the state equation:

$$p=\sum_{s=1}^{4}\rho_s R_s T+\rho_e R_e T_e \tag{10}$$

where R_s is molar gas constant of species s, T_e is the temperature of electron.

The initial and boundary conditions of Eq. (7) are determined by the results of target ablation.

The instantaneous laser energy deposition in the plasma S_{plasma} can be calculated as:

$$S_{plasma}(x,t)=I_0(t)\beta(t)\exp\left(-\int_{-\delta_p}^{x}\beta(t)dx\right) \tag{11}$$

where δ_p is the length of the ablation plasma.

In addition, the ionization and IB absorption are considered in the model. In the plasma plume, a local thermodynamic equilibrium stage is assumed. Meanwhile, the plasma plume is considered nonviscous and electrically neutral, containing five species (Al, Al^+, Al^{2+}, Al^{3+}, and e^-). Otherwise, the electron temperature is considered to be equal to the ions and neutrals. Hence, the partition function of species Q^s and Q^e can be expressed as follows:

$$\begin{cases} Q^s=V\left(2\pi m_s k_B T/h^2\right)^{3/2}\sum_{i=0}^{j^s} g_i^{(s)}\exp\left(-\Theta_{el,i}^{(s)}/T\right) \\ Q^e=V\left(2\pi m_e k_B T_e/h^2\right)^{3/2} \end{cases} \tag{12}$$

where m_s denotes the mass of species s, s = Al, Al^+, Al^{2+}, Al^{3+}. V is the plasma plume volume, h is Planck constant.

The species number density n_s can be solved in an equilibrium state [22]:

$$
\begin{aligned}
&\frac{n_{\mathrm{Al}^{+}} n_{\mathrm{e}}}{n_{\mathrm{Al}}}=\frac{Q^{\mathrm{Al}^{+}} Q^{\mathrm{e}}}{V Q^{\mathrm{Al}}} \exp \left(-\frac{I P_{\mathrm{Al}}}{k_B T}\right) \\
&\frac{n_{\mathrm{Al}^{2+}} n_{\mathrm{e}}}{n_{\mathrm{Al}^{+}}}=\frac{Q^{\mathrm{Al}^{2+}} Q^{\mathrm{e}}}{V Q^{\mathrm{Al}^{+}}} \exp \left(-\frac{I P_{\mathrm{Al}^{+}}}{k_B T}\right) \\
&\frac{n_{\mathrm{Al}^{3+}} n_{\mathrm{e}}}{n_{\mathrm{Al}^{2+}}}=\frac{Q^{\mathrm{Al}^{3+}} Q^{\mathrm{e}}}{V Q^{\mathrm{Al}^{2+}}} \exp \left(-\frac{I P_{\mathrm{Al}^{2+}}}{k_B T}\right) \\
&n_{\mathrm{Al}}+n_{\mathrm{Al}^{+}}+n_{\mathrm{Al}^{2+}}+n_{\mathrm{Al}^{3+}}+n_{\mathrm{e}}=n_T \\
&n_{\mathrm{Al}^{+}}+2 n_{\mathrm{Al}^{2+}}+3 n_{\mathrm{Al}^{3+}}=n_{\mathrm{e}}
\end{aligned}
\tag{13}
$$

Where IP_{Al}, $IP_{\mathrm{Al}^{+}}$ and $IP_{\mathrm{Al}^{2+}}$ are the first, second and third ionization energy of aluminum, respectively.

IB process is considered to be the most important mechanism of the plasma absorption, for nanosecond laser ablation of aluminum [22, 34, 36]. The IB absorption coefficient is given by the sum of the processes between electron and neutral atomic species, and between electron and ions, as follows [22, 37]:

$$
\begin{aligned}
&\beta=\beta^{\mathrm{IB}}=\beta_{\text{e-Al}}^{\mathrm{IB}}+\beta_{\text{e-i}}^{\mathrm{IB}} \\
&\beta_{\text{e-Al}}^{\mathrm{IB}}=\left[1-\exp \left(-\frac{h v_l}{k_B T}\right)\right] n_{\mathrm{e}} n_{\mathrm{Al}} Q_{\text{e-Al}} \\
&\beta_{\text{e-i}}^{\mathrm{IB}}=\left[1-\exp \left(-\frac{h v_l}{k_B T_{\mathrm{e}}}\right)\right] \frac{4 e^6 \lambda_l^3}{3 h c^4 m_e} \sqrt{\frac{2 \pi}{3 m_{\mathrm{e}} k_B T_{\mathrm{e}}}} n_{\mathrm{e}}\left(n_{\mathrm{Al}^{+}}+4 n_{\mathrm{Al}^{2+}}+9 n_{\mathrm{Al}^{3+}}\right)
\end{aligned}
\tag{14}
$$

where v_l and λ_l are the laser frequency and wave length. Where $Q_{\text{e-Al}}$ denotes the averaged cross section for a collision between electron and neutral atom (10^{-46} m^5) [22].

Obviously, the absorption coefficient of the plasma β and the length of the plasma plume δ_p are changing, during the laser ablation process. Therefore, it is necessary to calculate the target ablation coupled with the plasma expansion in each time step.

2.3. Numerical method

In the numerical calculation, the length of target and plasma plume are assumed to be 5 and 200 μm. Meanwhile, 400 and 1000 uniform meshes are utilized, respectively, in the target and plasma region. The time-step is taken to be $\Delta t = 10^{-12}$ s, respecting the CFL condition in the entire computational domain.

The heat conduction Eq. (6) is solved employing an explicit finite difference technique. And the gas-dynamical Eq. (8) is diverged utilizing AUSM+-up method [38, 39]. In the numerical calculation, the target ablation and plasma expansion are calculated simultaneously, and coupled in each time step. Since the plasma occurs due to the ablation of target, the ablation surface of target is the inlet of the plasma. Meanwhile, the instantaneous laser intensity reaching on the target surface is decided by plasma absorption.

3. Experimental setup

3.1. Measurement setup for discharge arcs

As shown in **Figure 3**, a ns-pulsed Nd:YAG laser (InnoLas Corp. SpitLight 600, wavelength: 1064 nm, pulse energy: ~600 mJ, pulse width: 7 ns) is used for laser ablation of propellants. The laser pulse is irradiated into a vacuum chamber (vacuum degree: <5.0 × 10^{-4} Pa) through a quartz window and focused on the propellant with a focusing lens (f = 400 mm). The thruster is ignited and initiated by the ns-pulsed laser. Discharge arcs formed between the electrodes of thruster. The discharge current is monitored with a current monitor (Rogowski Coil, CWT150, sensitivity: 0.2 mV/A, maximum current: 30.0 kA) and a digital oscilloscope (Tektronix DPO 4340). The discharge voltage between the electrodes is measured using a standard high-voltage probe (Tektronix P5100).

3.2. Description of thrust stand

The impulse bits in the six cases of **Table 1** are measured by a calibrated thrust stand. The calibrated thrust-stand utilizing inverted pendulum techniques is developed and applied to estimate μNs-class impulses. Schematic of experimental setup for impulse-bit measurement is given in **Figure 3**. The thruster was settled onto the inverted pendulum. As an ns pulsed Nd:YAG laser beam passes through a quartz glass window and lens, the thruster is initialized, and the plasma is produced. The inverted pendulum swung on a knife-edge immediately. The motion of the pendulum is attenuated by an electromagnetic damper device subsequently. And the pendular angular is measured by an optical non-contacting measurement method. A He-Ne laser beam penetrated through a quartz glass window and irradiated onto a moving mirror settled near the knife-edge. The mirror rotated along with the motion of pendulum. The laser beam eventually irradiates into a position sensitive detector (PSD) after passing through several mirrors as shown in **Figure 3**.

For measuring the impulse bits, the calibration procedures are conducted. Because of the zero drift phenomenon and other influencing factors that gradually change the equilibrium position

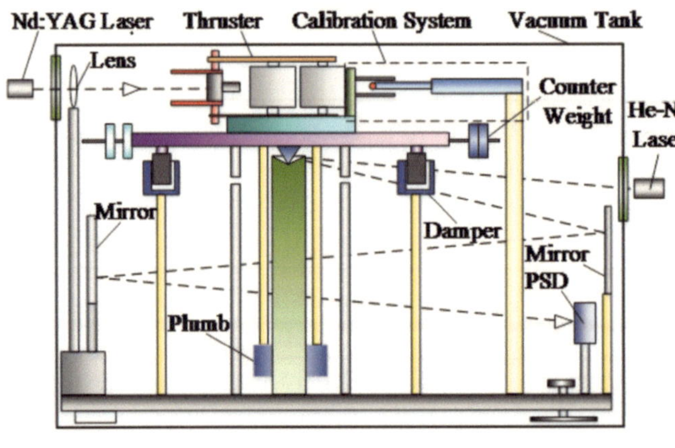

Figure 3. Schematic illustration of the impulse bit measurement system.

http://dx.doi.org/10.5772/intechopen.77511

Case	Propellants	d_1 (mm)	d_2 (mm)	A_0	B_0	A	B
I	Aluminum	20	10	4.8666E−3	5.6211E1	−6.5748E−4	2.2113E−2
II	Aluminum	30	10			−1.2289E−3	2.4758E−2
III	Aluminum	30	15			1.5489E−4	2.5930E−2
IV	Aluminum	30	20			1.0816E−3	2.8182E−2
V	Aluminum	40	20			1.1685E−3	2.8730E−2
VI	PTFE	20	10			3.5247E−3	2.3624E−2

Table 1. Six cases and coefficients obtained in calibration procedures.

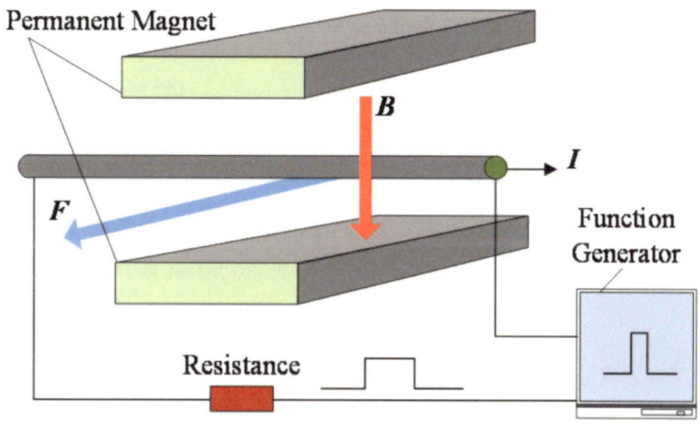

Figure 4. Schematic illustration of the electromagnetic calibration method.

of pendulum in experiments, especially in evacuation and gas charging procedure of the vacuum chamber, the measurement system should be calibrated before or after impulse bit measurements are implemented. The electromagnetic calibration subsystem is proposed as shown in **Figure 4** and described herein. A pulsed Ampere force is generated by a couple of parallel permanent magnet (NdFeB) plates and live wires as shown in **Figure 4**, and utilized to simulate the pulsed thrust of thruster. The pair of permanent magnet plates is settled behind the thruster, and the height of Ampere force action spot is set to be the same with the action spot of pulsed thrust of thruster. The distance between the couple of magnets is set 5 mm, and the crest value and duration of current in the conducting wires are controlled by a function generator as shown in **Figure 4**.

The detailed calibration procedure is described herein.

a. The relationship of Ampere force F and the volts d.c. output X of the function generator is tested by using an electronic balance (Mettler Toledo XS205, minimum mass 10 μg). The relationship can be expressed as:

$$F = A_0 + B_0 X \tag{15}$$

where the coefficients A_0 and B_0 are obtained after the calibration procedure.

b. The calibration devices are settled on the thrust stand with keeping the distance of permanent magnet plates and the relative location between the permanent magnet plates and the conducting wires. And the pendulum is propelled by the Ampere force with duration of t. Therefore, the relationship of PSD signals S and the volts d.c. output X of function generator can be obtained and written as:

$$S = g(X,t) \tag{16}$$

For the sake of simplification and considering that the pendular period of pendulum is generally longer than 5 s in the experiments, the duration of Ampere force t can be set 1 s by the function generator hereafter. Hence, the expression (16) can be expressed as:

$$S = g(X) \tag{17}$$

According to further study, it's indicated that the expression (17) can be simplified as:

$$S = A + BX \tag{18}$$

where, the coefficients A and B in expressions (18) could be obtained in the calibration procedures, as shown in **Table 1**. Besides, according to expression (15), the impulse bit I_b can be expressed as:

$$I_b = Ft = (A_0 + B_0 X)t \tag{19}$$

c. The relationship of impulse bit and PSD signal S can be obtained as:

$$I_b = (S - A)B_0 / B + A_0 \tag{20}$$

The uncertainty analysis is important and conducted herein. There are many sources of error for operating this thrust stand which must be identified, then quantified in order to evaluate the accuracy of the impulse bit measurement. It is shown in the expression (20) that the impulse bit is calculated from the voltage values of PSD signals and the coefficients in calibration procedures, A, A_0, B and B_0. Therefore, the uncertainties in the system data and results can be categorized into two main types. Firstly, the calibration procedures may generate errors for obtaining the coefficients A, A_0, B and B_0. Secondly, the impulse bit measurement is another source of error.

3.3. Measurement for mass shots

The mass shots are measured for the six cases by using an electronic balance (Mettler Toledo XS205, minimum mass 10 μg). It is apparent that the mass shots with different charged energy for a laser pulse energy of 600 mJ are approximately the same for the first five cases in **Table 1**, because of the isolation of laser ablation and discharge processes in the thruster. The typical number of pulse-shots to measure the mass shots is 100 times. The mass consumptions of propellants per shot are tested approximately as 15.7 and 7.8 μg for PTFE and aluminum, respectively. It is remarkable that the ablation of ceramic tube and ceramic insulator is much weak and not considered herein.

http://dx.doi.org/10.5772/intechopen.77511

A series of experiments are conducted to test the discharge characteristics and the performance parameters, such as impulse bit, and specific impulse. The experiments are conducted in six cases, as shown in **Table 1**.

4. Results and discussion

4.1. Plasma expansion and ionization

In order to describe the whole process of ablation plasma expansion and ionization, the properties of the plasma at several specific times should be investigated in detail, such as 3.5 ns (the ablation plasma produced), 10.5 ns (the phase explosion finished), 20 ns (the target ablation finished), 60.5 ns (the ablation plasma reaches the wall of the ceramic tube), 100 ns and 200 ns. By solving the gas-dynamical Eq. (8), the plasma velocity, temperature and electron number density fraction can be numerically calculated.

The spatial distribution of plasma velocity along the centerline of the ceramic tube for different times are presented in **Figure 5(a)**, and the variety of the peak plasma velocity with time is presented in the **Figure 5(b)**. It can be seen from **Figure 5(a)**, the peak plasma velocity appears at the front of the plasma. At 3.5 ns, the plasma has ejected from the target, and the peak velocity is about 2400 m/s. Afterwards, the phase explosion occurs at the target surface, and the ablation products carry out lots of heat from the target to the plasma. Hence the plasma velocity quickly increases, due to the phase explosion and the absorption of the laser energy. Thus at 10.5 ns, the peak plasma velocity increases to 4000 m/s. When the target ablation is finished (at 20 ns), the peak plasma velocity is around 4600 m/s. After finishing the target ablation, the plasma velocity still increases. But the increasing rate of the plasma velocity gradually decrease as shown in **Figure 5(b)**, due to the absence of the absorption of the laser energy and the injection of the ablation products.

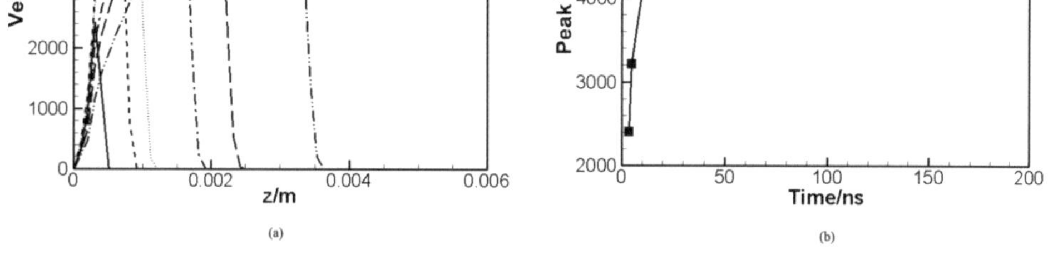

Figure 5. (a) Distribution of plasma velocity for different times. (b) Variety of the peak velocity with time.

The spatial distribution of plasma temperature along the centerline of the ceramic tube for different times are presented in **Figure 6(a)**. The variety of peak plasma temperature with time is presented in **Figure 6(b)**. As can be seen in **Figure 6(a)**, there are two peaks of the plasma temperature appears nearby the target surface (T_1) and at the front of the plasma (T_2). T_1 is due to the high-temperature ablation products ejection from the target to the plasma. T_2 is because of the plasma absorption of laser energy through IB absorption, which causes the plasma temperature to increase. At 3.5 ns, only T_1 exists and equals to 4500 K, due to the injection of high-temperature ablation products. However, afterwards, both T_1 and T_2 appear in the figure. The T_1 gradually increases to about 5700 K at 10.5 ns, and then decreases to below 2000 K at 200 ns. We can also conclude from **Figure 6(b)**, the variation tendency of T_1 is consistent with the target temperature. The T_1 decreases after 10.5 ns. However, the T_2 sharply increases during the laser pulse (from 3.5 to 20 ns), and then nearly keep constant after finishing the laser pulse (20 ns). Thus at 200 ns, only T_2 exists in the figure. What is coincident, the T_1 equals to T_2 at 15 ns, therefore the T_1 is bigger than T_2 before 15 ns and smaller than T_2 after 15 ns.

The spatial distribution of plasma electron number density fraction (η) along the centerline of the ceramic tube for different times are presented in **Figure 7(a)**. The variety of the fraction (ψ) of ionization area to the whole ceramic tube area with time is presented in **Figure 7(b)**. As it can be seen in **Figure 7(a)**, the ionization region mainly distributes at the front of the plasma, which is similar to the plasma velocity and temperature. The η increases during the laser pulse, and at 20 ns, the peak η approaches 0.5, which means the plasma is fully ionized at this region. Afterwards, the η keep constant at 0.5, but the area of the fully ionized region continuously increases. However, only primary ionization occurs in the plasma, due to the equilibrium relations of $\eta_{Al} + \eta_{Al^+} + \eta_{Al^{2+}} + \eta_{Al^{3+}} + \eta_e = 1$ and $\eta_{Al^+} + 2\eta_{Al^{2+}} + 3\eta_{Al^{3+}} = \eta_e$. We can also conclude from the **Figure 7(b)**, the ψ increases faster during the laser pulse, from 1% at 3.5 ns to 3% at 20 ns. Afterwards, the increasing rate of the ψ gradually decreases, at 200 ns, the ψ approaches 6%.

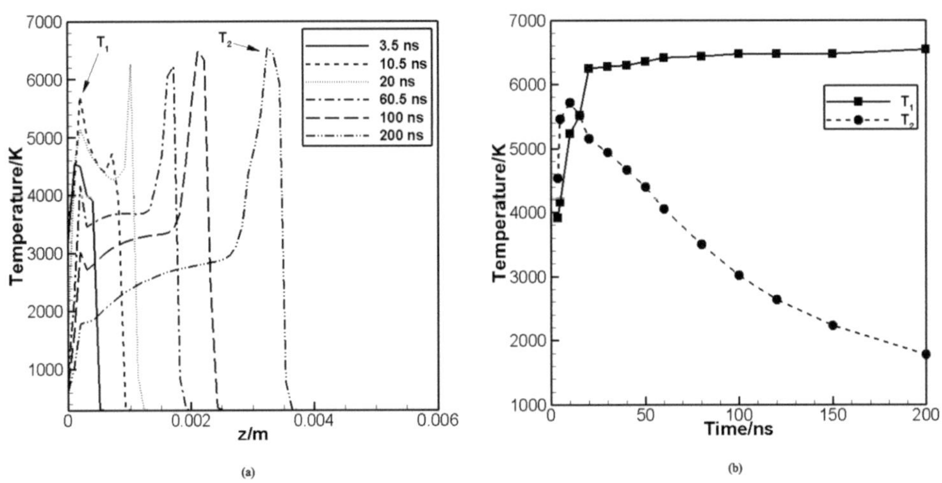

Figure 6. (a) Distribution of plasma temperature for different times. (b) Variety of peak temperature with time.

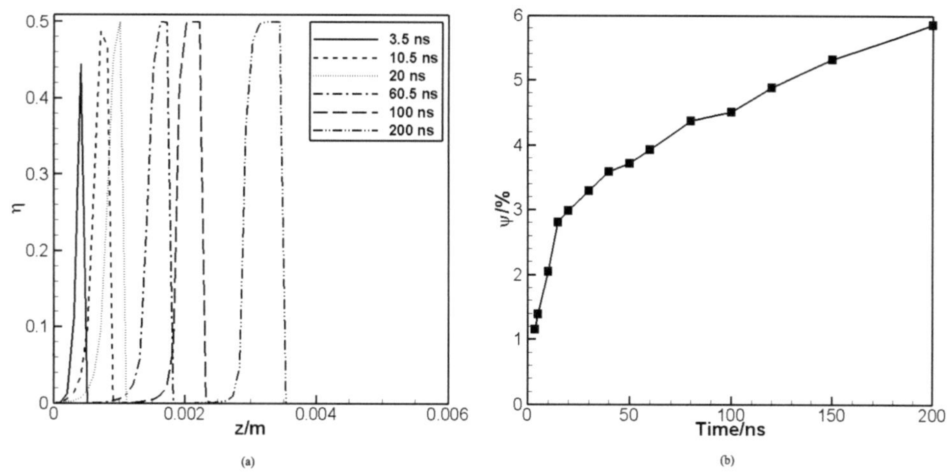

Figure 7. (a) Distribution of plasma electron number density fraction (η) for different times. (b) Variation of the fraction (ψ) of ionization area to the whole ceramic tube area with time.

4.2. Thrust performances

The impulse bits and discharge currents were measured with different parameters d_1 and d_2. Temporal variations of discharge current between the electrodes for charged voltage conditions from about 300 to 2000 V and a laser pulse energy of 600 mJ are shown in **Figure 8** respectively. The impulse bits in different cases are shown in **Figure 9**.

It is shown in **Figure 8** that the higher the charged energy, the higher the discharge current being achieved. In the case I, a maximal charged energy is 25.2 J, $d_1 = 20$ mm, $d_2 = 10$ mm and the propellant is aluminum. It is shown that the peak value of discharge current under the charged voltage of 2050 V is more than 20 kA. And it is proven that discharge current can be observed under very low charged voltage conditions (~316 V), in which case a discharge current raises up to 2.7 kA at 1.1 μs. Compared between **Figure 8(a)** and **(b)**, it can be concluded that the peak value of discharge current decrease as d_1 increasing for the situation of same charged voltage. Compared between **Figure 8(b)–(d)**, it can be concluded that the peak value of discharge current reaches maximum for the situation of $d_1 = 2d_2$, which means the propellant is placed in the middle of the electrodes.

The specific impulse and thrust efficiency could be calculated from the impulse bits and mass shots tested for the six cases. The curves of specific impulse for each working condition are illustrated as shown in **Figure 9**. In addition, the thrust efficiency of the thruster is defined as:

$$\eta = E_{kinetic} / E_{total} = 10^3 I_b^2 / (2mE_{total}) \tag{21}$$

where E_{total} is the total energy imported by the thruster, consisting of the single laser energy and the charged energy of capacitors, i.e., $E_{total} = E_{laser} + E_{charge}$. I_b represents the impulse bit, m is the mass ablated per shot and v is the mean velocity of plasma. And the factor 10^3 is required in the conversion of units. The thrust efficiency η was calculated and listed in **Figure 10**.

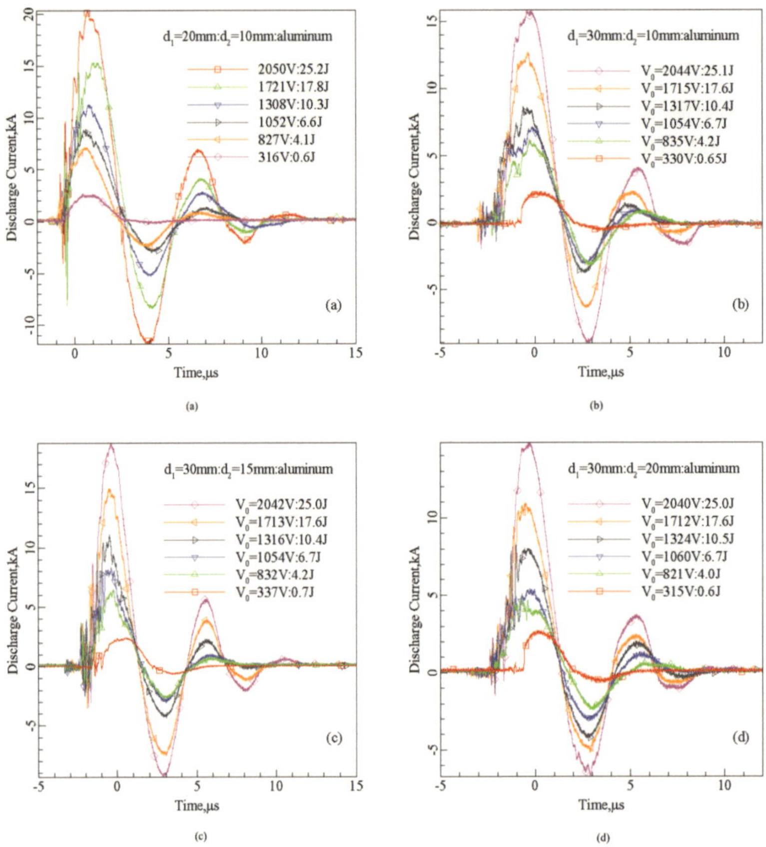

Figure 8. Discharge current between electrodes with different structural parameters: (a) case I, (b) case II, (c) case III and (d) case IV.

The impulse bits and specific impulses of thruster increase with the charged energy of capacitors, as shown in **Figures 9** and **10**. However, the thrust efficiency in case V decreases with the increment of charged energy in the range of 6–10 J, as shown in **Figure 11**. The rate of change of specific impulse or the thrust efficiency to charged energy is much smaller than at the higher energy levels.

Maximal specific impulse, impulse bit and thrust efficiency are achieved with a charged energy of above 25 J for all cases, as shown in **Figures 9–11**. For the case III, with a charged energy 25 J and the use of metal aluminum, a maximal impulse bit of 600 μNs, a specific impulse of approximate 8000 s and thrust efficiency of about 90% are obtained.

In addition, when the charged voltage is decreased to 0 V, the status can be called as the pure laser propulsion mode. And status with a charged voltage above zero is called as electromagnetic acceleration mode. The left ends of curves in **Figure 6** represent the impulse bit with low charged energy levels and that the discharge processes are weak, and the impulse bits equal approximately with the impulse bits in pure laser propulsion modes.

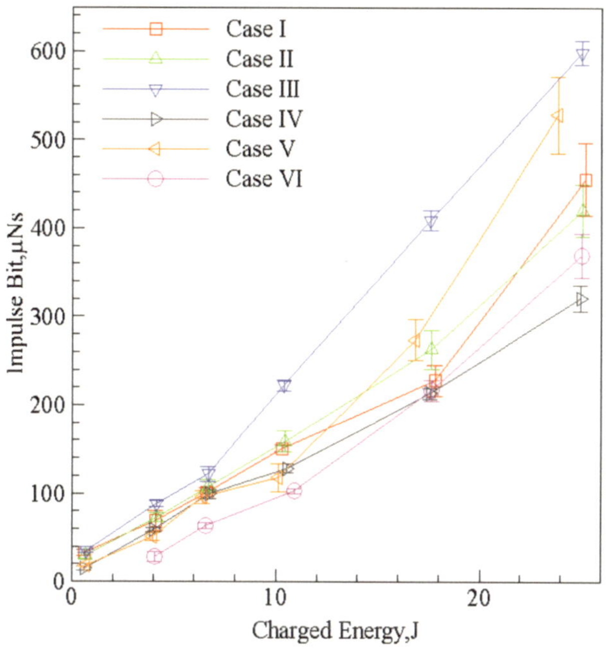

Figure 9. Variations of impulse bit with charged energy.

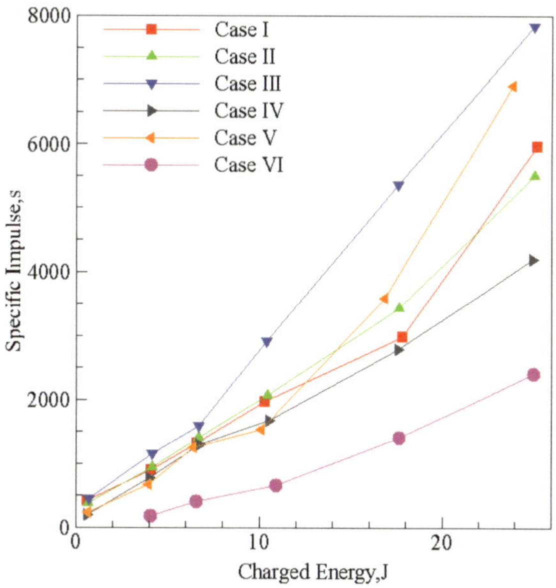

Figure 10. Variations of specific impulse with charged energy.

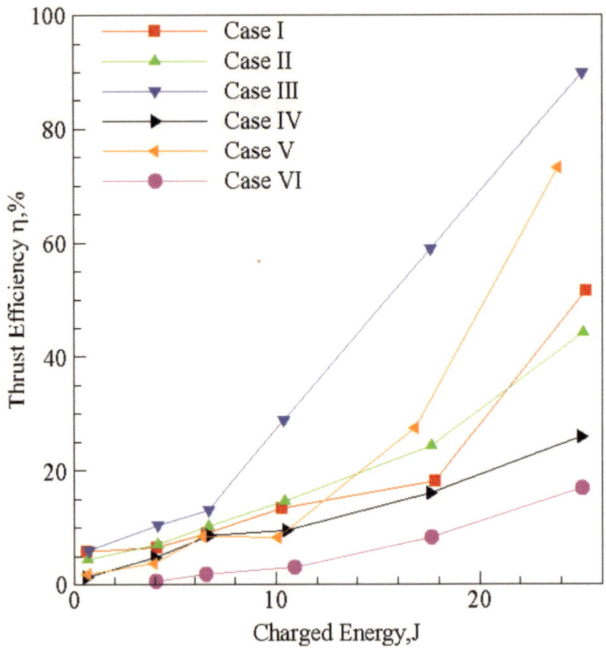

Figure 11. Variations of thrust efficiency with charged energy.

As shown in **Figure 9**, the impulse bit in pure laser propulsion mode can be estimated as 10–30 μNs with an approximate specific impulse of 100–300 s and thrust efficiency of 1–4%. In contrast with the pure laser propulsion mode, the thrust performance in electromagnetic acceleration mode is much better. It is apparent that the discharge processes do enhance and accelerate the laser plasma. And the thrust performance is proven to be improved by the electromagnetic acceleration.

5. Discussion

In this paper, a novel laser ablation plasma thruster with electromagnetic acceleration is presented. The plasma expansion and ionization induced by nanosecond pulsed laser in a ceramic tube are investigated by using the numerical method. The heat conduction model and fluid dynamics model are established to calculate the target ablation and plasma properties. Moreover, a rectangular electrode configuration with several capacitors and a ceramic tube is used, which allows the metals to be used as propellants. Preliminary experiments on the discharge characteristics and thrust performance are conducted and analyzed. Following conclusions are achieved and summarized.

1. Along the centerline of the ceramic tube, the peak values of plasma velocity, temperature and electron number density fraction distribute at the front of the plasma. The plasma velocity increases with time, but the increasing rate gradually decreases.

2. There are two peaks of the plasma temperature appear in the plasma along the centerline of the ceramic tube. The peak temperature nearby the target is due to the injection of the high-temperature ablation products. Hence its variation tendency is consistent with the target temperature. The peak temperature exist at front of the plasma is because of the IB absorption of the laser energy. Hence it increases during the laser pulse, and keep constant after finishing the laser pulse.

3. The plasma electron number density fraction increases during the laser pulse, and approaches 0.5 when the laser pulse finished, which means fully ionized in this region. After finishing the laser pulse, the plasma electron number density fraction keep constant, but the area of the fully ionized region continuously increases.

4. The thrust performance is proven to be improved by the electromagnetic acceleration. In contrast with the pure laser propulsion mode, the thrust performance of electromagnetic acceleration modes is much better.

5. The impulse bit and specific impulse increased with the charged energy. The optimal thrust performance of the thruster in experiments exists in large charged energy modes. With the charged energy 25 J and the use of metal aluminum, a maximal impulse bit of 600 μNs, a specific impulse of approximate 8000 s and thrust efficiency of about 90% are obtained. For the PTFE propellant, a maximal impulse bit of about 350 μNs, a specific impulse of about 2400 s, and thrust efficiency of about 16% are obtained.

Acknowledgements

The authors would like to thank National Natural Science Foundation of China for the financial assistance provided under the grant numbers 11772354 for this work.

Author details

Jianjun Wu*, Yu Zhang, Yuqiang Cheng, Qiang Huang, Jian Li and Xiaobin Zhu

*Address all correspondence to: jjwu@nudt.edu.cn

College of Aerospace Science and Engineering, National University of Defense Technology, Changsha, Hunan, China

References

[1] Pozwolski A. Electromagnetic propulsion of satellites. Acta Astronautica. 2008;**63**:575-577. DOI: 10.1016/j.actaastro.2008.03.018

[2] Silnikov MV, Kulakov KS, Kulakov SL, et al. Correction thruster development based on high-current surface discharge in vacuum. Acta Astronautica. 2015;**109**:177-181. DOI: 10.1016/j.actaastro.2014.10.021

[3] Zhang R, Zhang DX, Zhang F, et al. Deposition of fluorocarbon films by pulsed plasma thruster on the anode side. Applied Surface Science. 2013;**270**:352-358. DOI: 10.1016/j.apsusc.2013.01.029

[4] Hou DL, Zhao WS, Kang X. Operation analysis of pulsed plasma thruster. Acta Astronautica. 2008;**62**:404-409. DOI: 10.1016/j.actaastro.2008.01.004

[5] Burton RL, Turchi PJ. Pulsed plasma thruster. Journal of Propulsion and Power. 1998; **14**(5):716-735. DOI: 10.2514/2.5334

[6] Schonherr T, Nees F, Arakawa Y, et al. Characteristics of plasma properties in an ablative pulsed plasma thruster. Physics of Plasmas. 2013;**20**(033503):1-8. DOI: 10.1063/1.4794198

[7] Krejci D, Seifert B, Scharlemann C. Endurance testing of a pulsed plasma thruster for nanosatellites. Acta Astronautica. 2013;**91**:187-193. DOI: 10.1016/j.actaastro.2013.06.012

[8] Pottinger SJ, Krejci D, Scharlemann CA. Pulsed plasma thruster performance for miniaturised electrode configurations and low energy operation. Acta Astronautica. 2011;**68**: 1996-2004. DOI: 10.1016/j.actaastro.2010.11.011

[9] Horisawa H, Kimura I. Characterization of novel laser particle accelerators for space propulsion. In: 36th AIAA/ASME/SAE/ASEE Joint Propulsion Conference and Exhibit. Huntsville, Alabama; 16–19 July 2000. p. 36687

[10] Matsubara K, Hosokawa H, Akashi N, et al. A short-pulse laser-assisted pulsed plasma thruster. In: 51st AIAA/SAE/ASEE Joint Propulsion Conference; Orlando, FL; 27–29 July 2015; p. 4183

[11] Akashi N, Oigawa Y, Hosokawa H, et al. Plasma acceleration characteristic of a rectangular laser-electromagnetic hybrid thruster. In: 50th AIAA/ASME/SAE/ASEE Joint Propulsion Conference; Cleveland, OH; 28–30 July 2014; p. 3538

[12] Ono T, Uchida Y, Horisawa H, et al. Measurement of ion acceleration characteristics of a laser-electrostatic hybrid microthruster for space propulsion applications. Vacuum. 2009; **83**:213-216. DOI: 10.1016/j.vacuum.2008.03.098

[13] Ono T, Uchida Y, Horisawa H. Laser-electrostatic acceleration characteristics of a laser-electric hybrid thruster. In: 44st AIAA/ASME/SAE/ASEE Joint Propulsion Conference; Hartford, CT; 21–23 July 2008; p. 5008

[14] Horisawa H, Kawakami M, Kimura I. Laser-assisted pulsed plasma thruster for space propulsion applications. Applied Physics A. 2005;**81**:303-310. DOI: 10.1007/s00339-005-3210-8

[15] Horisawa H, Igari A, Uchida Y, et al. Electrostatic acceleration mode of laser-electric hybrid thrusters. In: 41st AIAA/ASME/SAE/ASEE Joint Propulsion Conference; Tucson, Arizona; 10–13 July 2005; p. 3697

[16] Horisawa H, Kimura I. Fundamental study on laser plasma accelerator for propulsion applications. Vacuum. 2002;**65**(3–4):389-396. DOI: 10.1016/S0042-207X(01)00447-X

[17] Zhang Y, Zhang D, Wu J, et al. A novel laser ablation plasma thruster with electromagnetic acceleration. Acta Astronautica. 2016;**127**:438-447. DOI: 10.1016/j.actaastro.2016.05.039

[18] Zhang Y, Zhang D, Wu J, et al. A thermal model for nanosecond pulsed laser ablation of aluminum. AIP Advances. 2017;**7**(075010):1-15. DOI: 10.1063/1.4995972

[19] Zhao XT, Tang F, Han B, et al. The influence of laser ablation plume at different laser incidence angle on the impulse coupling coefficient with metal target. Journal of Applied Physics. 2016;**120**(213103):1-8. DOI: 10.1063/1.4971247

[20] Liu T, Gao X, Hao Z, et al. Characteristics of plasma plume expansion from Al target induced by oblique incidence of 1064 and 355 nm nanosecond Nd:YAG laser. Journal of Physics D: Applied Physics. 2013;**46**(485207):1-6. DOI: 10.1088/0022-3727/46/48/485207

[21] Annemie B, Maryam A, David A, et al. Computer simulations of laser ablation, plume expansion and plasma formation. Advanced Materials Research. 2011;**227**:1-10. DOI: 10.4028/www.scientific.net/AMR.227.1

[22] Sakai T. Impulse generation on aluminum target irradiated with Nd:YAG laser pulse in ambient gas. Journal of Propulsion and Power. 2009;**25**(2):406-414. DOI: 10.2514/1.37767

[23] Cowpe JS, Pilkington RD, Astin JS, et al. The effect of ambient pressure on laser-induced silicon plasma temperature, density and morphology. Journal of Physics D: Applied Physics. 2009;**42**(165202):1-8. DOI: 10.1088/0022-3727/42/16/165202

[24] Adent M, Beyer E, Herziger G, et al. Laser-induced vaporization of a metal surface. Journal of Physics D: Applied Physics. 1992;**25**:57-65. DOI: 10.1088/0022-3727/25/1/008

[25] Gnedovets AG, Gusarov AV, Smurov I. A model for nanoparticles synthesis by pulsed laser evaporation. Journal of Physics D: Applied Physics. 1999;**32**:2162-2168. DOI: 10.1088/0022-3727/32/17/305

[26] Gusarov AV, Gnedovets AG, Smurov I. Two-dimensional gas-dynamic model of laser ablatin in an ambient gas. Applied Surface Science. 2000;**154-155**:66-72. DOI: 10.1016/S0169-4332(99)00389-X

[27] Gusarov AV, Gnedovets AG, Smurov I, et al. Simulation of nanoscale particles elaboration in laser-produced erosive flow. Applied Surface Science. 2000;**154-155**:331-336. DOI: 10.1016/S0169-4332(99)00451-1

[28] Colonna G, Casavola A, Giacomo A D, et al. Ablation of titanium metallic targets: A comparison between theoretical and experimental results. In: 33rd AIAA Plasmadynamics and Lasers Conference; Maui, Hawaii; 20–23 May 2002 p. 2159

[29] Tillack M, Blair D, Harilal SS. The effect of ionization on cluster formation in laser ablation plumes. Nanotechnology. 2004;**15**:390-403. DOI: 10.1088/0957-4484/15/3/028

[30] Bogaerts A, Chen Z. Effect of laser parameters on laser ablation and laser-induced plasma formation: A numerical modeling investigation. Spectrochimica Acta Part B. 2005;**60**:1280-1307. DOI: 10.1016/j.sab.2005.06.009

[31] Chen Z, Bleiner D, Bogaerts A. Effect of ambient pressure on laser ablation and plume expansion dynamics: A numerical simulation. Journal of Applied Physics. 2006;**99**(063304): 1-9. DOI: 10.1063/1.2182078

[32] Tan X, Zhang D, Li XA, et al. A new model for studying the plasma plume expansion property during nanosecond pulsed laser deposition. Journal of Physics D: Applied Physics. 2008;**41**(035210):1-8. DOI: 10.1088/0022-3727/41/3/035210

[33] Lu Q, Mao SS, Mao X, et al. Theory analysis of wavelength dependence of laser-induced phase explosion of silicon. Journal of Applied Physics. 2008;**104**(083301):1-7. DOI: 10.1063/1.2978369

[34] Bulgakova NM, Bulgakov AV. Pulsed laser ablation of solids: Transition from normal vaporizaiton to phase explosion. Applied Physics A. 2001;**73**:199-208. DOI: 10.1007/s003390000686

[35] Lide DR. CRC Handbook of Chemistry and Physics. 84th ed. CRC press; 2004

[36] Ghalamdaran S, Parvin P, Torkamany MJ, et al. Two-dimensional simulation of laser ablation with 235 nanosecond pulses. Journal of Laser Applications. 2014;**26**(1, 012009):1-8. DOI: 10.2351/1.4847795

[37] Root RG. Modeling of post-breakdown phenomena. In: Rakziemski LJ, Cremers DA, Dekker M, editors. Laser-Induced-Plasmas and Applications. New York; 1989. pp. 69-103

[38] Chang CH, Liou MS. A robust and accurate approach to computing compressible multiphase flow: Stratified flow model and AUSM$^+$-up scheme. Journal of Computational Physics. 2007;**225**:840-873. DOI: 10.1016/j.jcp.2007.01.007

[39] Liou MS. A sequel to AUSM, part II: AUSM$^+$-up for all speeds. Journal of Computational Physics. 2006;**214**:137-170. DOI: 10.1016/j.jcp.2005.09.020

Chapter 11

Physics of High-Density Radio Frequency Capacitively Coupled Plasma with Various Electrodes and Its Applications

Yasunori Ohtsu

Additional information is available at the end of the chapter

http://dx.doi.org/10.5772/intechopen.78387

Abstract

The radio frequency discharge plasma sources are widely utilized to prepare functional thin films and to etch insulated layers in semiconductor devices in microelectronic industry. Especially, a capacitively coupled plasma (CCP) is the most popular discharge because the equipment is very simple and almost maintenance free. However, there is a problem such as low-density plasma under low-gas pressure less than 10 Pa, that is, low processing rate. In this chapter, the production principle of conventional CCP and the special CCP with various electrodes and magnets is reviewed. The applications prepared by the special CCP system are also presented. Finally, the future plan including problems is described.

Keywords: RF plasma, capacitively coupled plasma, plasma processing, hollow cathode discharge, ring-shaped hollow plasma, magnetized plasma

1. Introduction

A large-scale integrated circuit (LSI), various sensors, and other semiconductor devices are essential for various electrical and electronic devices and automobile [1, 2]. In general, these devices are fabricated by plasma-processing techniques such as dry etching and plasma deposition using low-pressure plasma. The plasma processing is the most usually utilized chemical and physical processes in microelectronics fabrication for functional thin film preparation and dry etching of silicon-based films.

In the dry-etching processing, the silicon wafer patterned by the photoresist mask for LSI is exposed in plasma containing halogen molecules (e.g., CF_4, CHF_3, SF_6) [3–7]. In the case of these gases, the fluorine atoms dissociated by an electron impact collision with the molecules react with the silicon surface to generate a volatile etch product like SiF_4 so that the silicon wafer is etched with assistance of energetic ions produced in the plasma. For example, a high-density plasma sources have been developed to challenge patterning features less than 0.25 μm with high aspect ratios [8].

The plasma deposition has a sputtering deposition and a plasma-enhanced chemical vapor deposition (PECVD) method. The sputtering deposition is to impinge ions to the material target biased by a negative potential so as to sputter atoms from the target. The functional thin films such as transparent conductive oxide used in tableted computer and smartphone are deposited by sputtering method [9]. The sputtering depositions are prepared by DC magnetron and radio frequency (RF) magnetron plasma sources [10]. Various typed sputtering sources have been developed for the synthesis of high-quality thin films [10]. Especially, RF magnetron plasma source has an advantage that the insulated films can be deposited on the various substrates compared with DC magnetron plasma source. The PECVD is to dissociate molecule by electron impact so as to deposit radicals [8].

In these plasma-processing techniques, capacitively coupled plasma (CCP) with parallel plates was widely utilized. The physics of CCP has been studied by many researchers [11–16]. The sustaining mechanism of CCP is the electron heating process that the oscillating radio frequency sheath near the powered electrode plays an important role in electron acceleration as well as the collisional heating in the presence of electric fields and the emission process of secondary electrons from the electrode [11–16]. However, the electron heating process cannot produce high-density plasma.

The requirement of these semiconductor devices with high-speed operation and high performance is increasing annually. In order to perform the demand, high-speed plasma processing, that is, a high-density plasma production, is important. However, CCP does not attain high-density plasma. Thus, the high-density CCP is required. In this chapter, the production principle of conventional CCP and the special CCP with various electrodes and magnets is reviewed.

2. Production principle of conventional capacitively coupled plasma

In order to perform the high-speed processing of the LSI semiconductor, high-density plasma is needed for CCP. The power balance of CCP in the form of a global model [17] is expressed as the following equation:

$$P_{ab} = en_e u_B S(\varepsilon_e + \varepsilon_i + \varepsilon_c). \tag{1}$$

Here, P_{ab}, e, u_B, and S are the total absorbed power, the electronic charge, the Bohm velocity, and the surface area, respectively. ε_e and ε_i are the averaged energy loss per electron and ion lost to the electrodes. ε_c denotes the collisional energy loss per electron-ion pair created. The plasma density, n_e, can be determined from Eq. (1)

$$n_e = \frac{P_{ab}}{en_e S(\varepsilon_e + \varepsilon_i + \varepsilon_c)}. \tag{2}$$

When only thermal electrons are lost to the boundary surfaces [18], ε_e will be negligible. In general, the electron temperature decreases with increasing gas pressure [17]. Then, ε_c increases, while ε_i regularly decreases with increasing gas pressure, because the sheath is more collisional. For a fixed absorbed power, the increase in the plasma density is resulted with increasing gas pressure. Even for a relatively high-gas pressure of 50 Pa used in RF PECVD, however, the plasma density at a conventional driving frequency of 13.56 MHz is less than 10^{10} cm^{-3}.

In order to increase the plasma density at the typical frequency of 13.56 MHz, ingenious device is required. In the next section, the various ingenuities will be introduced.

3. Effect of high secondary electron emission oxide on high-density capacitively coupled plasma production

In general, plasma is generated by electrons with an energy higher than an ionization potential of target neutral gas [19]. According to the ionization cross section [20] for noble gases of He, Ne, Ar, and Xe, their ionization energy ranges from 10 to 30 eV, while the energy is a few 100 electron volts when the ionization cross section becomes maximum. These electrons are effectively possible to ionize neutral gases through inelastic collisions. Then, it is expected to produce high-density plasma. In CCP discharge, it is also easy to generate a high voltage of a few hundred volts between the powered and grounded electrodes, that is, CCP can generate secondary electron emission (SEE) from the powered electrode. It was reported that magnesium oxide (MgO) electrodes have a high SEE coefficient which are a few 10 times higher than that of conventional metal electrodes such as aluminum [21].

In this section, the effect of SEE as the acceleration mechanism of electrons is proposed to solve the serious problem of CCP density. The RF breakdown voltage and plasma density are studied experimentally. As show in **Figure 1(a)**, an RF power of 13.56 MHz was supplied to generate CCP between two electrodes of 20-mm diameter with a gap d_{gap} of 10 mm, which were mounted into the center of a cylindrical vessel of 160-mm diameter and 200-mm length. The back of the RF electrode is covered by a grounded metal enclosure to avoid additional discharge between the RF electrode and the grounded vessel. An MgO disk of 20-mm diameter and 2-mm thickness was connected to the Al metal electrode, as shown in **Figure 1(b)**. As a working gas, Ne or Ar gas was introduced in the vessel at 133 Pa.

Figure 2 shows RF breakdown voltage V_{Brf} characteristics. V_{Brf} is expressed as a peak-to-peak value measured by a high voltage probe and a digital oscilloscope. V_{Brf} characteristics exhibit a roughly U-shaped distribution like Paschen's curve of Townsend discharge [22] for both Ne

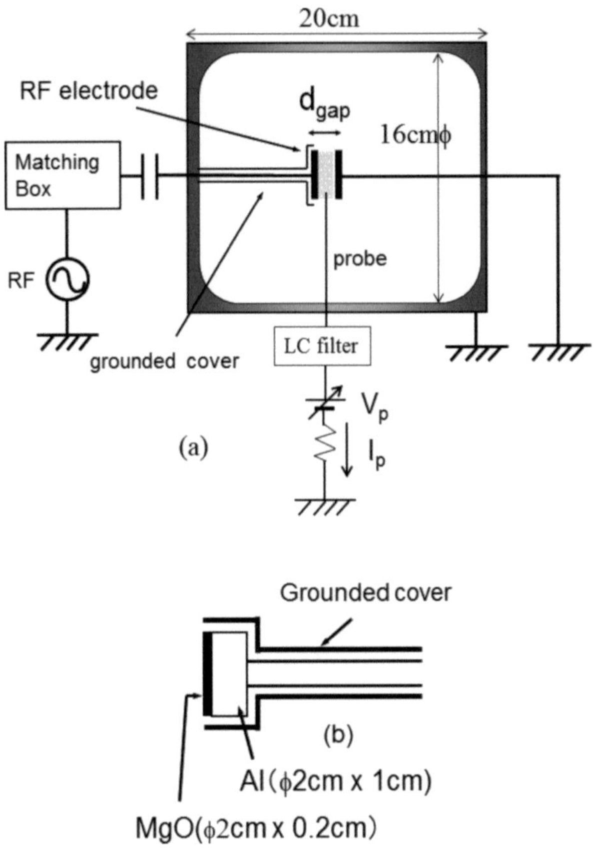

Figure 1. (a) Experimental apparatus and (b) electrode construction.

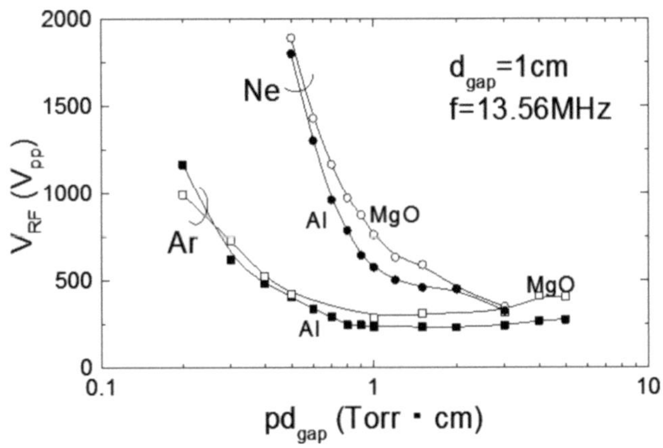

Figure 2. Breakdown voltage V_{RF} as a function of pd_{gap}. Here, p and d_{gap} denote gas pressure and gap distance, respectively.

and Ar gases. It is found that V_{Brf} for MgO electrode is higher than that for Al electrode for both Ne and Ar gases, whereas it is reported [21] that the coefficient of SEE for MgO is higher than that for Al. This is ascribed by the effect of voltage drop across MgO. It is also observed that V_{Brf} for Ar gas is lower than that for Ne gas because the ionization potential of Ar gas is lower than that of Ne gas [20].

Figure 3 shows plasma density j_{is} as a function of RF voltage V_{RF} applied between the RF electrode and the grounded vessel. The plasma density was estimated from an ion saturation current density j_{is} [23] measured by a tiny Langmuir probe to avoid the disturbance of the plasma by the probe. This is because the electron temperature was almost constant as a function of the RF voltage. The solid line shown in **Figure 3** corresponds to j_{is} when it is proportional to V_{RF}. The value of $j_{is} = 8 \times 10^{-3}$ A/cm^2 is almost an electron density of approximately 10^{10} cm^{-3}. As mentioned in **Figure 3**, the breakdown voltage V_{Brf} for Ne gas at both MgO and Al electrodes ranged from 500 to 1750 V_{pp}. For Al electrode, it is found that plasma density increases roughly proportionally to V_{RF}. The plasma density is only of the order of 10^9 cm^{-3} at even $V_{RF} = 1$ kV_{pp}.

According to the scaling law under the assumption that the stochastic heating without the SEE effect in the main discharge mechanism in CCP [17], the plasma density can be expressed as

$$n_e \propto \omega^2 V_{RF}/\varepsilon, \tag{3}$$

where ω, V_{RF}, and ε are the angular driving frequency, the RF voltage, and the energy needed to generate electrons and ions, respectively. Eq. (3) expresses that n_e is proportional to V_{RF} under the experiment that ω and ε are fixed. In the case of Al electrode, the experimental data almost correspond to the line of $n_e \propto \omega$. That is, the plasma production without SEE effect is predominant by stochastic heating where electrons accelerated by the oscillating RF sheath ionize neutral particles.

For MgO electrode, the plasma density is one order of magnitude higher than that for Al electrode. It is found that plasma density increases obviously with increasing V_{RF} for $V_{RF} > 400$ V_{pp}, and then attained over 10^{10} cm^{-3} for $V_{RF} > 800$ V_{pp} and then saturates. The SEE energy emitted from the MgO electrode corresponds to approximately 200 eV at $V_{RF} = 400$ V_{pp}. For electron-Ne

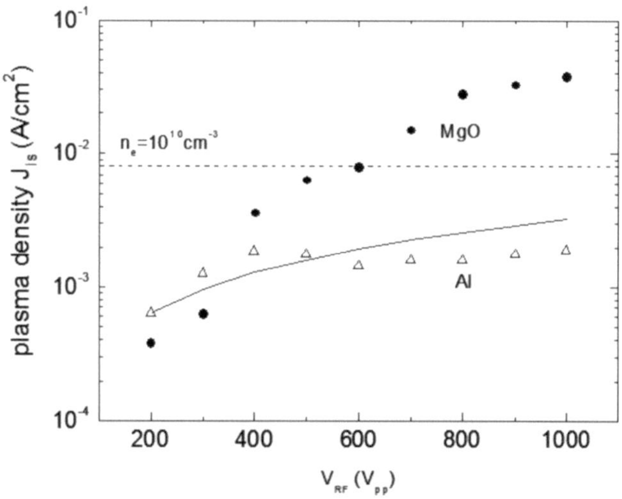

Figure 3. Plasma density j_{is} as a function of V_{RF} for Al and MgO electrodes. The solid line is the line of $n_e \propto \omega$.

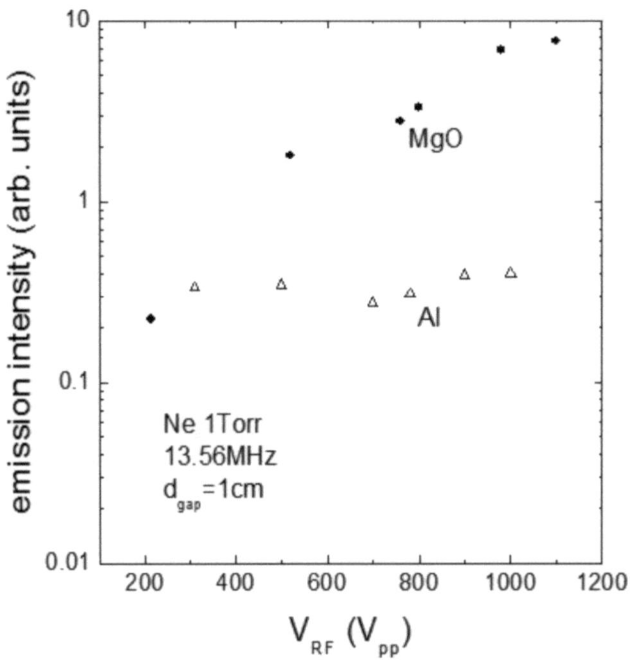

Figure 4. Optical emission intensity of He I line (588.1 nm) as a function of V_{RF}.

gas inelastic collision, this SEE energy of E_{SEE} = 200 eV has the maximum ionization cross section [20]. Thus, for V_{RF} > 400 V_{pp}, the ionization of Ne gas is effectively enhanced by the high SEE coefficient of MgO [21], which leads to the marked increase of plasma density. In fact, the RF power input to the plasma for MgO was higher than that for Al at a fixed applied voltage. Unfortunately, the power efficiency of the reactor could not be estimated because it was not easy to directly monitor current and phase difference between voltage and current.

The reason why the plasma density for MgO becomes constant for V_{RF} > 800 V_{pp} is considered as follows. The ionization cross section starts to decrease with increasing E_{SEE} for E_{SEE} > 400 eV [20]. Thus, the plasma density decreases with increasing V_{RF} for V_{RF} > 800 V_{pp}.

Figure 4 shows optical emission intensity of the Ne I line (588.1 nm) as a function of V_{RF}. Here, the optical emission intensity was measured with a spectrometer and a digital oscilloscope. The intensity increases markedly with increasing V_{RF}, in the same manner as plasma density. Therefore, the high-density plasma production is realized by using MgO with a high SEE coefficient in CCP. The effect of high secondary electron emission is one candidate to produce high-density plasma.

4. Effect of structured electrodes on high-density capacitively coupled plasma production

In this section, it is described that structured electrodes can produce the high-density capacitively coupled plasma. One of the structured electrodes is a hollow cathode. The hollow cathode discharge [24] is applied into a production mechanism of CCP to attain high-density plasma.

4.1. Multi-hollow electrode

The effects of a multi-hollow cathode discharge and a high SEE are applied to capacitively coupled plasma to produce high-density plasma [25, 26]. **Figure 5(a)** and **(b)** shows the experimental apparatus and construction of the multi-hollow electrode, respectively. As shown in **Figure 5(b)**, one plate has 35 holes with 5-mm diameter and 15-mm length, and these holes lay on a concentric circle. In order to emit secondary electron emission from the electrode facing the multi-hollow electrode, the other electrode is biased by the voltage of low frequency of 1 MHz. The plate is called as the substrate electrode.

Figure 6 shows plasma density and electron-neutral mean free path as a function of Ar gas pressure. Here, plasma density was estimated by ion saturation current density of a negatively biased probe because plasma density is proportional to ion saturation current [27]. The electron-neutral mean free path λ_{en} was also calculated by the following equation:

$$\lambda_{en} = (\sigma n)^{-1}, \tag{4}$$

where σ and n are electro-neutral cross section and Ar gas density, respectively. In **Figure 6**, as a reference value, an absolute value of the plasma density for Ar gas pressure 37.5 mTorr was estimated from electron saturation current density of probe and is approximately 8×10^{10} cm^{-3}. The plasma density drastically increases with an increasing gas pressure from 7.5 to 22.6 mTorr Pa and then varies by order, although the electron-neutral mean free path is inversely proportional to the gas pressure and reaches the hole size of 5 mm at the gas pressure of 113 mTorr. It is confirmed that the hollow cathode effect is achieved at the pressure range of 22.6–112.5 mTorr where the mean free path is comparable to the hole size of 5 mm.

The effect of SEE to attain high-density plasma production was examined by biasing the substrate electrode. **Figure 7** shows plasma density as a function of substrate biasing voltage V_b for axial position $z = 5$ mm and radial position $r = 0$ where the origins of z and r are the surface and the center of the powered electrode, respectively. The dashed line denotes the absolute value of plasma density for $V_b = 0$, which was estimated to be approximately 10^{10} cm^{-3} from probe characteristics. It is seen that plasma density increases with the substrate-biased voltage V_b and approached approximately 10^{11} cm^{-3} at $V_b = -800$ V. This suggests that the increase in V_b performs to ionize neutral atoms by inelastic collisions of secondary high-energy electrons. The film preparation is tried by using methane gas. **Figure 8** shows the deposition rate as a function of substrate biased voltage V_b. It is found that the deposition rate exponentially increases with increasing V_b in the range of $100 < V_b < 500$ V and then saturates. This is because the hybrid discharge combines the hollow cathode discharge and the secondary electron emission. Therefore, the drastic increase in the deposition rate is ascribed by realizing this hybrid discharge. For $V_b > 300$ V, the deposition rate exceeds the previous reported maximum value which is approximately 30 nm/min [28]. The deposition rate of 200 nm/min is attained at $V_b > 500$ V.

4.2. Ring-shaped hollow electrode

In the previous subsection, the effect of multi-hollow electrode on high-density capacitively coupled plasma was described [29]. In this subsection, the ring-shaped hollow electrode was tried to produce high-density capacitively coupled plasma. The high-density plasma in the

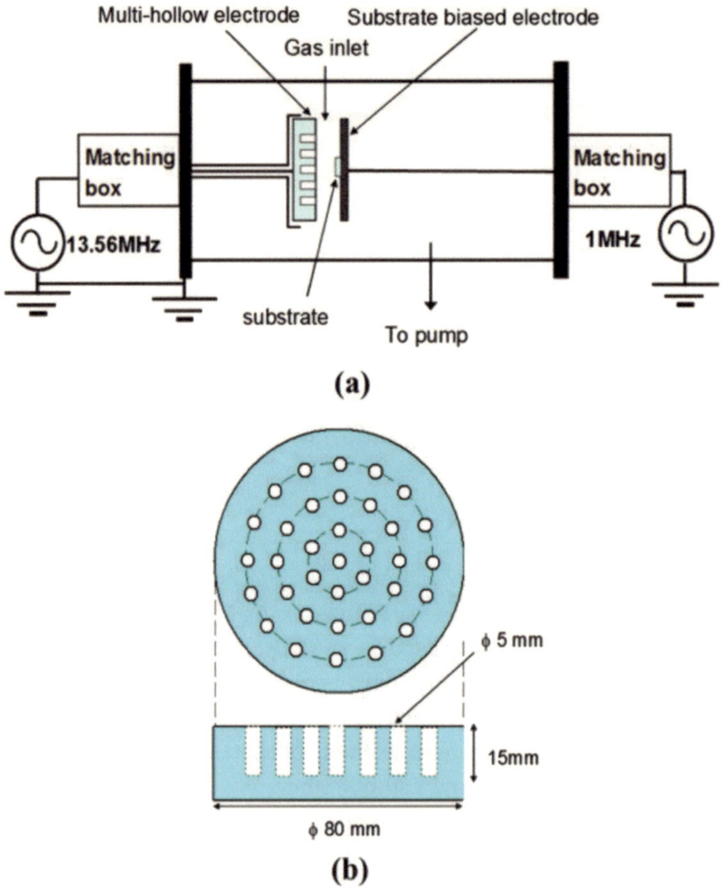

Figure 5. (a) Experimental apparatus and (b) construction of multi-hollow electrode.

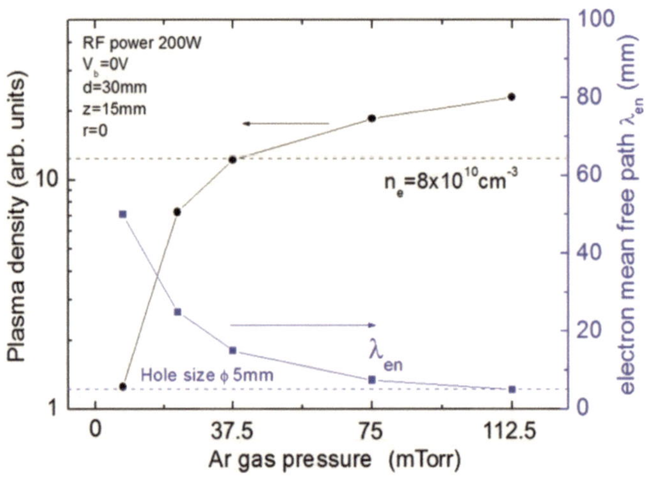

Figure 6. Plasma density and electron-neutral mean free path as a function of Ar gas pressure.

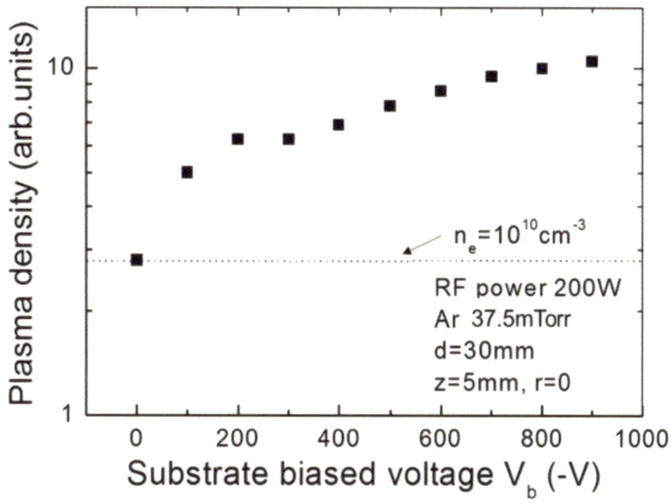

Figure 7. Plasma density as a function of RF-biased voltage.

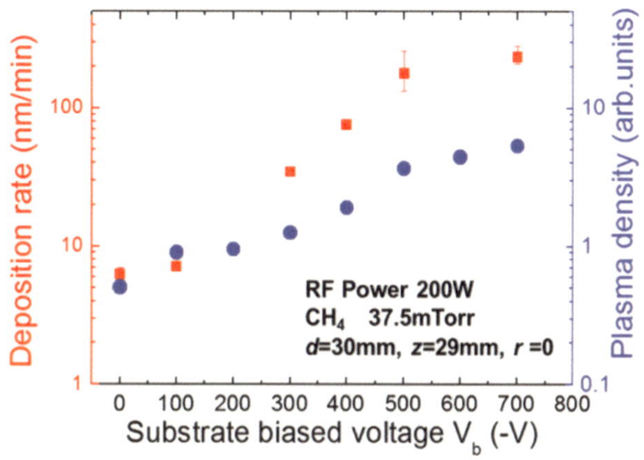

Figure 8. Deposition rate of thin films and plasma density as a function of RF-biased voltage.

trench diffuses toward the downstream region, and then the radial profile of plasma density becomes uniform at a certain axial position. The influence of trench width and gas pressure on plasma density and its profile is examined, comparing the case of a conventional flat electrode.

It is very important to accelerate electrons in the trench by moving RF cathode sheath for producing the high-density plasma with a hollow cathode effect. To satisfy the hollow cathode effect, it is required that the hollow trench width W be twice as long as the sheath thickness d_s as shown in **Figure 9(a)**. The depth of the hollow trench must be larger than the sheath thickness. In order to determine the trench width W, the sheath thickness d_s can be estimated by the following equation:

$$d_s = \frac{\sqrt{2}}{3} \lambda_D \left(\frac{2 e V_{sh}}{T_e} \right)^{\frac{3}{4}}, \quad (5)$$

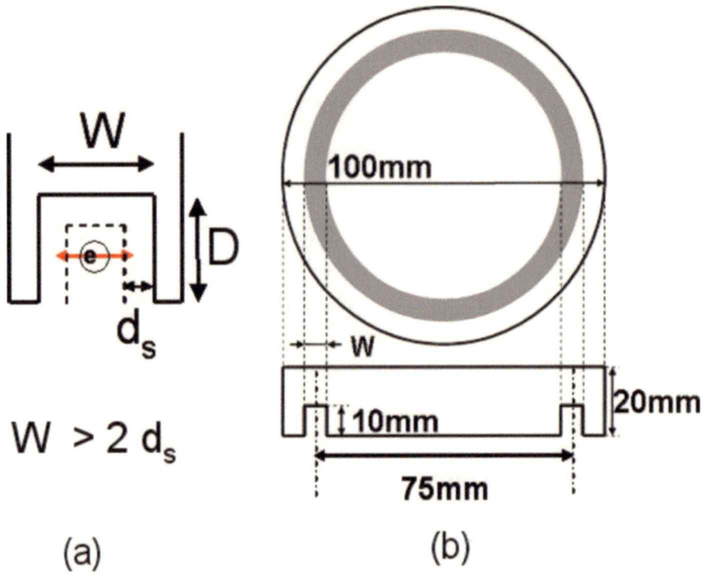

Figure 9. (a) Cross section near the ring-shaped trench and (b) structure of the RF hollow powered electrode.

where λ_D, T_e, and V_{sh} denote the Debye length, the electron temperature, and the time-averaged sheath voltage of the RF electrode which was used as the typical self-biased voltage, respectively. Eq. (1) is derived from the Child-Langmuir Law [1, 11]. The sheath thickness d_s can be estimated to be approximately 2 mm by Eq. (4) using $T_e = 4$ eV, $n_e = 10^{10}$ cm^{-3} and $V_{sh} = 200$ V. In this experiment, D is set at 5 and 10 mm. **Figure 9(b)** shows the cross section of the ring-shaped hollow electrode with 100-mm diameter and 20-mm thickness.

Figure 10(a) and **(b)** show typical images of plasma emission near the RF electrode for the ring-shaped hollow electrode and the flat electrode, respectively. As shown in **Figure 10(a)**, a high intensity of plasma emission is observed near the ring-shaped hollow trench for the ring-shaped hollow electrode. The hollow cathode effect is attained in the trench. On the other hand, the conventional flat electrode shows a uniform glow plasma on the whole electrode.

Figure 11 shows plasma density n_e as a function of Ar gas pressure p for the hollow and the flat electrodes. The position is fixed at $z = 8$ mm and $r = 38$ mm. For the hollow electrode, the plasma density raises rapidly with raising Ar gas pressure p and has a peak at $p = 180$ mTorr and then remains almost constant. At $p = 350$ mTorr, the plasma density decreases discontinuously, and then at $p > 350$ mTorr, the plasma density is independent of the gas pressure. The electron-neutral mean free path λ_{en} decreases with increasing gas pressure. λ_{en} at $p = 350$ mTorr is approximately 1 mm which is of the order of the magnitude of the hollow trench size. The maximum plasma density is approximately 2×10^{11} cm^{-3}. It is seen that the high-density plasma with the order of magnitude of 10^{11} cm^{-3} is achieved in a wide range of gas pressure from 100 to 300 mTorr. In the flat electrode, the plasma density shows a similar tendency with the hollow electrode. The maximum value of the plasma density is one order of magnitude lower than that of the hollow electrode.

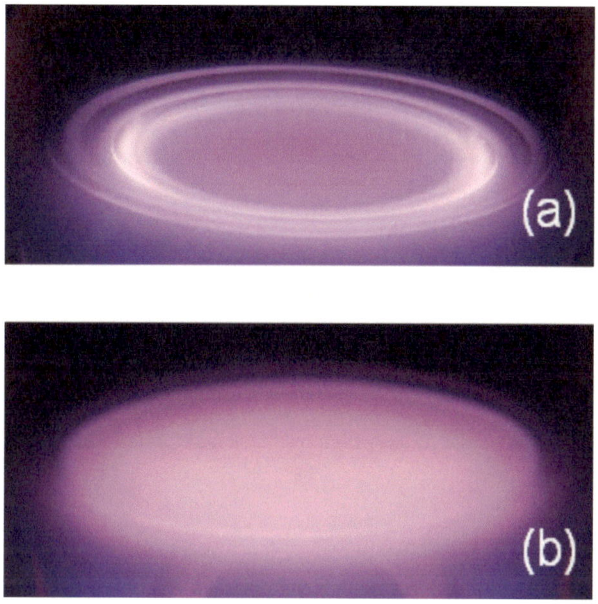

Figure 10. Typical images of plasma structure for (a) the hollow and (b) the flat electrodes.

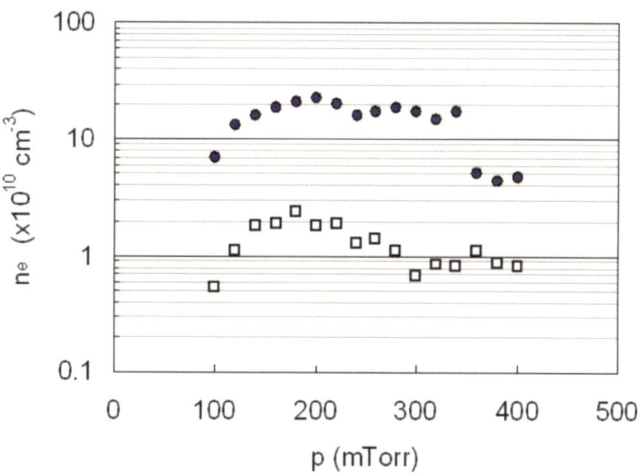

Figure 11. Plasma densities as a function of Ar gas pressure for the hollow (solid circles) and the flat electrodes (squares). Here, the ring-shaped trench width is $W = 5$ mm.

Figure 12 shows plasma density as a function of Ar gas pressure for $W = 5$ and 10 mm. Here, the measured position is at $z = 12$ mm and $r = 38$ mm. It is seen that the critical pressure when the plasma density attained 10^{11} cm^{-3} decreases with increasing the trench width. In fact, the critical values are 140 and 100 mTorr at $W = 5$ and 10 mm, respectively. The sheath thickness d_s was calculated to be approximately 1.7 mm at $n_e = 10^{11}$ cm^{-3} at the critical gas pressures. The condition of $W > 2d_s$ is satisfied at the critical gas pressure.

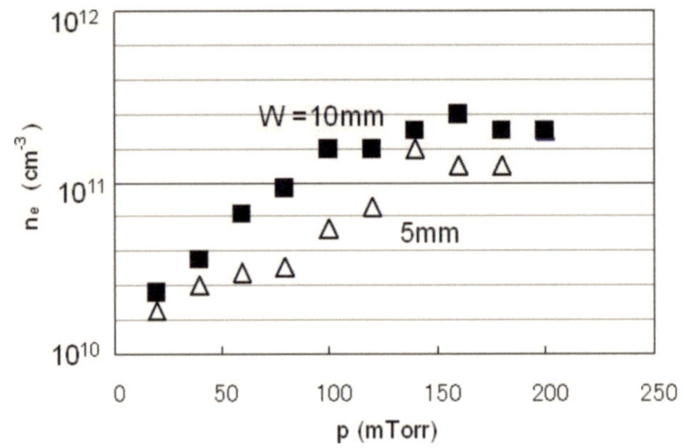

Figure 12. Plasma density as a function of Ar gas pressure at the hollow electrode for W = 5 and 10 mm under lower pressure conditions.

4.3. Magnetized ring-shaped hollow cathode discharge

In a low pressure of 1 Pa, it is difficult to produce plasma using only hollow cathode discharge [30]. In this subsection, in order to attain high-density plasma in the low-gas pressure, the combination of hollow cathode discharge and magnetic confinement with magnets is proposed. The addition of a magnetic field is one candidate for performing discharge in the low-gas pressure. It is easy to magnetize electrons under the low-gas pressure.

Figure 13 shows the construction of a ring-shaped hollow cathode for (a) the NS-NS arrangement, (b) the NS-SN arrangement, and (c) the NS arrangement of permanent magnets. The neodymium magnets were used. Three arrangements of permanent magnets were investigated. The NS-Ns arrangement is that two NS magnets with 7 × 10 mm^2 in cross section and 10 mm in length are positioned at both walls of a hollow trench and six couples of the NS-NS magnets are used as shown in **Figure 13(a)**. In the NS-SN arrangement, NS and SM magnets are mounted as shown in **Figure 13(b)**. As shown in **Figure 13(c)**, six NS magnets with 10 × 15 mm^2 in cross section and 6 mm in length are set at the bottom of a hollow trench for the NS arrangement. The outside of magnets is covered by iron yokes with 1 mm in thickness as shown in **Figure 13**.

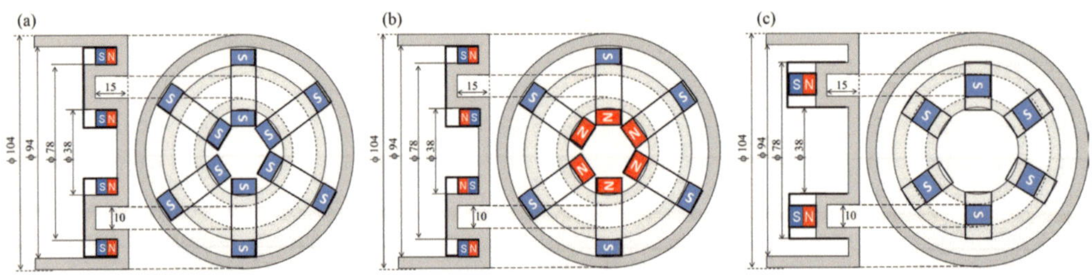

Figure 13. Constructions of a ring-shaped hollow cathode for (a) the NS-NS arrangement, (b) the NS-SN arrangement, and (c) the NS arrangement of permanent magnets. The neodymium magnets were used.

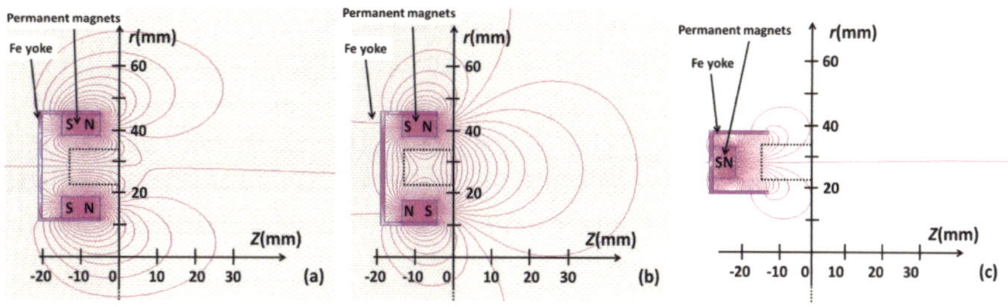

Figure 14. Two-dimensional distributions of magnetic field lines near the hollow cathode for (a) the NS-NS arrangement, (b) the NS-SN arrangement, and (c) the NS arrangement of the permanent magnets.

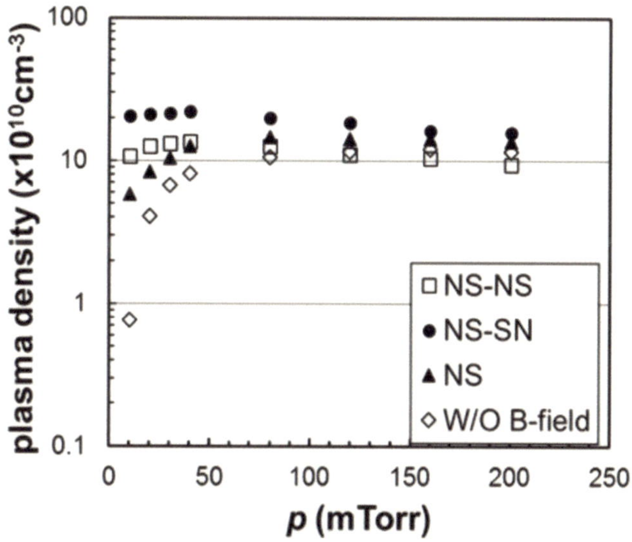

Figure 15. Plasma density as a function of Ar gas pressure at r = 29 mm and z = 23 mm at various magnet arrangements. Here, the data without magnetic field are also included for the comparison.

Figure 14 shows two-dimensional distributions of magnetic field lines at (a) NS-NS, (b) NS-SN, and (c) NS arrangements of permanent magnets, respectively. Here, the area enclosed by dashed lines is the hollow trench. As shown in **Figure 14(b)**, for the NS-SN arrangement, it is clear that the profile of the magnetic field lines is quite different from the other arrangement. The magnetic field lines show a cusp profile in the trench. The NS-SN arrangement is the best profile for magnetic confinement of electrons.

Figure 15 shows plasma density as a function of gas pressure at r = 29 mm and z = 23 mm at various arrangements. For the case without magnet, it is found that plasma density decreases from approximately 10^{11} cm^{-3} to 8 × 10^{9} cm^{-3} with a decreasing gas pressure less than 80 mTorr and saturates for a gas pressure region of 80–200 mTorr. It is difficult to sustain high-density plasma without magnet at a low-gas pressure. It is noticeable that the addition of magnets with hollow cathode discharge is effective for producing high-density plasma under the low-gas pressure. For the case of NS-SN arrangement, high-density plasma over 10^{11} cm^{-3} is achieved at all gas pressures.

In these chapters, in order to improve the plasma density in CCP, various typed CCP discharges have been presented. In Section 3, it is indicated that RF electrode with a high secondary electron emission oxide of MgO is effective to produce high-density capacitively coupled plasma. The plasma density for MgO electrode increases drastically with increasing RF voltage compared with the metal electrode of Al. In Section 4, it is described that the structured electrode plays an important role to improve the plasma density. This mechanism is ascribed by the hollow cathode effect.

5. Conclusions

The radio frequency capacitively coupled plasma source is widely utilized in the semiconductor fabrications. However, the source has a serious problem, although it has some merits such as simple structure and maintenance free. In this chapter, some solutions are introduced by adding simple methods. The first method is the high secondary electron emission electrode. The second method is multi-hollow and ring-shaped electrodes. The third method is magnetized ring-shaped electrode. All methods attained high-density plasma production. Especially, the ring-shaped electrode with magnets performed high-density plasma with 10^{11} cm^{-3} under a low-gas pressure. These methods will be useful to advance capacitively coupled plasma for microelectronic technology.

Author details

Yasunori Ohtsu

Address all correspondence to: ohtsuy@cc.saga-u.ac.jp

Faculty of Science and Engineering, Saga University, Saga, Japan

References

[1] Sugawara M, Stanfield BL, Handa S, Fujita H, Watanabe S, Tsukamoto T. Plasma Etching. Oxford University Press; 1998

[2] Roth JR. Industrial Plasma Engineering. Vol.2. Institute of Physics; 2001

[3] Heinecke RAH. Control of relative etch rates of SiO_2 and Si in plasma etching. Solid-State Electronics. 1975;**18**:1146

[4] Ephrath LM. Selective etching of silicon dioxide using reactive ion etching with CF_4-H_2. Journal of the Electrochemical Society. 1979;**126**:1419

[5] Coburn JW. In situ Auger electron spectroscopy of Si and SiO_2 surfaces plasma etched in CF_4-H_2 glow discharges. Journal of Applied Physics. 1979;**50**:5210

[6] Maruyama K, Sakai A, Goto T. Measurement of the CF_3 radical using infrared diode laser absorption spectroscopy. Journal of Physics D. 1993;**26**:199

[7] Maruyama K, Ohkouchi K, Ohtsu Y, Goto T. CF_3, CF_2 and CF Radical Measurements in RF CHF_3 Etching Plasma Using Infrared Diode Laser Absorption Spectroscopy. Japanese Journal of Applied Physics. 1994;**33**:4298

[8] Chen FF, Chang JP. Lecture Notes on Principles of Plasma Processing. Springer US; 2003

[9] Minami T, Nanto H, Tanaka S. Highly conductive and transparent zinc oxide films prepared by rf magnetron sputtering under an applied external magnetic field. Applied Physics Letters. 1982;**41**:958

[10] Musil J. Low-pressure magnetron sputtering. Vacuum. 1998;**50**:363

[11] Godyak VA, Piejak RB. Abnormally low electron energy and heating-mode transition in a low-pressure argon rf discharge at 13.56 MHz. Physical Review Letters. 1990;**65**:996

[12] Graves DB. Fluid model simulations of a 13.56-MHz rf discharge: Time and space dependence of rates of electron impact excitation. Journal of Applied Physics. 1987;**62**:88

[13] Goedde CG, Lichtenburg AJ, Lieberman MA. Self-consistent stochastic electron heating in radio frequency discharges. Journal of Applied Physics. 1988;**64**:4375

[14] Vender D, Boswell RW. Numerical modeling of low-pressure RF plasmas. IEEE Transactions on Plasma Sciences. 1990;**PS-18**:725

[15] Okuno Y, Ohtsu Y, Komatsu C, Fujita H. Measurement of electron energy distribution function in an asymmetric radio frequency discharge plasma. Journal of Applied Physics. 1993;**73**:1612

[16] Ohtsu Y, Okuno Y, Fujita H. Observation of radio-frequency discharges at various frequencies. Journal of Applied Physics. 1993;**73**:2155

[17] Lieberman MA, Lichitenberg AJ. Principle of Plasma Discharge and Materials Processing. 2nd ed. New Jersey: Wiley; 2005

[18] Liu YX, Gao F, Liu J, Wang YN. Experimental observation of standing wave effect in low-pressure very-high-frequency capacitive discharges. Journal of Applied Physics. 2014;**116**:043303

[19] Ohtsu Y, Fujita H. Production of high-density capacitively coupled radio-frequency discharge plasma by high-secondary-electron-emission oxide. Applied Physics Letters. 2004;**85**:4875. DOI: 10.1063/1.1827353

[20] Brown SC. Basic Data of Plasma Physics. 2nd ed. Revised. Cambridge: MIT; 1966

[21] Choi EH, Lim JY, Kim YG, Ko JJ, Kim DI, Lee CW, Cho GS. Secondary electron emission coefficient of a MgO single crystal. Journal of Applied Physics. 1999;**86**:6525

[22] Raizer YP, Shneider MN, Yatsenko NA. Radio-Frequency Capacitive Discharges. New York: CRC Press; 1995

[23] Chen FF. Plasma Diagnostic Techniques. New York: Academic; 1965

[24] Roth JR. Industrial Plasma Engineering. Principles. Vol. 1. Bristol: IoP; 1995

[25] Ohtsu Y, Fujita H. Production of high-density capacitive plasma by the effects of multihollow cathode discharge and high-secondary-electron emission. Applied Physics Letters. 2008;**92**:171501

[26] Ohtsu Y, Nakamura C, Misawa T, Fujita H, Akiyama M, Yukimura K. Production of high-density capacitively coupled plasma with RF multi-hollow cathode and/or high secondary electron emission for DLC film preparation. Plasma Processes and Polymers. 2009;**6**:S458-S461

[27] Hershkowitz N. How Langmuir Work. Plasma Diagnostics. Vol. 1. San Diego: Academic; 1989

[28] Robertson J. Diamond-like amorphous carbon. Materials Science and Engineering. 2002;**R37**:129

[29] Ohtsu Y, Urasaki H. Development of a high-density radio frequency plasma source with a ring-shaped trench hollow electrode for dry etching. Plasma Sources Science and Technology. 2010;**19**:045012 (6 pp)

[30] Ohtsu Y, Eguchi J, Yahata Y. Radio-frequency magnetized ring-shaped hollow cathode discharge plasma for low-pressure plasma processing. Vacuum. 2014;**101**:46-52

Chapter 12

Space Plasma Interactions with Spacecraft Materials

Daniel P. Engelhart, Elena A. Plis, Dale Ferguson,
W. Robert Johnston, Russell Cooper and
Ryan C. Hoffmann

Additional information is available at the end of the chapter

http://dx.doi.org/10.5772/intechopen.78306

Abstract

Spacecraft materials on orbit are subjected to the harsh weather of space. In particular, high-energy electrons alter the chemical structure of polymers and cause charge accumulation. Understanding the mechanisms of damage and charge dissipation is critical to spacecraft construction and operational anomaly resolution. Energetic particles in space plasma break molecular bonds in polymers and create radicals that can act as space charge traps. These electron-induced chemical changes also result in changes to the spectral absorption profile of polymers on orbit. Radicals react over time, either recreating identical bonds to those in the pristine material, leading to material recovery, or creating new bonds, resulting in a new material with new physical properties. Lack of knowledge about this dynamic aging is a major impediment to accurate modeling of spacecraft behavior over its mission life. This chapter first presents an investigation of the chemical and physical properties of polyimide films (PI, Kapton-H®) during and after irradiation with high-energy (90 keV) electrons. Second, the deleterious effects of space plasma on a spacecraft component level are presented. The results of this physical/chemical collaboration demonstrate the correlation of chemical changes in PI with the dynamic nature of spacecraft material aging.

Keywords: space plasma, geosynchronous earth orbit, polyimide, electrical charge transport, material optical signature, satellite arcing, reflectivity

1. Introduction

When a solid dielectric interacts with plasma, it is subject to a number of different destructive and non-destructive processes. A charged particle impinging on a surface with sufficient energy will penetrate the surface and lose energy into the bulk of the material. The deposition of this kinetic energy will result in electronic and vibrational excitations. If sufficient energy is deposited into a single bond to excite it beyond its dissociation energy, the chemical bond can be broken. The dominant mechanism of energy deposition depends strongly on the mass of the particle. Massive particles such as heavy ions will impart large amounts of ballistic energy over a relatively small depth, displacing nuclei in the solid, exciting phonons, and vibrational transitions sufficient to rupture any chemical bond and create radicals. A less massive particle such as an electron can be expected to deposit energy primarily in the form of electronic excitation. Sufficient electronic energy deposition will also rupture bonds and create radicals, but the damage will be deposited over a longer trajectory and the chemical damage will be more bond-selective. Ions, electrons, and photons incident on a surface will also eject secondary electrons, initiating charge imbalance near the surface. After the kinetic energy of a charged particle has been exhausted, the ion or electron can imbed itself into the bulk, creating a local charge imbalance at the penetration depth of the particle. When considering the interaction of space plasma with solid dielectrics, it is important to distinguish between energy deposition and charge deposition [1, 2].

The following chapter will focus on the interaction of electrons, which comprise the most damaging species in the Geosynchronous Earth Orbit (GEO) environment in terms of energy deposition [3–5], with dielectric materials such as polymers and solar array cover glass [6]. The results of this plasma/material interaction are characterized in terms of modification of the materials' optical and electrical properties. It will be shown that these materials change dramatically in the plasma environment and that these changes can have profound effects on the performance of spacecraft components.

The thermal plasma in GEO (6.6 Earth radii, or R_E) is predominantly quasi-neutral plasma consisting of atomic hydrogen ions (replenished mainly by the solar wind and ionized by Lyman-alpha radiation from the Sun) and electrons. It is typically at a very low-density (0.1–1 cm^{-3}) but high plasma temperature (typically 4–10 keV). During geomagnetic storm conditions, the plasma temperature can increase dramatically (to 16–30 keV), as magnetic reconnection in the magnetotail accelerates trapped electrons, and the accompanying ions. The second population of particles exists at GEO in the highly non-thermal outer radiation belt. These highly energetic electrons (0.1–10 MeV) form a toroidal belt extending out to about 60,000 km (10 R_E) from Earth, with a maximum intensity typically at about 3–5 R_E at the equator. Thus, although GEO orbit is beyond the distance of maximum intensity, it is well within the limits of the outer belt, and satellites orbiting within it are subjected to significant penetrating electron fluxes, as can be seen in **Figure 1** [3–5].

The exterior of spacecraft is typically comprised of materials that regulate temperature by reflecting visible sunlight and radiating infrared radiation. Among the most common of these materials is multi-layer insulation (MLI), which usually consists of many thin layers of aluminized (or silvered) polymers such as Kapton® (polyimide or PI). Tenets of modern spacecraft design dictate that radiation-sensitive electronics be positioned in the bulk of the spacecraft in a shielded conductive "Faraday cage". These Faraday cages can only be

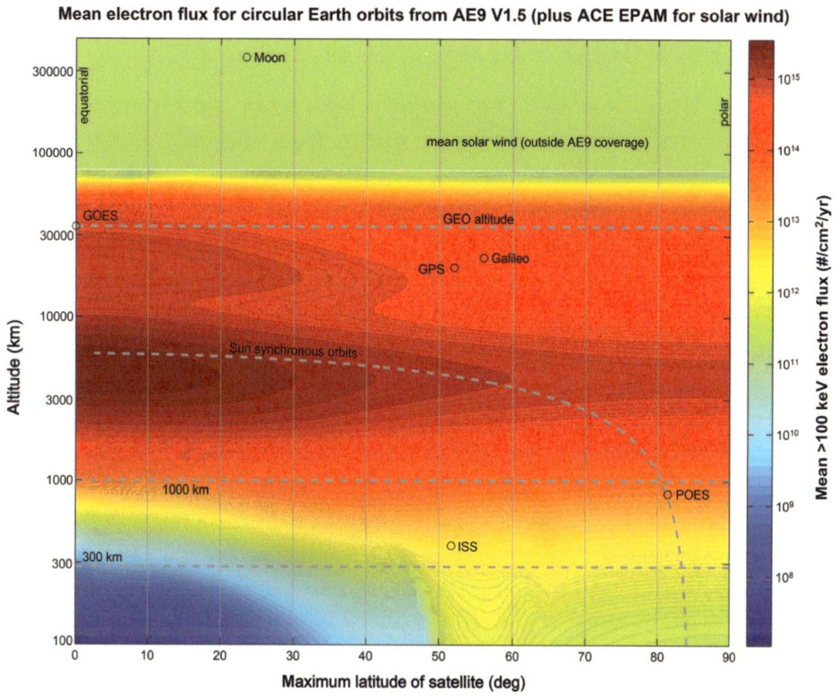

Figure 1. Plot of mean annual electron flux (>100 keV electrons) orbit averaged, experienced by a satellite in a circular geocentric orbit as a function of altitude and maximum latitude (inclination for prograde orbits, a supplement of inclination for retrograde orbits). Representative orbits are shown as dashed lines for reference, as are the positions of the moon, International Space Station (ISS), and several other satellites. Figure based on AE9 V1.5 mean radiation belt model, [7] with mean solar wind values outside AE9 coverage from data collected by the Electron, Proton, and Alpha Monitor on the Advanced Composition Explorer (ACE EPAM) [5, 8].

penetrated by electrons of energy > 2 MeV, while electrons of > 0.25 MeV energy can fully penetrate the MLI layers [3]. However, surface materials are subject to deterioration by electrons of even 0.1 MeV, which have the greatest flux due to the steep outer belt electron spectrum, and are typically stopped within the outermost polymer layer [5]. In addition to damage caused by energy deposition, non-penetrating electrons deposit charge in the material.

The material properties of greatest interest to spacecraft engineers are those that influence temperature and differential charging. Surface temperature is dependent on the selectivity of the surface, defined as the ratio of incident light absorbed (absorptivity, α; heating) to radiated infrared flux (emissivity, ε; cooling). Any change in color or surface roughness will change one or both of these quantities, and lead to a change in equilibrium temperature. Unfortunately, both quantities may be changed by polymer degradation due to radiation.

Optical properties of materials are also important for surveillance and health-monitoring purposes. Satellites at GEO are not spatially resolvable with ground-based optics, even with adaptive optics techniques. Therefore, the observable quantities are brightness, position, color (or reflected spectrum), and polarization. Any change in these quantities may lead to satellite misidentification, or be indicative of a change in satellite health.

Spacecraft charging is the potential a spacecraft or spacecraft component will assume as a result of the balance of incoming charged particles from the environment and particles removed via material conduction or electron emission. Interaction with space plasma can result in potential gradients between different spacecraft components (differential charging) of thousands or even tens of thousands of volts, which can seriously affect the operation of the spacecraft. Differential charging is most pronounced in GEO where the environment is dominated by electrons and the flux of mitigating ions is relatively low. Material properties also affect the charging behavior of a spacecraft. For instance, secondary electron emission and photoelectron emission tend to discharge surfaces, so spacecraft charging is very sensitive to these material properties. Also, dielectric materials can lose surface charge to interior chassis "ground" by electron conduction; this property also changes as the dielectric interacts with the ambient plasma. This is important due to the fact that in the presence of space plasma, electric arcs can jump from a negative surface to a positive surface if the potential difference between them is great enough. Depending on the amount of energy stored in the "surface capacitance," these arcs can lead to contamination of adjacent surfaces, radio frequency interference, sharp current transients and electronic upsets, and in the worst case can develop into "sustained arcs" that can destroy entire solar arrays or other electrical circuits.

In general, charge buildup anywhere on or in the spacecraft can lead to arcing. Changes in surface and bulk conductivity, secondary electron emission, and photoemission will lead to changes in the local electric fields and therefore, changes in the susceptibility of spacecraft to arcing. Unfortunately, although these electrical properties are known to change with radiation exposure, the magnitude of the changes is poorly characterized at present.

In the laboratory materials degradation tests reported here, the incident electrons are 90 keV electrons unless otherwise noted. This energy is chosen because the outer belt non-thermal electron flux is highest at lower energies, so they are most important from an energy deposition standpoint. Finally, 90 keV electrons can easily be produced by commercial electron guns and are less dangerous from an X-ray production standpoint than higher energy electrons. High-energy ions can also produce material degradation, but the fluxes of these particles in GEO are much lower than for electrons [5, 7].

2. Space weather simulation facility

A spacecraft on orbit interacts with the ambient plasma environment, comprised of electrons, photons, and ions in a vacuum, in a number of different ways. For example, the surface of a spacecraft can develop a large potential relative to the spacecraft chassis due to non-penetrating charged particle deposition. Additionally, highly energetic photons [ultraviolet (UV) and vacuum ultraviolet (VUV)] and charged particles deliver high levels of radiation with sufficient energy to break chemical bonds to the craft, surface materials in particular. This energy deposition leads to chemical bond breakage and reformation and radical formation [9]. These chemical changes manifest themselves as changes to a number of different physical properties including absorptivity and emissivity, electrical conductivity, and mechanical properties of the materials. Because it is not experimentally convenient to study on orbit spacecraft *in situ*, the spacecraft charging and instrument calibration laboratory in the space vehicles directorate of the US air force research laboratory have constructed a space weather simulation chamber, nicknamed Jumbo, in which electron and VUV photon damage can be inflicted on a variety of spacecraft materials and the effects of this damage can be quantified. The results of these simulated space-aging experiments can be used to refine existing models for spacecraft differential charging [10], predict, and characterize anomalous satellite behavior, identify space debris, and even be used to diagnose the health of satellites on orbit from ground-based observations.

The Jumbo space weather simulation chamber consists of a 6′ long, 6′ diameter cylindrical vacuum chamber, energetic particle sources, and various probes used for material characterization [11]. Pressures of < 10^{-6} Torr can be achieved from the atmosphere in ~1 h using a combination of mechanical pumps and a turbo- and/or a cryopump. In GEO, the electromagnetic radiation most likely to damage polymer materials is the Lyman-α line of hydrogen at 121.6 nm. These VUV) photons have enough energy (~10 eV) to break bonds in polymer materials. In order to mimic Lyman-α line of hydrogen, the chamber is equipped with four Krypton lamps (Resonance Ltd.) with emission lines at 123.6 and 116.5 nm, providing approximately 10 suns of VUV radiation. The prime electron source is a Kimball physics EG8105-UD electron flood gun with a range of 1–100 keV. This electron gun is used both for material aging (high-energy, deeply penetrating electrons) and for charging of materials (low-energy, shallowly penetration electrons) in order to perform SPD experiments from which material bulk conductivity/resistivity can be derived [12].

In addition to the particle sources, which are used to age common spacecraft materials; jumbo is equipped with a number of probes, which are used to characterize physical characteristics of the chosen material before, during, and after irradiation. The primary characterization tools used *in vacuo* are an integrating sphere with fiber optic coupled external light source and spectrometer, which is used to characterize optical changes and a non-contact electrostatic voltmeter, which is used to investigate material charge transport characteristics using the SPD method.

Optical transmission/reflectance spectra are recorded using an in-vacuum integrating sphere so that the measured quantity is the directional hemispherical reflectance (DHR). For each measurement, an ASD FieldSpec Pro spectroradiometer operating over a 350–2500 nm range collects first a white reference spectrum from a piece of in-vacuum calibrated Spectralon®. The electron beam is then temporarily extinguished and the motion system moves the integrating sphere to a sample carousel where it measures samples as they are rotated into the measurement position [13].

The SPD method utilizes charge injection via a low-energy electron beam to induce an electric potential near the surface of the material. To perform an SPD experiment, the front surface of the material is dusted by a beam of 5 keV electrons for 1–2 s immediately, after which the non-contact voltmeter is positioned 1–2 mm from the surface and begins to record the surface potential of the dielectric. After the charged body induced by the beam has reached the grounded backplane, the dissipation of surface potential is primarily determined by the loss of electrons from the material and is directly proportional to the material conductivity. SPD measurements were performed in darkness to eliminate the possibility of optically enhanced conductivity and photoemission. Since SPD is measured immediately following the charging electron beam, persistent radiation-induced conductivity (RIC) is still active [14]. However, it is only in effect between the surface and the penetration depth of the electrons, which for the case of 5 keV electrons in polyimide is less than a micron [3]. This minimizes the effect of RIC, which is assumed to be negligible for the bulk of the material in this measurement.

In addition to in-vacuo characterization, a portable vacuum pumped window is used to transport aged materials to a commercial UV/visible transmission/reflection spectrometer (Perkin-Elmer Lambda 950) and a Fourier transform infrared reflection spectrometer (Surface Optics SOC-400T). The portable vacuum window was designed to enable characterization of air sensitive materials using existing bench-top instruments without subjecting the space-weathered samples to unnecessary air exposure. Post-irradiation air exposure has been shown to modify certain materials' chemistry extensively and on a very short time scale (minutes) [15].

3. Effect of space plasma on satellites

3.1. Electrostatic charge and discharge

Interaction of space plasma with materials on orbit has been shown to drastically and permanently change spacecraft materials' charging properties. There is strong evidence that Galaxy 15, a GEO communications satellite, was incapacitated for 8 months due to arcing caused by the space environment, [16] even though in its 5 years of previous flight it had experienced similar environments several times with no ill effects. Charging in the auroral streams or when satellites exit eclipse can cause arcing on solar arrays, single event upsets in electronics, or as in the case of ADEOS-2, arcing in power lines leading to a complete loss of power [17].

Different parts of a satellite have different susceptibilities to charge accumulation. For example, a space-like test of a four-cell GPS-like solar array showed that most arcs occurred on the SCV10-2568 room-temperature vulcanization rubber (RTV) used to glue the cells and wiring to the Kapton substrate or near solar cell edges or corners. Each arc discharged about the coverglass area of the array. From these results, it is hypothesized that at least some of the arcing that occurs on GPS satellites on orbit are from the RTV and that some of the contamination that degrades the GPS solar array performance over time comes from arcs on the RTV as well as from the arcs from silver interconnects [18].

The likelihood of electronic discharge on a satellite on orbit is a function of the differential potentials that develop throughout the craft. That potential is dictated by the balance of incoming charge from the environment and charge that is removed via conduction and secondary electron emission. Since incoming charge cannot be controlled, the only way to predict what electric potential a material will adopt is to control the rate that charge is removed.

Figure 2 is a plot of the SPD rate as a function of polyimide resistivity [12]. When that decay rate exceeds the orbital period (1 day for GEO) the material can gather charge for the entire mission and the likelihood of discharge increases. The red and green stars in **Figure 2** show the resistivity of pristine polyimide and polyimide that has been radiation damaged, respectively. The irradiated material has become far less susceptible to charge accumulation, which is important to take into account in the design of spacecraft and the characterization of anomalous spacecraft behavior.

Charge transport characteristics of materials exposed to space plasma are defined by their fundamental material properties; in particular for insulators, density, and energy distribution of electron trap states within the band gap. Under the approach described by Dennison and Hoffmann, [20] as materials are bombarded with a flux of penetrating high-energy radiation, energy is shared with many bound (valence) electrons within the material, which are excited into energy levels scattered in the conduction band. These excited electrons quickly thermalize to shallow localized trap states just below the conduction band edge. Next, electrons can, among other processes, (i) be thermally re-excited into the conduction band, leading to thermally assisted charge transport, and termed radiation-induced conductivity (RIC); [21] or (ii) hop to an adjacent trap, termed thermally assisted hopping conductivity or dark current (DC) conductivity [22]. In addition, as a material age, more traps are generated [23].

Figure 3 Presents a representative charge–discharge curve of electron-irradiated PI material during and after bombardment with non-penetrating electrons. This curve can be divided into three regions: (I) charging section driven by the balance of electron deposition, secondary electron generation, RIC and DC conductivity; (II) pre-transit discharge section dominated by the RIC and DC conduction, as is the time regime when the deposited charge body is traversing the material but has not made contact with the grounded backplane; (III) post-transit discharge section dominated DC conduction.

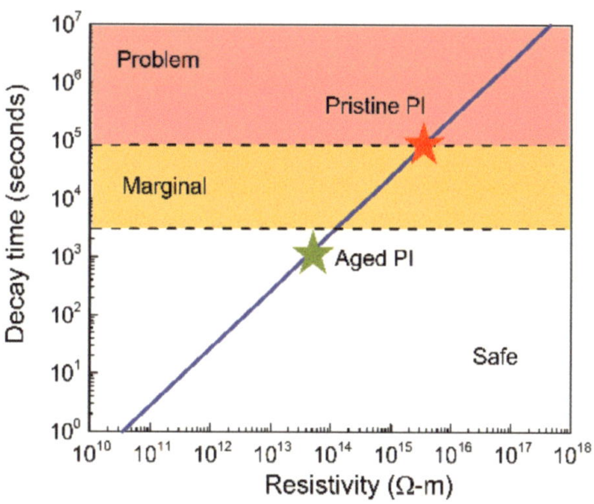

Figure 2. The plot of charge decay time versus resistivity value for PI. The red star indicates the resistivity of pristine PI and the green star that of laboratory-aged Kapton-H®. Figure adapted from [19].

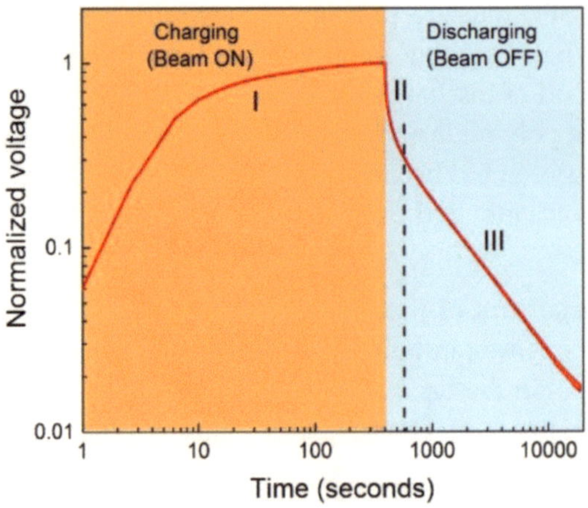

Figure 3. Representative charge/discharge curve of PI material bombarded with a non-penetrating electron. Shaded areas represent the three regions of the charge/discharge curve. See text for further details.

Theoretical models developed over the past several decades allow the extraction of many material parameters from a material's charge/discharge curve, including the density of trapped states (region I), trapping and de-trapping rates, and effective electron mobility (region II), and dark resistivity and conductivity of the material (region III) [24–28].

In particular, the charging part of the charge/discharge curve (region I) may be modeled with the following equation developed by Sim [24]:

$$V_s(t) = \frac{q_e d N_t}{\varepsilon_0 \varepsilon_r}\left\{1 - \frac{R(\varepsilon_b)}{d}\right\} R(\varepsilon_b)\left[1 - exp\left\{\frac{s_c J_b \tau_{onset}(1-\sigma_{yield})}{q_e(1-m)}\right\}\left[1 - \left(1 + \frac{t}{\tau_{onset}}\right)^{1-m}\right]\right] \tag{1}$$

Here q_e is the charge of an electron, C; ε_0 and ε_r are the permittivity of free space and relative permittivity of the chosen material, respectively; J_b is the electron beam flux, nA/cm^2; d is a sample thickness, cm. The secondary electron yield, σ_{yield}, the number of electrons emitted per incident electron, may be estimated based on the measurements and models of Song et al. [29]. The range, R (ε_b), is the maximum distance an electron of a given incident energy can penetrate through the material before all kinetic energy is lost and the electron comes to rest. Free parameters for Eq. (1) are the density of states, N_t; capture cross-section s_c; characteristic onset time for the current decay to occur, τ_{onset}; and a power parameter m, with $0 < m < 1$.

The pre-transit discharge section (region II) may be described using a model based on original work of Toomer and Lewis [25] supplemented by Aragoneses *et al* [26].

$$\frac{V(t)}{V_0} = 1 - \frac{V_0 \mu_0}{2 d^2 R}\left(r_t t + \frac{r_t}{R}\{1 - e^{-Rt}\} + \frac{r_r a_1^2}{2\alpha}\{1 - e^{-2\alpha t}\}\right) - \frac{V_0 \mu_0}{2 d^2 R}\left(\frac{r_t a_1^2}{R + 2\alpha}\{1 - e^{-(R+2\alpha)t}\}\right) - \lambda t^{\beta} \tag{2}$$

where d is the sample thickness, μm; V_o is the initial surface potential; R is the parameter describing charge transport dynamics; r_r and r_t are the probabilities of charge per unit time to be released from the trap and to be re-trapped in different trapping center, respectively, s^{-1}; λ is a dispersive term, taking into account the disordered structure of polyimide; μ_0 is the carrier mobility between traps, $m^2\ V^{-1}\ s^{-1}$; a_1 and β are fitting parameters. It is assumed that a fraction a_1 of the charges are initially placed in the surface traps, from where they move to the bulk at a certain rate α, s^{-1}. The remainder of the deposited charge ($1-a_1$) is injected directly into the bulk immediately after the discharge.

The post-transit region of the surface potential decay curve (region III) starts when a charged body has reached the grounded back plane. This can take place within a fraction of a second for high conductivity polymers or years in the case of low conductivity polymers such as PTFE (Teflon®). After the front of the charge, body has reached the grounded backplane the dissipation of charge is primarily determined by the loss of electrons from the material. This region is fit by Eq. (3), from which a decay time and the dark resistivity of the material may be derived [30]:

$$V_s(\tau_{decay}) = m\,\tau_{decay}^{b} \qquad \varrho_{dark} = \frac{\tau_{decay}}{\epsilon_0\,\epsilon_r} \tag{3}$$

where τ_{decay} is charge decay time, seconds; m and b are fitting parameters; ε_0 and ε_r are permittivity of free space and relative permittivity of the PI material, respectively. The conductivity of the material is then calculated as inversely proportional to the resistivity of the material, $1/\Omega\cdot cm$.

3.2. Material changes

Interaction of PI with highly energetic particles of space plasma will modify its chemical structure. The extent of this modification is a function of several simultaneous kinetic processes, namely, damage (interaction of material with highly energetic particles, resulting in broken chemical bonds), healing (formation of bonds identical to those damaged, returning the material to its pristine state), and scarring (formation of new chemical bonds in the damaged material, which are different from those in the pristine material) [15]. Often, macroscopic properties are measured making it difficult to distinguish healing from scarring; hence we refer to the sum of healing and scarring as recovery.

Figure 4 Shows the photographs of a radiation-damaged PI sample taken immediately after irradiation with 90 keV mono-energetic electron flood gun to the dose of 5.6×10^7 Gy and a pristine PI control sample. This energetic dose is equivalent to that experienced by PI during approximately 8 years in GEO orbit [3, 5, 7]. The damaged sample has a deep brown color, which differs from the characteristic amber color of pristine PI. From **Figure 4**, it is clear both electron damage and subsequent exposure to air have a significant effect on the optical properties of PI in the visible spectrum.

The effect of damaging radiation on the optical properties of the irradiated material is evident from the transmission spectra of radiation-damaged PI, as shown in **Figure 5**. The red-shift of the absorption edge indicates an effective shrinking of the PI's "band gap" to ~1.8 eV in the damaged material due to the emergence of radiation-induced electronic states. Compared to the measured "band gap" of Kapton of 2.3 eV, these states are energetically shallow [31].

Figure 4. Photographs of pristine reference Kapton (left) and a radiation damaged Kapton sample (right) taken after an electron dose of 5.6×10^7 Gy. Right picture was taken after 2 min of air exposure.

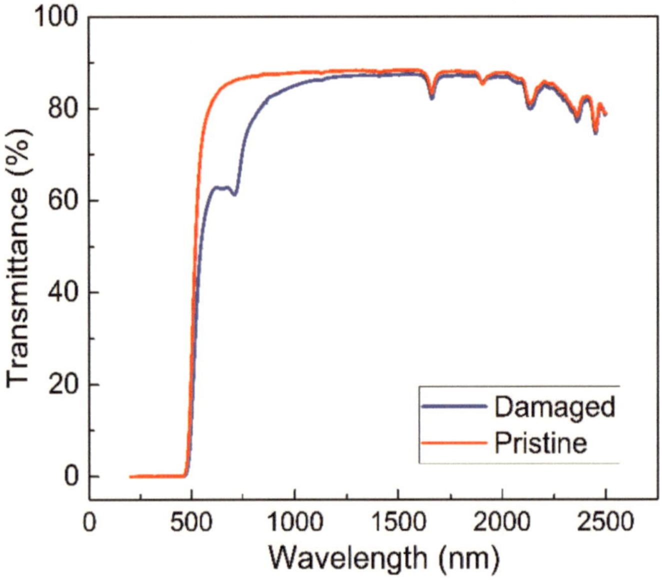

Figure 5. Transmittance spectra of pristine and radiation damaged PI samples.

Fourier-Transform Infrared (FTIR) spectroscopy was used to understand the underlying chemistry of radiation-induced damage in PI material. FTIR probes chemical bonding by exciting vibrational transitions within the polymer. Changes in the position and intensity of the IR absorption "fingerprint" of damaged material will offer insights into what chemical bonds are being modified during the radiation-induced degradation. PI has a complex IR signature with each peak corresponding to a specific vibration within the monomer. Several characteristic vibrational assignments have been identified [32–34] and are summarized in **Table 1**.

Figure 6 presents FTIR spectra of pristine and radiation-damaged and subsequently air exposed polyimide. Measurements were made in a portable vacuum sealed CaF_2 window, with an absorption cutoff at 1200 cm^{-1}. Comparison of the FTIR spectra reveals two interesting radiation-induced changes in the IR fingerprint of the damaged film. First, the absorption at the wavelength associated with the carbonyl out-of-phase stretch increased after electron bombardment. This suggests first that existing carbonyl moieties were not preferentially broken due to the electron bombardment. A significant reduction in the absorption associated

Assignment	Absorption (cm^{-1})	Characterization
δ(phenyl)	1004	Phenyl ring deformation
ν(C-N-C)	1117	Imide stretch
ν(C-O-C)	1261	Bridging C-O-C stretch
δ(C-N-C)	1380	Imide stretch
ν(phenyl)	1465	Phenyl ring C-C stretch
C=C	1515	Aromatic C=C stretching
ν(phenyl)	1601	Phenyl ring C-C stretch
ν(C=O)	1675	Out-of-phase carbonyl stretch
ν(C=O)	1753	
Cyclic anhydride	1890–1940	Cyclic anhydrides, presented in not fully cured polymer

Table 1. Vibrational assignments of polyimide.

with phenyl ring C-C stretch after electron bombardment suggests that ether breakage is accompanied by rupture of the phenyl rings in the monomer, possibly leading to the formation of a new pi-bonded carbon structure containing the new carbonyl. It is important to note that because this sample had been exposed to air after damage but before the investigation, the damage products evident from the FTIR spectra result from the sum of both damage and material recovery [35].

To evaluate the effect of space plasma on the charge transport properties of PI, bulk conductivities of radiation-damaged samples were evaluated. **Figure 7** compares bulk conductivities of PI samples irradiated with a dose of 5.6×10^7 Gy and recovered in the air (top panel) and under vacuum (bottom panel). The initial post-irradiation conductivity of the two damaged PI samples was nearly the same [5×10^{-17} $(\Omega\cdot cm)^{-1}$] and [8×10^{-17} $(\Omega\cdot cm)^{-1}$]. However, air exposure of radiation-damaged PI resulted in rapid recovery, within 3 h, of the irradiated material's conductivity from its initial value to that of pristine Kapton [2×10^{-20} $(\Omega\cdot cm)^{-1}$], whereas the vacuum recovery of radiation-damaged PI retained the same value for over 3 weeks (504 h) after the damaging process.

From **Figure 7** it is obvious that air exposure has a significant effect on charge transport properties of the radiation-damaged PI material. This illustrates the necessity of in-vacuum characterization techniques with minimized air exposure to the irradiated material. Still, observation of air recovery process of radiation-damaged PI may provide some insights into the chemistry driving the aging and recovery process.

The conductivity of two radiation damaged PI samples (dose of 5.6×10^7 Gy) as a function of cumulative air exposure time is plotted in **Figure 8**. After irradiation, Samples 1 and 2 were stored under vacuum and only exposed to air during conductivity measurements. Both samples recovered to nearly the conductivity of pristine Kapton after 250 and 400 h (vacuum and air exposed), respectively. However, when conductivity is plotted purely versus air

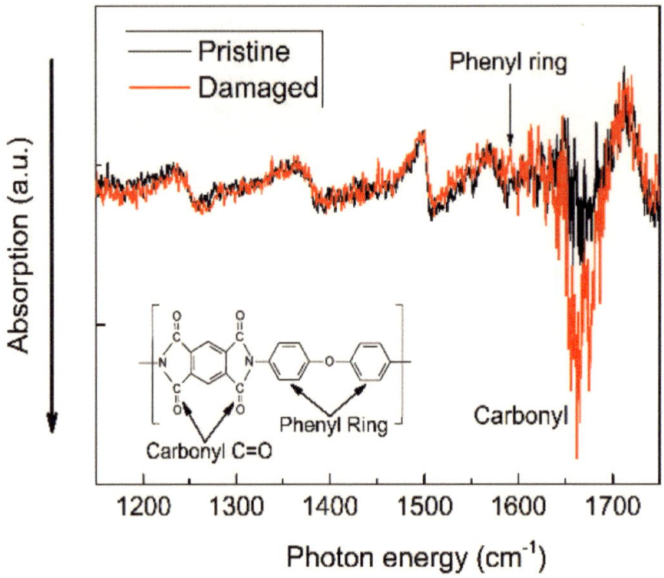

Figure 6. Absorption spectra of reference (pristine) and radiation-damaged with a dose of 5.6 × 10^7 Gy PI samples. Lower values on the ordinate indicate more absorbed light in the polymer. Notice increased absorption at the carbonyl stretching frequency (1675 cm^{-1}) and decreased absorption at the phenyl ring C-C stretch (1435–1570 cm^{-1}) in the damaged sample. The inset shows chemical structure of polyimide with relevant moieties identified.

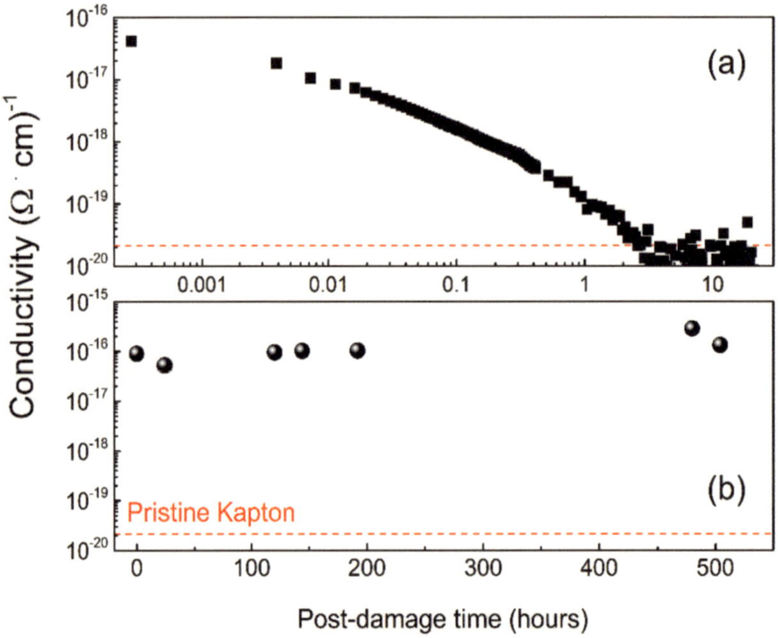

Figure 7. Comparison of (a) air- and (b) vacuum-recovered conductivities of PI irradiated with a dose of 5.6 × 10^7 Gy. The dashed line indicates the conductivity of pristine Kapton-H®.

exposure, as shown in **Figure 8**, it is apparent that the healing process proceeds primarily under exposure to the atmosphere. It has since been reported that the enhanced conductivity of electron-irradiated PI is stable under vacuum conditions [35].

To further investigate the effect of damaging radiation on PI, including the concentration and nature of free radicals, EPR measurements were performed on electron irradiated PI material. Radical concentrations are reported as arbitrary units and scaled according to relative peak-to-peak intensity. A reference sample (pristine PI) showed no EPR signal indicating that the number of unpaired electrons was below the detection limit in pristine PI, as was expected. However, a strong initial EPR signal was measured in the damaged material that decayed with exposure to air (**Figure 8**).

The fact that the conductivity of radiation-damaged PI decays on the same time scale as the concentration of radicals suggests that these properties are interconnected. These corresponding time scales are also suggestive that the concentration of radicals plays a critical role in the transport of electrons through the bulk of the material. It is reasonable to assume that creation and decay of radicals in the material will modify the density and energetic distribution of electron trap states in the bandgap of PI [36–38]. This is further supported by the UV/Vis spectroscopy that shows the development of energetically shallow traps in the bandgap of damaged PI, as seen in **Figure 5**.

Moreover, charge transport in disordered materials like PI occurs via incoherent hopping among transport sites [22, 38, 39]. Bulk conductivity is influenced by both energetic and geometric disorder. That is to say, the facility with which an electron can travel through a disordered material in the presence of a strong electric field is dependent on both the energetic distribution of transport sites within the material and the geometric distribution of the transport sites. The latter dependency arises due to the variation of intersite electronic

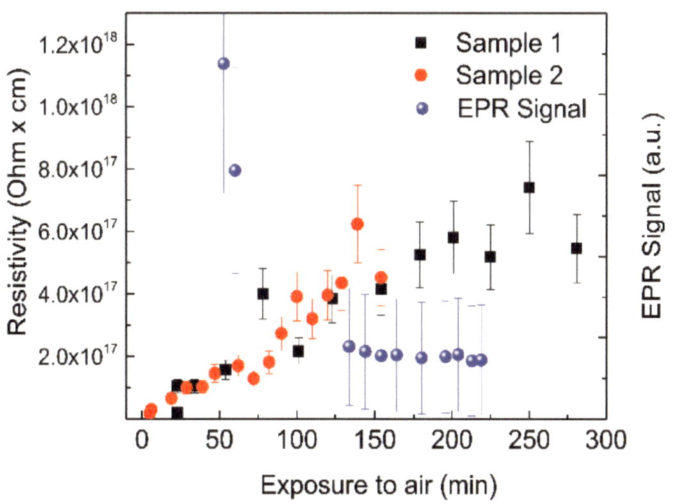

Figure 8. Resistivity (inverse of conductivity) for two samples and EPR signal of radiation-damaged PI film plotted as a function of cumulative air exposure time.

wavefunction overlap arising from the positional and orientational distribution of these hopping sites [38]. It is our hypothesis that the radical sites created due to bond-specific rupture during electron bombardment can act as electron hopping sites, which are not present in the pristine material.

Finally, it has been commonly assumed that exposure to air would be deleterious to understanding how materials recover in a vacuum. However, since PI is very stable under normal conditions a small amount of air exposure is accepted as necessary and largely unavoidable in the majority of studies that have been published [40–42]. Data presented here show that air exposure dominates the post-irradiation chemistry of PI and that even limited air exposure (less than 10 min) will cause dramatic and unwanted effects that will obscure experimental studies. This fact illustrates the necessity of in-vacuum characterization methods as well as a careful examination of material handling techniques when reviewing the literature.

4. Deleterious effects of plasma interactions with spacecraft

While interactions with space plasma can sometimes change material properties in a beneficial way (see succeeding section), these interactions also often have deleterious effects on spacecraft performance, often related to a change of electrical potentials on either internal or surface materials. If the local electric field becomes larger than some threshold value, electrostatic discharges (sometimes called arcs) will occur. Spacecraft arcs have been seen to have the following effects:

1. Large and rapid current spikes, which can cause latchups or even destroy sensitive electronic components.
2. Transients in spacecraft power.
3. Electromagnetic interference (EMI), which can interfere with communications, and so on.
4. Contamination of surrounding surfaces, which can affect optical transparency (usually thermal, but maybe solar cell current output) and electrical properties.
5. Short-circuits between individual solar cell junctions, causing the loss of power from individual cells.
6. Sustained arcing between adjacent solar array strings or from solar array strings to spacecraft ground, which can lead to permanent short-circuits and power loss in one or all strings (total loss of power).

Due to these and other issues, efforts are usually made to prevent arcs by choosing appropriate materials and power system engineering. Elaborate computer codes are used to see where and when spacecraft charging is likely to occur and to adjust surface and interior materials where arcing susceptibility is found. If material properties change dramatically due to the space radiation, however, these models are invalidated, as they rely on pristine material characteristics, and unwanted surprises are likely to occur.

Unexpected, unexplained slow and pernicious power loss has been consistently observed in GPS satellites [43]. Although the loss was quickly determined to be due to the contamination of the solar array surfaces, a seemingly exhaustive search for sources of the contamination turned up no suspects. Confronted with the challenge, engineers decided to oversize the solar arrays by 25%, such that the deteriorated solar arrays would still provide adequate power at the end-of-life. This was an expensive and difficult solution because it led to increased spacecraft weight and volume, important launch considerations. Recently, it has been discovered that GPS solar arrays have been undergoing extensive arcing and gradually contaminating their own surfaces, decreasing the amount of sunlight that can reach the active parts of the solar cells [18, 43]. The arcing had gone undetected because of heavy filtering of electrical transients in the power system. Ground-based testing has shown that the power loss can be explained by the contamination produced by thousands of solar array arcs seen on orbit. In this case, a change in material properties due to the space plasma environment led to an unexpected and deleterious result.

As with changes in human DNA, changes in spacecraft materials properties may be beneficial but are usually not. If, however, the materials properties changes can be quantified and predicted, engineering can achieve a survivable spacecraft throughout its anticipated life. In the case of GPS, understanding the cause of solar array power degradation is leading to designs that can prevent the cause of the degradation, and in the end, provide a more reliable, cheaper, lighter, and smaller satellite that can still fulfill its mission.

5. Modification of material optical signatures after exposure to a space plasma environment

In general, it has been found that the space environment fundamentally changes spacecraft materials [15, 44–46]. The nature of particles primarily responsible for the damage is dependent on the orbit; in GEO, high-energy electrons are the primary damaging species in terms of energy deposition into the bulk of the material [5]. This energy deposition breaks chemical bonds within the material, resulting in the creation of new electronic transitions in the material. These new electronic transitions manifest themselves in altered optical and physical properties, such as color change and electrical conductivity.

Over the last decade, astronomical reflectance spectroscopy has been proposed to characterize the material properties of artificial space objects, such as satellites, rocket bodies, and human-made debris [47, 48]. In this application, reflected light is collected from a remote target illuminated by a continuous source, the Sun. The reflected light produces a spectrum whose shape and absorption features are indicative of a specific material composition. Since each material has a unique spectral fingerprint, it may be unambiguously identified if its spectral features are well differentiated.

One example of applied optical characterization of materials in space is the assessment of high area-to-mass ratio (HAMR) objects, consisting primarily of spacecraft MLI. Several layers of MLI have comprised of aluminum sputtered polyimide. It has been shown that after electron irradiation the hemispheric reflectance spectrum, the sum of all diffuse and specular reflectance of a surface under diffuse illumination, of this aluminized PI changes considerably. This change in reflectance is due in part to the change in absorbance of the polymer portion of the material. Since the PI is backed by a thin reflective aluminum layer, the light that would otherwise pass through the PI is instead reflected back into the spectrometer. Pristine PI is fairly transparent between 500 and 700 nm the absorption edge at 500 nm is what gives PI its characteristic amber color. The damaged material's new electronic structure results in resonant absorption of lower energy photons, which results in a shift of the absorption band edge to around 730 nm. This phenomenon manifests itself as a darkening of the material in the visible spectrum [15, 49].

Because the change in optical absorption is greater at some wavelengths than others, the ratio of one spectral band to another, known as a color–color plot, can be used to characterize the extent of space plasma exposure experienced by spacecraft material. **Figure 9** shows the evolution of the color–color plots of two different aluminum sputtered PI layers of MLI after various electron exposures. The color–color plots were generated using the passbands associated with the r′ and i′ filters defined by the Sloane Digital Sky Survey [50–53]. For a more detailed description of the experiment, refer to reference [13].

Conceivably, the magnitude of change of these color–color indices is great enough to enable identification and/or characterization of space debris clouds. Knowledge of the origin of these debris clouds can be used to facilitate space debris remediation [54, 55].

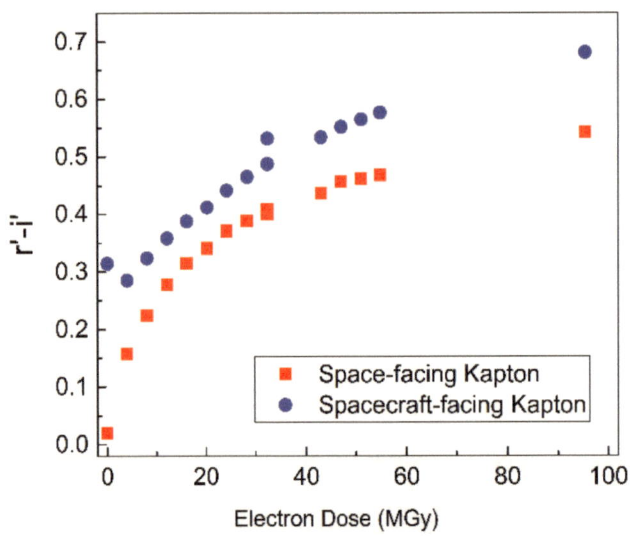

Figure 9. r′-i′ plots for space- and space-craft facing Kapton. The magnitude of change for all materials is comparable. However, the absolute magnitude of the dose differs due to the material composition.

6. Conclusion

As the population of man-made objects in space grows ever more rapidly, understanding the interaction of space plasma with commonly used materials becomes only ever more important. Clearly, knowledge of the different facets of this interaction can be used to the advantage of aerospace scientists either as a diagnostic tool or as guidelines that can result in more efficient and robust spacecraft design. Further, detailed knowledge of how space plasma interacts with materials in Earth's orbit will guide the development of next-generation spacecraft materials.

While the information presented in this chapter is not exhaustive, it serves to illustrate the broad variety of chemical and physical changes that can occur when a given material interacts with space-like plasma. The examples presented here were chosen due to the ubiquity of polyimide and solar arrays on existing spacecraft. The simulated GEO environment (high-energy electrons) used to study these interactions is a first-order approximation for the rather more complicated GEO environment. Although electrons are the primary damaging species in GEO in terms of energy deposition, a more representative space weather simulation would include protons and UV photons in order to study synergistic damage processes that occur during the interaction of materials with a more complicated charged particle environment [5].

Acknowledgements

We would like to acknowledge support from the Air Force Office of Scientific Research, Remote Sensing and Imaging Physics Portfolio (Dr. Stacy Williams). Grant 17RVCOR414.

Conflict of interest

No conflict of interest is declared.

Author details

Daniel P. Engelhart[1]*, Elena A. Plis[1], Dale Ferguson[2], W. Robert Johnston[2], Russell Cooper[2] and Ryan C. Hoffmann[2]

*Address all correspondence to: afrl.rvborgmailbox@us.af.mil

1 Assurance Technology Corporation, Carlisle, MA, USA

2 Air Force Research Laboratory, Space Vehicles Directorate, USA, Kirtland AFB, Albuquerque, NM, USA

References

[1] Marletta G, Iacona F. Heat-induced versus particle-beam-induced chemistry in polyimide. Nuclear Instruments & Methods in Physics Research Section B-Beam Interactions with Materials and Atoms. 1993;**80-1**:1045-1049

[2] Marletta G. Chemical reactions and physical property modifications induced by keV ion beams in polymers. Nuclear Instruments and Methods in Physics Research Section B: Beam Interactions with Materials and Atoms. 1990;**46**(1-4):295-305

[3] Berger MJ, Coursey JS, Zucker MA, Chang J. ESTAR, PSTAR, and ASTAR: Computer Programs for Calculating Stopping-Power and Range Tables for Electrons, Protons, and Helium Ions (version 1.2.3). Gaithersburg, MD: National Institute of Standards and Technology; 2005. Available: http://physics.nist.gov/Star [2018, May 30]

[4] Jursa AS, Laboratory USAFG. Handbook of Geophysics and the Space Environment: Air Force Geophysics Laboratory. USA: Air Force Systems Command, U.S. Air Force; 1985

[5] Ginet G, O'Brien T, Huston S, Johnston W, Guild T, Friedel R, et al. AE9, AP9 and SPM: New models for specifying the trapped energetic particle and space plasma environment. Space Science Reviews. 2013;**179**(1-4):579-615

[6] Mateo-Velez JC, Sicard-Piet A, Lazaro D, Inguimbert V, Sarrailh P, Hess S, et al. Severe geostationary environments: Numerical estimation of spacecraft surface charging from flight data. Journal of Spacecraft and Rockets. 2016;**53**(2):304-316

[7] O'Brien TP, Johnston WR, Huston SL, Roth CJ, Guild TB, Su YJ, et al. Changes in AE9/AP9-IRENE Version 1.5. IEEE Transactions on Nuclear Science. 2018;**65**(1):462-466

[8] Space Weather Prediction Center. US National Oceanographic and Atmospheric Admini stration; [Cited 2018]. Available from: https://www.swpc.noaa.gov/products/ace-real-time-solar-wind

[9] Mehnert R. Materials modification using electron beams. In: Misaelides P, editor. Application of Particle and Laser Beams in Materials Technology. Nato Advanced Science Institutes Series, Series E, Applied Sciences. Vol. 2831995. pp. 557-580

[10] Mandell MJ, Davis VA, Cooke DL, Wheelock AT, Roth C. Nascap-2k spacecraft charging code overview. IEEE Transactions on Plasma Science. 2006;**34**(5):2084-2093

[11] Cooper R, Hoffmann R. Jumbo space environment simulation and spacecraft charging chamber characterization. Air Force Technical Report; 2015. AFRL-RV-PS-TP-2015-0012

[12] Plis EA, Engelhart DP, Likar J, Hoffmann RC, Cooper R, Ferguson D. Electrical behavior of carbon-loaded Kapton for spacecraft applications. Journal of Spacecraft and Rockets. 2017:1-2

[13] Engelhart DP, Cowardin H, Maxwell J, Plis E, Ferguson D, Barton D, Schiefer S, Hoffmann R. Space Weathering Experiments on Spacecraft Materials. Advanced Maui Optical and Space Surveillance Technologies Conference; Maui, HI, USA; 2017

[14] Tyutnev A, Saenko V, Pozhidaev E, Ikhsanov R. Experimental and theoretical studies of radiation-induced conductivity in spacecraft polymers. IEEE Transactions on Plasma Science. 2015;**PP**(99):1

[15] Engelhart DP, Plis E, Humagain S, Greenbaum S, Ferguson D, Cooper R, et al. Chemical and electrical dynamics of polyimide film damaged by electron radiation. IEEE Transactions on Plasma Science. 2017;**45**(9):2573-2577

[16] Ferguson D, Denig W, Rodriguez J. Plasma Conditions During the Galaxy 15 Anomaly and the Possibility of ESD from Subsurface Charging. 49th AIAA Aerospace Sciences Meeting Including the New Horizons Forum and Aerospace Exposition. Orlando, FL, USA: Aerospace Sciences Meetings: American Institute of Aeronautics and Astronautics; 2011

[17] Hosoda S, Kim J, Cho M, Toyoda K, Kawakita S, Kusawake M, et al. Ground investigation of sustained arc phenomena in power cables on ADEOS-II satellite. Journal of the Japan Society for Aeronautical and Space Sciences. 2006;**54**(633):427-433

[18] Ferguson DC, Hoffman RC, Engelhart DP, Plis EA. Voltage threshold and power degradation rate for GPS solar array arcing. IEEE Transactions on Plasma Science; 2017

[19] Dennison J, Frederickson A, Swaminathan P. Charge Storage, Conductivity and Charge Profiles of Insulators As Related to Spacecraft Charging. All Physics Faculty Publications; 2003 (Paper 1485). http://digitalcommons.usu.edu/physics_facpub/1485

[20] Hoffmann R, Dennison JR, Thomson CD, Albretsen J. Low-fluence electron yields of highly insulating materials. IEEE Transactions on Plasma Science. 2008;**36**(5):2238-2245

[21] Dennison J, Gillespie J, Hodges J, Hoffmann R, Abbott J, Hunt AW, et al. Radiation Induced Conductivity of Highly-Insulating Spacecraft Materials. In: 10th Spacecraft Charging Technology Conference; 2009. Biarritz: France

[22] Brunson J. Hopping Conductivity and Charge Transport in Low Density Polyethylene. Logan, UT, USA: Utah State University; 2010

[23] Blaise G. Charge localization and transport in disordered dielectric materials. Journal of Electrostatics. 2001;**50**(2):69-89

[24] Sim A. Unified Model of Charge Transport in Insulating Polymeric Materials. Logan Utah: Utah State University; 2013

[25] Toomer R, Lewis T. Charge trapping in corona-charge polyethylene films. Journal of Physics D: Applied Physics. 1980;**13**(7):1343

[26] Aragoneses A, Mudarra M, Belana J, Diego JA. Study of dispersive mobility in polyimide by surface voltage decay measurements. Polymer. 2008;**49**(10):2440-2443

[27] Meyza X, Goeuriot D, Guerret-Piécourt C, Tréheux D, Fitting H-J. Secondary electron emission and self-consistent charge transport and storage in bulk insulators: Application to alumina. Journal of Applied Physics. 2003;**94**(8):5384-5392

[28] Hoffmann R, Dennison J. Measurement methods of electron emission over a full range of sample charging. IEEE Transactions on Plasma Science. 2012;**40**(2):298-304

[29] Song Z, Ong C, Gong H. Secondary and backscattered electron yields of polymer surface under electron beam irradiation. Applied Surface Science. 1997;**119**(1):169-175

[30] Dennison JR, Brunson J, Swaminathan P, Green NW, Frederickson AR. Methods for high resistivity measurements related to spacecraft-charging. IEEE Transactions on Plasma Science. 2006;**34**(5):2191-2203

[31] Yohei Komiyama SS, Miyake H, Tanaka Y, Takada T. Observation of surface discharge phenomena on dielectric films under low pressure using pockels effect. In: Jacob Kleiman MT, Kimoto Y, editors. Protection of Materials and Structures from the Space Environment. Amsterdam, Netherlands: Springer Verlag; 2012. p. 456

[32] Nurmukhametov R, Likhachev DY, Lavrov S, Kardash JY. Features of electronic absorption spectra of aromatic polyimides and polyisoimides. Polymer Science USSR. 1989; **31**(2):434-440

[33] Ortelli E, Geiger F, Lippert T, Wokaun A. Pyrolysis of Kapton® in air: An in situ DRIFT study. Applied Spectroscopy. 2001;**55**(4):412-419

[34] Li R, Li C, He S, Di M, Yang D. Damage effect of keV proton irradiation on aluminized Kapton film. Radiation Physics and Chemistry. 2008;**77**(4):482-489

[35] Engelhart DP, Plis E, Ferguson D, Cooper R, Hoffmann R. Optical and chemical characterization of polyimide in a GEO-like environment. Technical Paper: 2016 Advanced Maui Optical and Space Surveillance Technologies Conference; 2016

[36] Silva HS, Dominguez IF, Perthue A, Topham PD, Bussiere PO, Hiorns RC, et al. Designing intrinsically photostable low band gap polymers: A smart tool combining EPR spectroscopy and DFT calculations. Journal of Materials Chemistry A. 2016;**4**(40):15647-15654

[37] Ennis CP, Kaiser RI. Mechanistical studies on the electron-induced degradation of polymethylmethacrylate and Kapton. Physical Chemistry Chemical Physics. 2010;**12**(45): 14902-14915

[38] Gartstein YN, Conwell EM. Off-diagonal disorder and activation-energy of high-field hopping motion. Physical Review B. 1995;**51**(11):6947-6952

[39] Noro S, Yamadaya T. ESR and electrical-conduction in PYROLYZED polyimides. Journal of Polymer Science Part B-Polymer Physics. 1991;**29**(3):359-363

[40] Srinivasan R, Hall R, Allbee D. Generation of electrically conducting features in polyimide (KaptonTM) films with continuous wave, ultraviolet laser radiation. Applied Physics Letters. 1993;**63**(24):3382-3383

[41] Tahara H, Kawabata T, Zhang LL, Yasui T, Yoshikawa T. Exposure of spacecraft polymers to energetic ions, electrons and ultraviolet light. Nuclear Instruments & Methods in Physics Research Section B-Beam Interactions with Materials and Atoms. 1997;**121**(1-4):446-449

[42] Engelhart DP, Ferguson D, Cooper R, Hoffmann R. Optical and Chemical Characterization of Polyimide in a GEO-like environment. Advanced Maui Optical and Space Surveillance Technologies Conference; Maui, USA; 2016

[43] Ferguson D, Crabtree P, White S, Vayner B. Anomalous GPS power degradation from arc-induced contamination. Journal of Spacecraft and Rockets. 2016

[44] Wu Y, Sun C, Xiao J, Li R, Yang D, He S. A study on the free-radical evolution and its correlation with the optical degradation of 170 keV proton-irradiated polyimide. Polymer Degradation and Stability. 2010;**95**(7):1219-1225

[45] Sun YM, Zhu ZY, Jin YF, Liu CL, Wang ZG, Liu J, et al. The effects of high electronic energy loss on the chemical modification of polyimide. Nuclear Instruments & Methods in Physics Research Section B-Beam Interactions with Materials and Atoms. 2002;**193**:214-220

[46] Choi HS, Lee J, Cho KS, Kwak YS, Cho IH, Park YD, et al. Analysis of GEO spacecraft anomalies: Space weather relationships. Space Weather-the International Journal of Research and Applications. 2011:9

[47] Aberkromby KJ, Hamada K, Guyote M, Okada J, Barker E. Remote and ground truth spectral measurement comparisons of FORMOSAT III. Advance Maui Optical and Space Surveillance Technologies Conference; Maui, USA; 2007

[48] Bedard MD. Using a physics-based reflection model to study the reddening effect observed in spectrometric measurements of artificial space objects. Advance Maui Optical and Space Surveillance Technologies Conference; Maui, USA; 2007

[49] Plis E, Engelhart DP, Barton D, Cooper R, Ferguson D, Hoffmann R. Degradation of polyimide under exposure to 90keV electrons. Physica Status Solidi B-Basic Solid State Physics. 2017;**254**(7)

[50] Sloan Digital Sky Survey [cited 2017 June]. Available from: http://www.sdss.org/dr12/algorithms/ugrizvegasun/

[51] Fukugita M, Ichikawa T, Gunn JE, Doi M, Shimasaku K, Schneider DP. The sloan digital sky survey photometric system. Astronomical Journal. 1996;**111**(4):1748-1756

[52] Astrophysical Institute Potsdam [cited 2017 June]. Available from: http://www.aip.de/en/research/facilities/stella/instruments/data/sloanugriz-filter-curves

[53] Astrodon Photometrics [cited 2017 June]. Available from: http://www.astrodon.com/store/p12/Astrodon_Photometrics_Sloan_Filters.html

[54] Cowardin H, Seitzer P, Abercromby K, Barker E, Schildknecht T. Characterization of Orbital Debris Photometric Properties Derived from Laboratory-Based Measurements. Advanced Maui Optical and Space Surveillance Technologies Conference; Maui, USA; 2010

[55] Schildknecht T, Musci R, Ploner M, Beutler G, Flury W, Kuusela J, et al. Optical observations of space debris in GEO and in highly-eccentric orbits. Advances in Space Research. 2004;**34**(5):901-911

Chapter 13

Repetitive Nanosecond Volume Discharges under Airflows

Jingfeng Tang, Liqiu Wei and Daren Yu

Additional information is available at the end of the chapter

http://dx.doi.org/10.5772/intechopen.81919

Abstract

Atmospheric pressure discharges are widely used in active airflow control, material synthesis, and air treatment. The key to an optimal application performance lies in how to generate stable and diffuse plasma especially in a large volume and in high-speed airflows. This chapter presents the study of repetitive nanosecond volume discharges under high-speed airflows. The volume discharge strongly depends on the airflows, and the corresponding discharge modes vary from filament to diffuse modes with addition of airflows. The role of airflows provides negative effects on discharge currents as well as discharge densities. Moreover, a type of discharge device with upstream and downstream structure is proposed to demonstrate that charged particles produced by the upstream discharge are transported to the downstream zone and play a pre-ionization and enhanced effect to the downstream discharges.

Keywords: repetitive nanosecond discharges, volume discharges, airflows, upstream and downstream flow

1. Introduction

Recently, atmospheric pressure discharge plasma has been considered for many applications, such as airflow control [1–20], material modification [21–31], air purification [31–49], and so on. With the difference from low-pressure discharge plasmas, atmospheric pressure discharge plasmas are usually operated under an open environment, and the collision between ions and electrons is very frequent. Such strong elastic and/or inelastic collision produces a significant chemical effect and corresponding thermal effect. In addition, air as the media of the discharge process is usually in a flow state, such with an airflow state with different velocities. The various particles in the plasmas always perform a macro-overall movement, and the exchange of energy and momentum occurs between plasmas and airflows. Therefore, the atmospheric pressure discharge plasma is actually in a typical multi-field couple system, and the coupling interaction between airflows and discharges is of extensive concerns [1].

As the couple interaction between plasmas and airflows, the plasmas macroscopically exhibit a fluid state property, the distribution of plasma particles is influenced by the heat and mass transfer from airflows, and discharge modes and discharge intensities are also changed. As a simultaneous inverse role, the energy release by discharge can cause impulsive interference and thermal effect on airflows, and a change of airflow field distribution can be generated. The airflow transport effect determines the distribution of uncharged particles, and such distribution provides an ionization condition, thus affecting the discharge breakdown. The transfer of heat and mass from airflows provides a new factor on the plasma diffusion, and the discharge energy dissipation and discharge plasmas also provide an active control of airflow distribution. To be sure, discharge plasmas under airflows have undergone a fundamental change. With the presence of airflows, discharge plasmas are more dominant with a strong interaction between plasmas and airflows.

The couple between plasmas and airflows has been considered as the interaction between discharge plasma dynamics and gas dynamics, from a view of time and space scales. For discharge plasmas, the topical time scales include as follows: the establishment of electric field and the generation of plasma (nanoseconds), the dissipation and the quenching of plasma (nanoseconds and/or microseconds), and the life of charged particles (seconds and/or hours). For airflows, the typical time scale is mainly determined by airflow conditions, such as the transport time of airflow (microseconds and/or milliseconds) and the convection heat transfer time (milliseconds and/or seconds) [2]. For discharge plasmas, the typical space scale is about a small size (micrometers and/or millimeters), such as the mean free path and the thickness of plasma sheath. For airflows, the typical space scale is about a big size (millimeters and/or meters), such as the thickness of boundary layer [3]. Under such couple with a multi-time and a multi-space scale, it is necessary to recognize the phenomenon and mechanism of discharges under airflows.

2. Background for discharges under airflows

Plasma flow control is a type of active flow control technology based on discharge plasma technologies, which is advantageous of little power, quick response, and perfect actuation. Russians, Americans, and other research groups [4–20] have done an in-depth study on plasma flow control, as well as the interaction between plasmas and airflows, to improve the aerodynamic characteristics and promote the scientific basis for efficiency. Discharge plasmas applied to flow control mainly include surface discharges [4–14] and volume discharges [15–20]. Surface discharges are used to flow separation control by DBD discharges with a momentum exchange to neutral airflows, which generate a complex pattern of quasi-planar and spherical compression wave [4–8], as well as which are related to a strong demand on stable discharges within the flow boundary layer [7–14]. Volume discharges are applied in MHD flow control to achieve the acceleration and deceleration of airflow, which require high-intensity volume discharges under airflows condition, as well as which essentially need discharge enhancement methods under MHD airflow environment [15–19].

Plasma material processing is promising for material modification and its industrial applications. Efficient discharges are important demands to realize the surface modification and functional structure construction [21–31]. The moderate power density and the uniform energy distribution are beneficial to material modification. However, under a gas flowing condition, discharges can easily transit from a stable state into an unstable state, which can cause a disaster to the industrial application of plasma material processing. In order to obtain a uniform and stable discharge, discharges under airflows are employed to excite plasmas. Work groups [29] illustrate the surface modification of polyimide films by the discharge under airflow, it is found that the plasma at a homogeneous DBD is evenly distributed than at a filamentary DBD, and by the more efficient introduction of atomic oxygen to the PP surface in the case of homogeneous DBD.

Plasma air purification attracts widespread attention in recent years, mainly related with corona discharges under an air supply channel. Plasma air purification has been developed in many applications, including electrostatic precipitation [32–40], industrial gas exhaust treatment [41–44], and indoor air purification [45–49]. Work groups [35] illustrate that the electric power and the energy loss of corona discharges highly depend on airflow velocities, and corona discharge modes are also related to airflow conditions.

In addition, discharges under airflows are of complicated technical challenges, and the mechanism and its characteristics need a deep and wide investigation.

3. Atmospheric pressure volume discharges under airflows

In order to get a better understanding of airflow effects on volume discharge characteristics, this chapter presents the study of nanosecond pulse volume discharges in high-speed airflows.

3.1. Discharge mode characteristics under airflows

The experimental system is shown in **Figure 1**, which includes an air wind tunnel driven by a fan, a nanosecond pulse generator, discharge system, and measurement system. By changing the speed of the fan, the flow velocity at the end of the wind tunnel can be adjusted with a maximum value of up to 200 m/s. A pitot tube is used to measure the flow velocity of the airflow. The plate-plate electrodes are set in a horizontal and parallel manner. The two electrodes are composed of stainless steel plates with a thickness of 2 mm. The electrode edges were fully polished in order to avoid the point discharge occurring at the electrode edges. The two dielectrics are made of mica with a permittivity $\varepsilon r = 6$ and a thickness of 1 mm. The discharge system is installed at the downstream of subsonic wind tunnel exit with the flow direction perpendicular to the electrode surface. The applied voltage has repetitive pulses with a fixed pulse width of 5 ns and a maximum amplitude of 50 kV with a rise time of 5 ns, corresponding to the frequency ranged from 100 Hz to 3.5 kHz, respectively. The voltage and

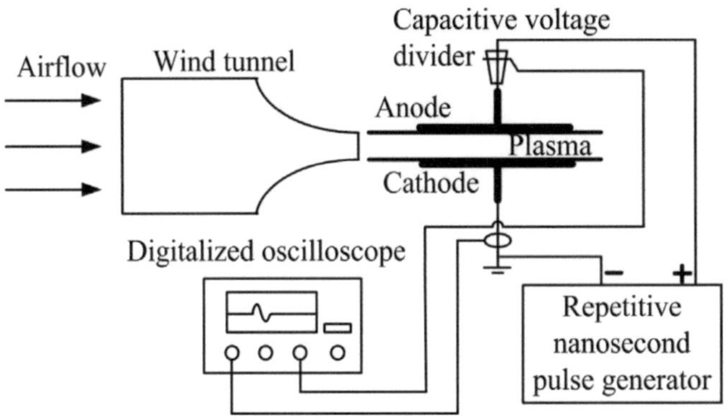

Figure 1. Schematic of experimental setup.

Figure 2. Discharges at different airflow speeds. (a) Flow speed $v = 0$ m/s, (b) $v = 10$ m/s, (c) $v = 35$ m/s, (d) $v = 50$ m/s, and (e) $v = 100$ m/s. Exposure time is 1/1250 s.

current are measured by using a high-voltage probe and a Rogowski coil with a response time of less than 1 ns. The voltage and current signals are recorded by a digitalized oscilloscope with a bandwidth of 1 GHz.

The typical luminous discharge images under airflows are shown in **Figure 2**. In the quiescent air (i.e., the flow velocity is 0 m/s), a multichannel and inhomogeneous violet discharge is present in the discharge volume. The discharge filaments are straight, and the filament foots are randomly and extensively distributed on the dielectric surface, as shown in **Figure 2(a)**. Increasing the flow velocity to 10 m/s, the number of the bright filaments is slightly reduced, but the change of glow component cannot be clearly observed, as shown in **Figure 2(b)**. When the flow velocity varies from 10 to 20 m/s, the filament number is gradually reduced, and the change of discharge luminance and distribution are relatively small. When the airflow with a speed of 35 m/s is introduced into the volume discharge, as shown in **Figure 2(c)**, interestingly, a diffuse discharge in a large volume is promoted. The unsteady nature of the filamentary part of the discharge cannot be easily observed with the naked eye in the discharge volume. The diffuse and homogeneous discharge mostly occurs at the middle region, and the filament discharge occurs mainly at the inlet region and partly at exit region of the channel. The filament channels on the two sides may be connected to the electric field concentration occurred at electrode edges. With the increasing of airflow speeds, the volume discharge modes vary from filament to diffuse modes. Further improving the airflow speed to 50 m/s, as shown in **Figure 2(d)**, a diffuse discharge also occurs at the middle discharge region, as well as with a reduction of luminous intensity.

Moreover, several excitation conditions are selected for the detailed investigation of airflow effects, in which the applied voltage amplitude is chosen from 10 to 30 kV; PRF is selected from 100 to 3800 Hz, and airflow speed is changed from 0 to 200 m/s, respectively. In a quiescent air or under a low speed, it presents filament and inhomogeneous violet discharges in a large volume, as likely shown in **Figure 2(a)**. At such airflow speeds, the number of the bright filaments is slightly increased with only increasing the applied voltage, as a process of the pinch of several filaments. However, the change of discharges cannot be clearly observed with only changing PRFs. With a speed less than 30 m/s, there are less filaments in the volume, and the change of discharge luminance is relatively small with changing the applied voltages and PRFs. With the airflow speed increasing higher, for example, a speed of 50 m/s as shown in **Figure 2(d)**, the relative uniform discharge in a large volume is promoted. Importantly, the discharge becomes unstable and almost fades away at a PRF less than 220 Hz.

3.2. Discharge density characteristics under airflows

As a nanosecond pulse is applied to a plate-plate gap, initial electrons are accelerated by the electric field, and an avalanche process is followed under the electron multiplication of collision ionization. Fast electrons with high energy can run away from the head of the critical avalanche and dominate in the subsequent development of the critical avalanche. When the head of the critical avalanche reaches near to the anode, a discharge bridge is built up between the two dielectrics as well as electrodes, and a discharge current runs through the gap space, which can be represented as the first pulse current. When the applied pulse is gone, more electrons accumulate in the dielectric surface near the anode, interact with interaction of accumulated positive particles near the cathode, and build up a strong electric field imposed on the discharge space. Such space electric field will induce another avalanche process between the two electrodes, and the second pulse currents arise in our experiments.

With an applied pulse voltage of 18 kV and a pulse repetitive frequency of 1800 Hz, the applied voltage and current waveforms under different airflows are shown in **Figure 3**. It is provided that the plate-plate DBD discharge is characterized by a series of two-stage pulse currents. In contrast with the unipolar pulse of applied voltage, the discharge current behaves bipolar and consists of both positive and negative pulses. The discharge current distributes irregularly, which can be attributed to the random nature of breakdown and complicated dynamics in the air gap.

There is a series of two-stage pulse currents for each nanosecond pulsed discharge. Even considering the existence of discharge delays, the first pulse currents always occur with a same breakdown voltage. Furthermore, it can be drawn from **Figure 4** that the first pulse currents increase first and then reduce with airflow speeds. With the flow speed increasing

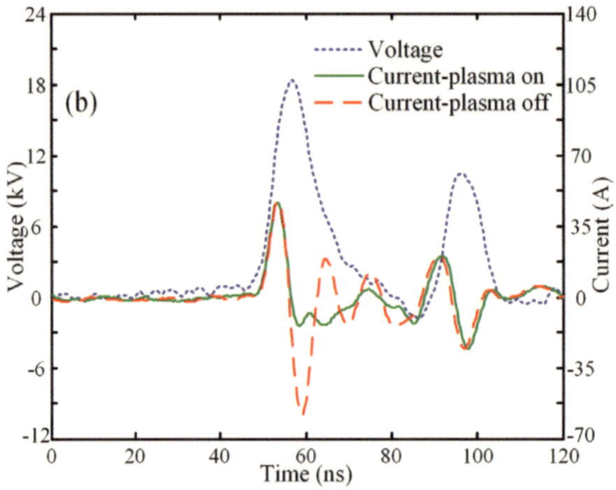

Figure 3. Applied voltage and current waveforms for discharges under airflows.

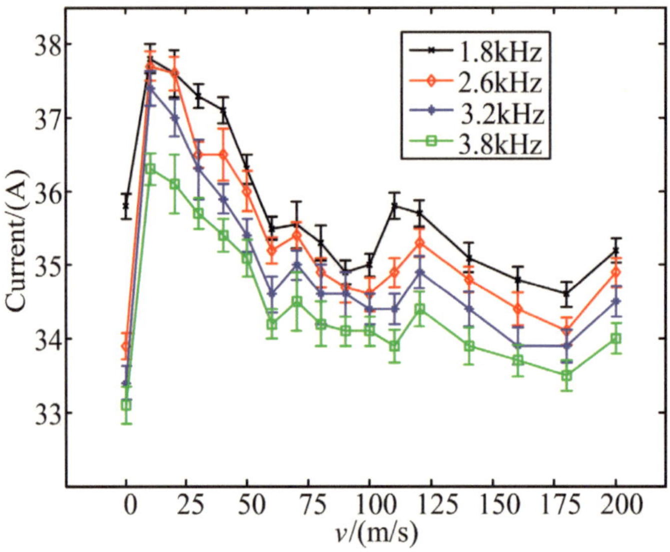

Figure 4. First pulse currents under airflows at different PRFs.

from 0 to 20 m/s, the amplitude of the current rapidly increases from 33.1 to 36.3 A at a PRF of 3800 Hz and then slowly decreases to about 34 A. It goes across a reverse "V"s curve with a turning point at the flow velocity of 20 m/s. The change of first pulse current values imposed by airflows is 4 A. Importantly, there are some factors leading to measurement errors in this works, which mainly include the pressure instability of air supplied to the wind channel, the measurement errors of airflow speed with the pilot tube, and the high frequency noises concerning with nanosecond discharges.

It is clearly shown in **Figure 5** that the second pulse currents are influenced by the addition of airflows to discharge zone. The second pulse currents almost occur at a same time point. At a PRF of 1800 Hz, the second pulse current has a rapid but small change from −34 to −32 A

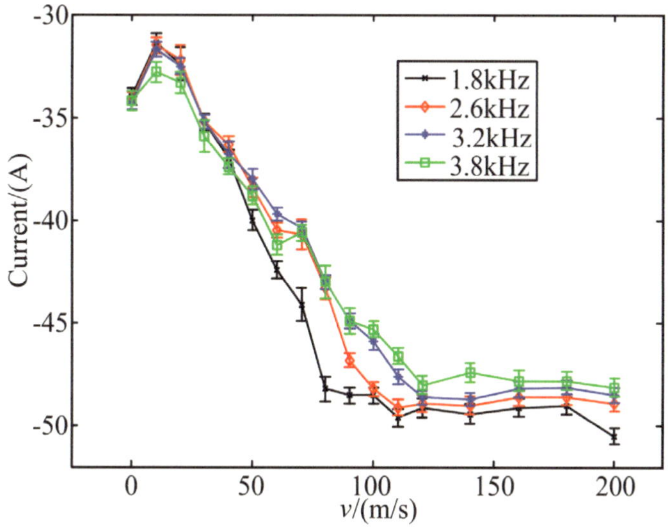

Figure 5. Second pulse currents under airflows at different PRFs.

with the airflow speed increasing from 0 to 20 m/s, follows with a rapid and big change from −32 to −47 A, and then becomes gradually relative stable as the airflow speed is bigger than 80 m/s. It is mostly concerned that the difference between second pulse current values is about 20A for different airflows.

With the airflow added into the gap space, the volume discharge modes vary from filament to diffuse modes, and the induced two-stage pulse currents are also influenced. These behaviors may be attributed to the combined action of two effects. On one hand, the airflows breathe more species into the discharge volume from upstream to downstream, especially the excited metastable state, which causes a decrease in the number of initial electrons. On the other hand, the airflows remove accumulated charges on the surface of the dielectric, which is favorable for the development of discharge.

As the first pulse current is arising, a bigger voltage pulse with a 5 ns rise time, as well as a stronger electric field, is being imposed on the electrodes as one discharge driver. Since the pulse action time is too small, airflows in this period can be considered as "frozen" and without any flow mobility. Therefore, the first pulse currents nearly maintain consistency for different airflow speeds. However, at the second pulse current arising, accumulated particles near the electrodes induce a space electric field as another discharge driver. Since this driver always acts for several milliseconds, airflow mobility must be considered in this period. As indicated in **Figure 5**, the second pulse currents for quiescent air are bigger than that of airflows with speeds less than 40 m/s. Since airflows breathe more species, especially under excited metastable states, out of the discharge volume, the loss of heavy particles under the force of airflows is unfavorable for the development of discharge. Under airflows with a speed of more than 40 m/s, the second pulse currents are always bigger than that of quiescent air. It is concerned with the distribution change of the accumulated charges on the surface of the dielectrics by the force of airflows, which is favorable for the development of discharge. With the speed increasing to more than 80 m/s, the second pulse currents become gradually relatively stable, which seems unaffected by airflows. The detailed effects under such high airflow speeds are very interesting and are also taken into considerations in the future works.

3.3. Conditional boundary characteristics under airflows

The boundary conditions of DBD volume discharges under airflows velocities are investigated. With the addition of airflows, no matter for any voltage, both the initiation and the extinction pulse repetition frequency need to be kept to a specified value to maintain the presence of discharge. Under a particular applied voltage and when the pulse repetitive frequency is rather low, any discharge does not occur; nevertheless, a current is detected due to the existence of the displacement electric field. With the growth of pulse repetitive frequency, the discharge can be established with an obvious change of discharge current.

The initiation and extinction boundaries are identified by making a subtraction of detected currents between discharge on and off, as shown in **Figures 6** and **7**. To improve the accuracy, the airflow and the applied voltage are kept constant as soon as possible while with an adjustable PRF, and all data are acquired with 10 times measuring. As the airflow velocity accelerates, the initiation and extinction value of PRFs is increased. In a still air, when the applied voltage reaches to 17 kV, the corresponding PRFs needed to reach about 1 kHz, which is almost an order of magnitude larger than the value under 27 kV. The lower is the applied voltage, the larger are the needed PRFs as well as the difference of PRFs. The initiation and extinction PRFs are almost inversely proportional to the applied voltage value under a certain airflow velocity. Once the airflow velocity attains 100 m/s under 17 kV, the discharge is not detected even though the PRF is up to the maximum value of applied value. As a special case of contrast, the initiation and extinction PRFs show a slight dependence on the velocities

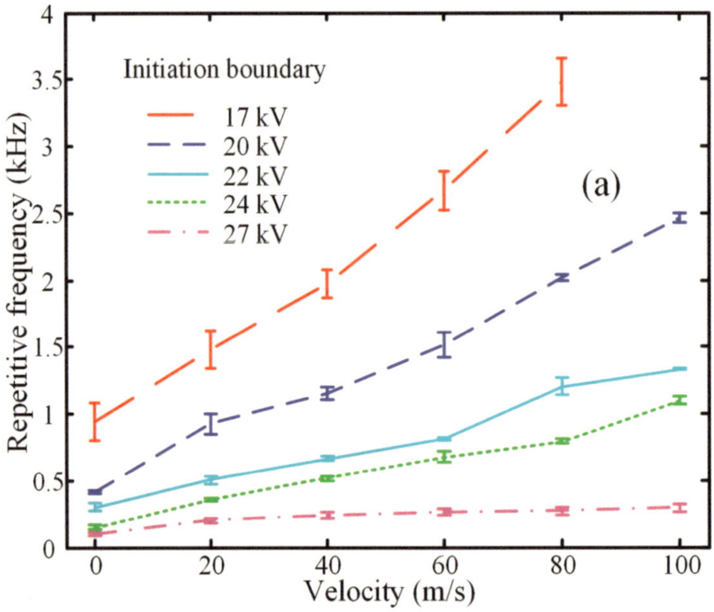

Figure 6. Initiation frequencies versus airflow speeds.

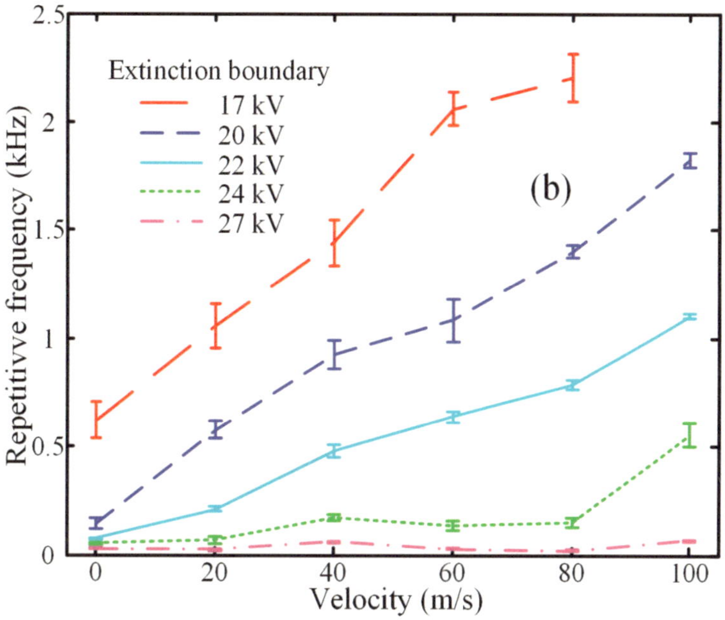

Figure 7. Extinction frequencies versus airflow speeds.

under an applied voltage of 27 kV. It can be concluded that the boundary PRFs are decided by the combined action between the applied voltages and the injected airflows.

In still air, mostly free electrons move rapidly toward the anode under the applied electric field. Many ions and metastable particles are produced by the complicated collisions between various particles in the discharge zone. Considering the interval time between two voltage pulses, only metastable particles with a lifetime of the order of several seconds can survive as residual seeds, while most ions have quenched. So, the increase of PRF can raise the concentration of residual seeds, which make the discharge breakdown more easily to occur. With the addition of the airflows, a large number of seeds are blown away from the discharge region, which makes a difficulty on the next pulsed discharge as well as leads to a growth of initiation PRFs. However, with the applied voltage growing, the free electrons obtain more energy from the electric field, and seed particles deposit on the dielectric plate. Owing to the existence of the viscous boundary layer, the seed particles are hardly moved away by the airflows. As a result under 27 kV, only a slight dependence of the initiation and extinction PRFs is observed with the airflow velocities.

4. Atmospheric pressure volume discharges under upstream and downstream airflows

The previous chapters demonstrate that the volume discharge mode is influenced by airflows and the discharge intensity decreases with the increase of airflow velocities. In this chapter, a type of discharge device with an upstream and downstream structure is provided to investigate the interactions between airflows and discharges. The upstream and downstream discharges under airflow include the generation and transport of charged particles.

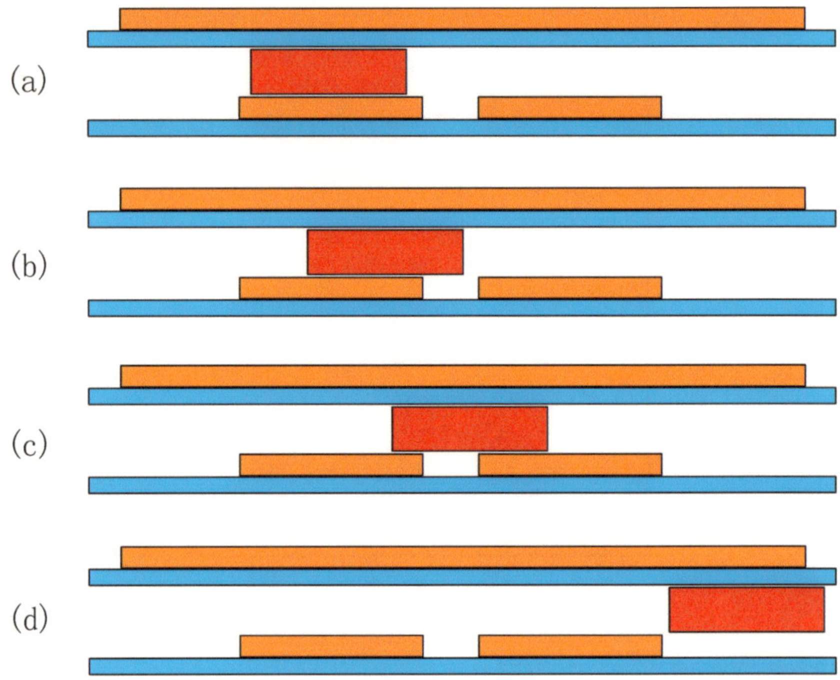

Figure 8. Schematic of discharges under upstream and downstream constructure with different airflow velocities.

The generation of charged particles is related to external pulse excitation, and the transport of particles can be influenced by the effect of airflow. Furthermore, the upstream and downstream discharge in the airflow channel is a multi-time scale problem, including the transport time of airflow and the time of charged particle generation. The characteristic time of airflow is related to the movement path and the airflow velocity, and the generation of particles is adjusted by the repetition frequency of the discharge.

By matching the transport effect of airflow and the pre-ionization of charged particles, some of the particles generated by the upstream discharge are transported to the downstream region, and those particles play a pre-ionization role in the downstream discharge, which causes the enhancement of the downstream discharge intensity. For the purpose of further recognizing the relationship between upstream and downstream discharge under airflow, the behavior of the particles were analyzed as schematically shown in **Figure 8**. In the quiescent air, the discharge products always stay only in the upstream zone, as shown in **Figure 8(a)**; as the airflow is injected, the discharge products produced by upstream zone will be transported to the downstream zone, as shown in **Figure 8(b)** and **8(c)**; as the further increasing of airflow velocity, the discharge product will be blow out the downstream zone, as shown in **Figure 8d**.

The time range of air flow transport time t_f is derived as Eq. (1). By properly controlling the pulse repetition interval time t_p and the air flow transport time t_f, the charged particles could be transported from upstream region to the downstream region and enhance the downstream discharge intensity:

$$L/v <= t_f <= (L + 2^* S)/v \tag{1}$$

where the pulse interval time is denoted as t_p, the distance between the upstream and downstream regions is L, the width of the electrode plate is S, and the airflow velocity is v (m/s).

4.1. Discharge mode characteristics under upstream and downstream airflows

With a pulse repetitive frequency of 1000 Hz, the applied voltage and current waveforms of the DBD volume discharge can be detected, which has been descripted in Section 3.1. By changing the air velocity, the discharge mode characteristics under different airflow velocities are shown in **Figure 9**.

For discharge in a static condition, as shown in **Figure 9(a)**, the upstream and downstream discharge exhibits a filament discharge mode. In the middle region of the electrode, the filaments exhibit a vertical distribution, and the path of the filaments close to the edge of the electrode shows a curved state because of the edge effect of the metal electrode.

For $t_f > t_p$, the discharge filaments move along the direction of flow, the filament path changes from the original vertical state to the curved state along the direction of airflow, and some parts of the discharge area become uniform. The result is shown in **Figure 9(b)**.

For $t_f \leq t_p$, with increasing airflow velocity, most of the discharge area exhibits a uniform state, and the paths of the discharge filaments at the edge of the metal electrodes are attracted by the central discharge region. The discharge filaments shrink to the inside region of discharge, and the discharge region exhibits a pinched state, as shown in **Figure 9(c)** and **(d)**.

4.2. Discharge spectrum characteristics under upstream and downstream airflows

The normalized emission spectra for the upstream and downstream discharges are provided as **Figure 10**. Due to the weak luminescence intensity of volume discharges, the spectrometer exposure time was set to 2 s, which means that the emission intensity was averaged temporally and spatially for 2000 cycles.

With increasing airflow velocity, the intensity of the upstream discharge emission spectrum decreases, which corresponds to a decrease in discharge intensity. As for downstream

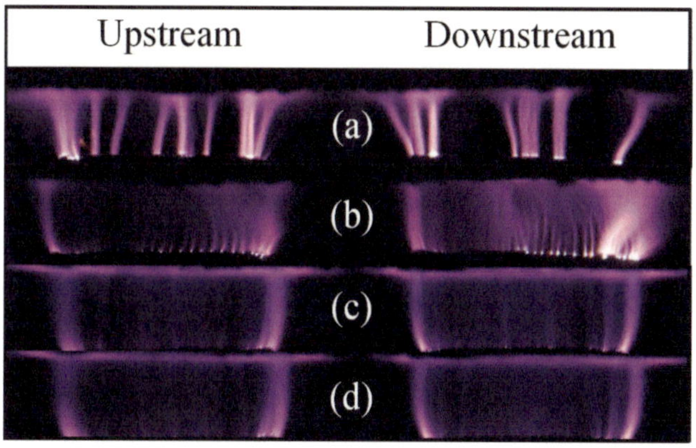

Figure 9. Discharge at different airflow velocities: (a) static air, (b) low speed ($t_f > t_p$), (c) medium speed ($t_f = t_p$), and (d) high speed ($t_f < t_p$).

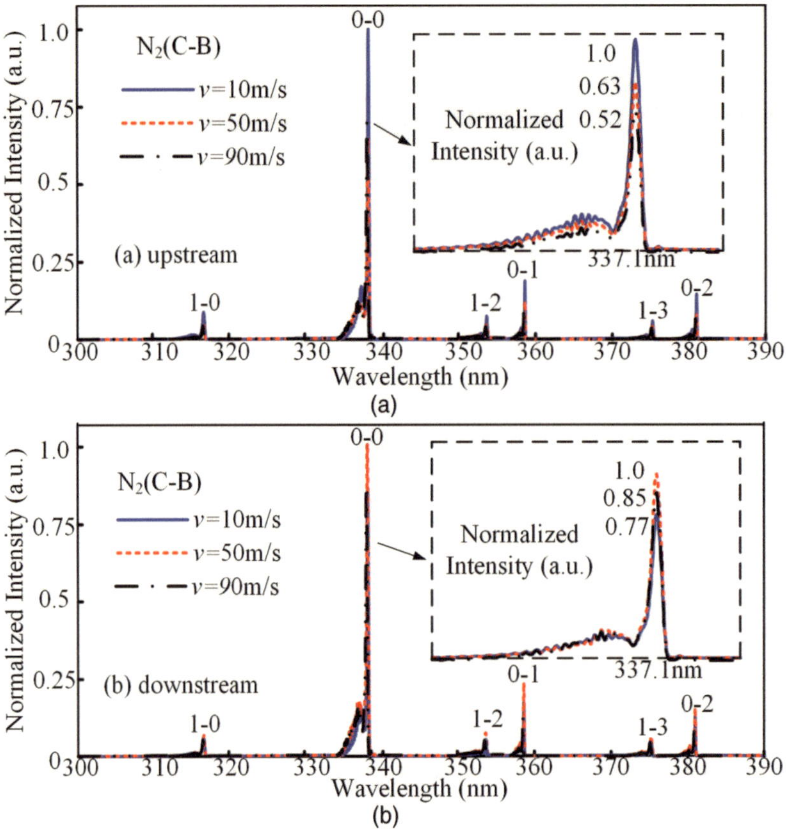

Figure 10. Normalized intensity discharge at different velocities: (a) upstream and (b) downstream.

discharge, when $t_f = t_p$, the intensity of the emission spectrum increases, which proves that the concentration of charged particles in the downstream region increased. At $t_f < t_p$ and $t_f > t_p$, the corresponding spectral intensity gradually decreases.

The dependence of rotational temperature of the upstream and downstream discharges on airflow rates is shown in **Figure 11**. It can be seen that the rotational temperature is approximately 390 K when the discharge is excited in the static air. The rotational temperature decreases to approximately 320 K as the airflow velocity increases gradually to 80 m/s. The low gas temperature may be attributed to two reasons. One is that the duty cycle of the pulse power supply is low; another is that more energy is delivered to the energetic electrons. The gas temperature is substantially the same in the upstream and downstream regions. This result shows that the gas temperature under airflow is not the key factor that causes the difference between the upstream and downstream discharge intensity.

Under the condition of reasonable matching between t_f and t_p, the mass transfer effect of airflow plays a dominant role in the upstream and downstream discharges. The particles generated by the upstream discharge can be transported to the downstream discharge region. The combined effect of flow transport and pre-ionization of charged particles enhances the downstream discharge.

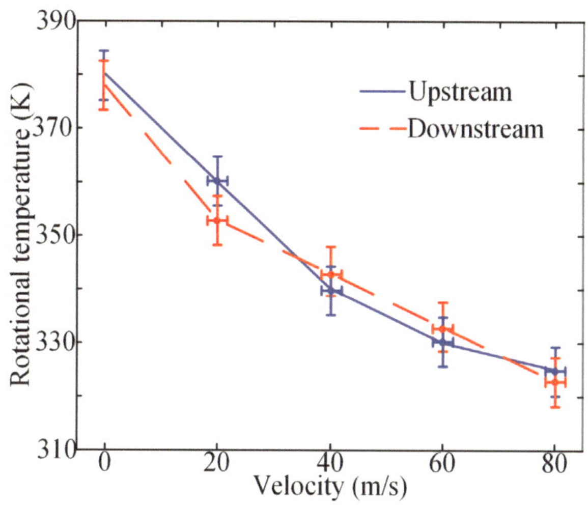

Figure 11. Rotational temperature at different velocities.

4.3. Discharge difference between upstream and downstream region

4.3.1. Various airflow velocities

The comparison of discharge currents between upstream and downstream discharges is shown in **Figure 12**. The current peak is used to illustrate the discharge intensity under different flow velocities.

For $t_f > t_p$, the amplitude of the upstream discharge decreases from 55 to 53 A, and the downstream discharge fluctuates between 38 and 36 A with increased airflow speed, and for the condition of $t_f = t_p$, the amplitude of the upstream discharge gradually decreases from 53 to 45 A. For the downstream discharge, the amplitude exhibits an opposite trend, changing from 36 to 46 A. With increasing airflow velocity, the amplitude of the upstream discharge current continues to decrease, and the increasing rate of the downstream discharge current tends to be zero.

4.3.2. Various PRFs

The pulse frequencies were adjusted to $f = 2$ kHz and $f = 200$ Hz, and the results are shown in **Figure 13**. When the pulse frequency is 2 kHz, for $t_f > t_p$, the current amplitudes of the upstream discharge gradually decrease from 59 to 57 A, and the downstream discharge fluctuates between 45 and 48 A. When the airflow velocity increases until $t_f = t_p$, the amplitude of the upstream discharge gradually decreases from 57 to 49 A. The amplitude of the downstream discharge exhibits an opposite trend and increases from 48 to 56 A.

When the discharge frequency was adjusted to 200 Hz, the pulse interval time t_p was 5 ms. Because of the low pulse repetition frequency, the charged particles in the space completed the diffusion and recombination process within the pulse interval time, causing the reduction in discharge intensity. In the meantime, at a lower airflow velocity ($t_f > t_p$), the amplitude of

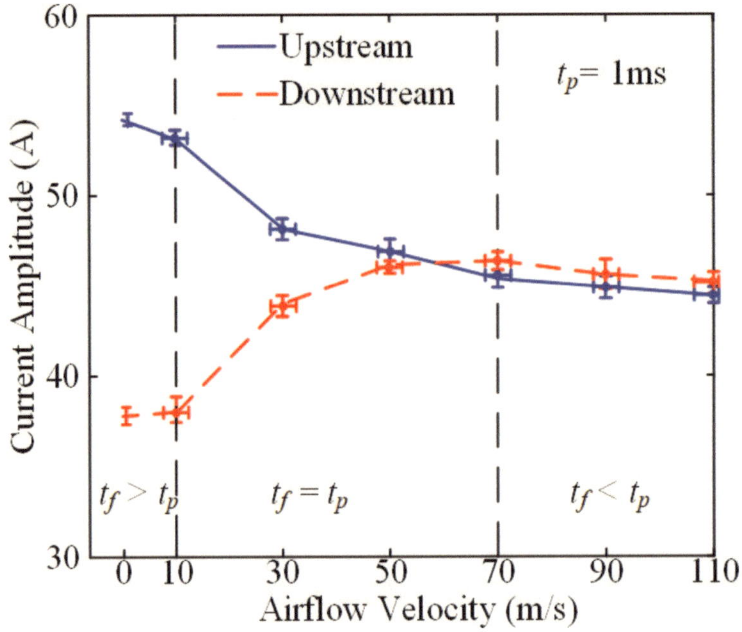

Figure 12. Discharge current amplitude of the upstream and downstream regions.

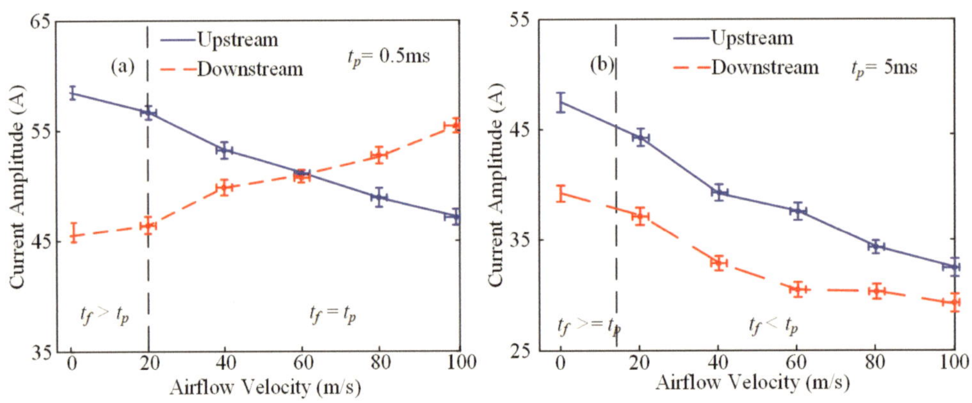

Figure 13. Discharge current amplitude at different PRFs: (a) PRF = 2 kHz and (b) PRF = 200 Hz.

the discharge current in the upstream region attenuated from 44 to 35 A, and the downstream discharge current weakened from 38 to 30 A.

4.3.3. Various distances between upstream and downstream

The conditions of $L = 20$ mm and $L = 50$ mm were studied. The experimental results for $f = 1$ kHz are shown in **Figure 14**. When the discharge electrode spacing was adjusted to $L = 20$ mm, for $t_f > t_p$, the charged particles generated by the upstream discharge could not transit to the downstream discharge region. The peak value of the upstream discharge current decayed from 54 to 50 A, and the amplitude of the downstream discharge current fluctuated between 37 and 39 A. During the

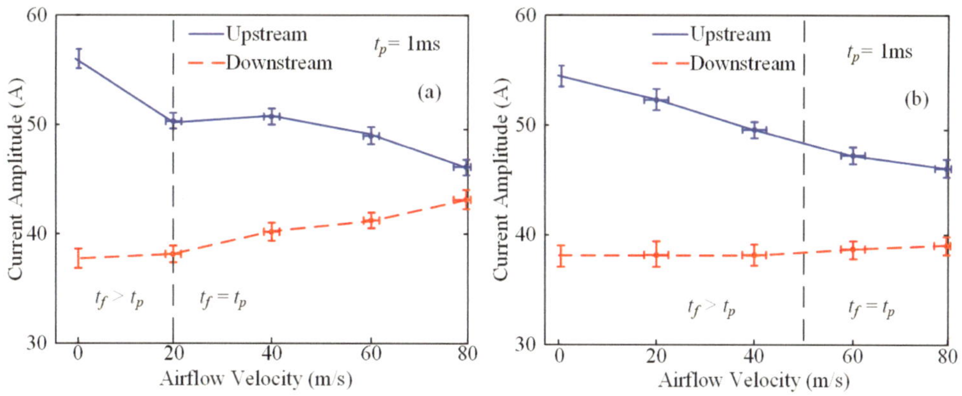

Figure 14. Discharge current amplitude at different distances: (a) $L = 20$ mm and (b) $L = 50$ mm.

process of increasing airflow velocity until $t_f = t_p$, particles generated in the upstream discharge region were blown to the downstream discharge region. The peak value of the upstream discharge current decayed gradually from 50 to 45 A, and the corresponding downstream discharge current amplitude gradually increased to 43 A. When the electrode spacing was adjusted to 50 mm, with the gas velocity reaching 50 m/s (i.e., $t_{f1} = t_p$), the particles generated in the upstream discharge region could be transported to the downstream discharge region. The upstream and downstream discharge currents are almost enhanced with the increased gas flow velocity.

4.4. Interaction between upstream and downstream discharges

According to the process of pulse discharge, the electrons move to the anode under the effect of pulse voltage, and the ions remain nearly static in the time scale of nanoseconds; the discharge current is formed by electron migration. The electrons cover the surface of the dielectric, and a large amount of ions are accumulated in the discharge space. At the falling edge of the voltage pulse, when the applied voltage potential is lower than the potential generated by the positive charge space, some electrons move toward the cathode and form a reverse current. During the movement, the electrons are neutralized with some positive ions, and the concentration of positive charges declines in the space.

Because of the different time scales of pulse discharge and airflow, the airflow cannot directly affect the process of pulse discharge. The rising time of the pulse discharge takes place in a few nanoseconds, and the duration of current is in the tens nanoseconds. For particles, the diffusion and recombination of neutral and charged particles take place on a micro- and millisecond time scale, and the lifetime of the metastable particles produced by the discharge in short gaps is thought to have a significant high concentration of some hundreds of microseconds to milliseconds. Therefore, the airflow can affect only the discharge distribution of particles during the pulse interval times.

When the airflow is applied in the discharge space, the distribution and movement of particles in the discharge region could be influenced by the airflow; a schematic of the particle motion paths is shown in **Figure 15**, and the analysis is as follows.

Metastable particles are transported by the drag force of the airflow and move from the upstream discharge area to the downstream discharge area, which particles play the pre-ionization role to

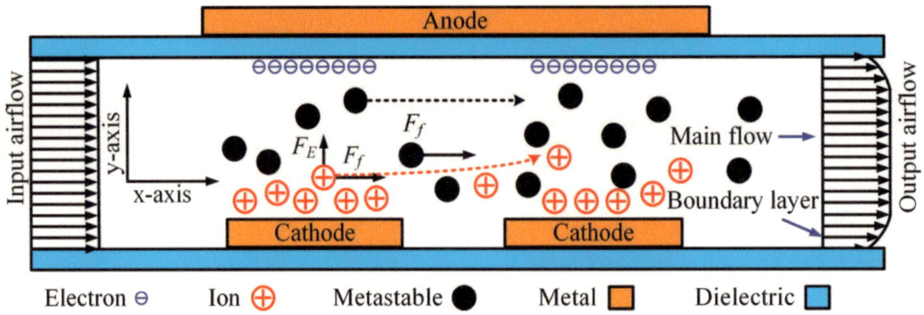

Figure 15. Schematic of particle motion paths.

the downstream discharge. For the charged ions in space, the ions move along the airflow direction mainly under the effect of the drag force of airflow and the electrostatic force from the surface electrons [15, 16]. According to the literature, during a pulse interval time of approximately 1 ms, the electric field strength between the surface electrons and the space charges is several kV/cm, and the ion density n_i is approximately 10^{12} cm^{-3}. The estimated electric field force of the ions is approximately 1.6×10^{-2} N. In terms of the bulk flow, the conservation of momentum of the bulk fluid is described by the Navier-Stokes equation. The air density is approximately 1.29 kg/m^3, the volume of the discharge area is approximately 5 mm × 30 mm × 30 mm, and the pressure difference between inlet and outlet is nearly 0.5% of the atmospheric pressure. The estimated flow field force to the ions is approximately 1.2×10^{-1} N, and the ratio of the flow force and the electrical field force is approximately 8:1. For the ions in the gap, the distance of ions moved along the y-axis is approximately 2 mm during a pulse interval of 1 ms, which is less than the gap distance. The estimation illustrates that the ions are blown to the downstream area during the pulse interval time under the coupling force of the airflow and electric field.

Furthermore, the space charges have different velocities at different regions in the airflow channel. In the region near the dielectric, the space charges have a lower velocity because of the drag force in the boundary layer and the larger electrostatic force from the surface charges. However, the space charges in the middle region of the gap have the fastest velocities and the smaller electrostatic force, and the drag force from the airflow plays a dominant role compared with the electrostatic force of the surface electrons. The curved discharge channel under the low-speed condition also illustrates that the effect of the drag force from the mainstream area is stronger than the electrostatic force between the surface electrons and the space charges; this phenomenon is also discussed in the literature [15, 22]. Therefore, both the estimation results and the discharge phenomenon illustrate that some particles, such as metastable particles and ions, could transit to the downstream discharge area during the pulse interval time; these particles act as seed electrons to the downstream discharge, and the downstream discharge intensity is enhanced.

Acknowledgements

Parts of this chapter source from authors' recent work [50, 51]. The authors thank to PhD student Zhou Desheng and Tang Miao to provide the experimental materials for the book. The authors would like to acknowledge the support of the National Natural Science Foundation of China (no. 51437002).

Author details

Jingfeng Tang, Liqiu Wei* and Daren Yu

*Address all correspondence to: weiliqiu@hit.edu.cn

Harbin Institute of Technology, Harbin, People's Republic of China

References

[1] Starikovskiy A, Aleksandrov N. Nonequilibrium Plasma Aerodynamics. Rijeka: In Tech; 2011. pp. 55-96

[2] Zheng JG, Zhao ZJ, Li J, et al. Numerical simulation of nanosecond pulsed dielectric barrier discharge actuator in a quiescent flow. Physics of Fluids. 2014;**26**(3):036102

[3] Kriegseis J, Grundmann S, Tropea C. Airflow influence on the discharge performance of dielectric barrier discharge plasma actuators. Physics of Plasmas. 2012;**19**(7):073509

[4] Adamovich IV, Little J, Nishihara M, et al. Nanosecond pulse surface discharges for high sped flow control, AIAA 2012 3137. Reston: AIAA; 2012

[5] Bletzinger P, Ganguly BN, van Wie D, et al. Plasmas in high speed aerodynamics. Journal of Physics D: Applied Physics. 2005;**38**(4):33-57

[6] Moreau E. Airflow control by nonthermal plasma actuators. Journal of Physics. D, Applied Physics. 2007;**40**(3):605-636

[7] Boeuf JP, Pitchford LC. Electrohydrodynamic force and aerodynamic flow acceleration in surface dielectric barrier discharge. Journal of Applied Physics. 2005;**97**(10):103307

[8] Font GI, Morgan WL. Recent progress in dielectric barrier discharges for aerodynamic flow control. Contributions to Plasma Physics. 2007;**47**(12):103-110

[9] Corke TC, Enloe CL, Wilkinson SP. Dielectric barrier discharge plasma actuators for flow control. Annual Review of Fluid Mechanics. 2010;**42**:505-529

[10] Li YH, Wu Y, Song HM, et al. Plasma Flow Control. Rijeka: InTech; 2011. pp. 21-54

[11] Nie WS, Cheng YF, Che XK. A review on dielectric barrier discharge plasma flow control. Advances in Applied Mechanics. 2012;**42**(6):722-734

[12] Riherd M, Roy S. Serpentine geometry plasma actuators for flow control. Journal of Applied Physics. 2013;**114**(8):083303

[13] Wu Y, LiY H, Jia M, et al. Influence of operating pressure on surface dielectric barrier discharge plasma aerodynamic actuation characteristics. Applied Physics Letters. 2008;**93**(3):031503

[14] Che XK, Shao T, Nie WS, et al. Numerical simulation on a nanosecond pulse surface dielectric barrier discharge actuator in near space. Journal of Physics. D, Applied Physics. 2012;**45**(14):145201

[15] Murakamia T, Okuno Y. Experiments and numerical simulations on high-density magnetohydrodynamic electrical power generation. Journal of Applied Physics. 2008; **104**:063307

[16] Lee CH, Lu HY. Quasi-one-dimensional parametric study for MHD generator in MHD bypass scramjet system. In: AIAA Paper 2007-644

[17] Bityurin VA. A feasibility study and experimental evaluation on MHD acceleration for application to advanced propulsion and hypervelocity ground testing. In: AIAA Paper 2000, 2301

[18] Su CB, Li YH, Cheng BQ. Experimental investigation of MHD flow control for the oblique shockwave around the ramp in low-temperature supersonic flow. Chinese Journal of Aeronautics. 2010;**22**:22-32

[19] Fujino T, Sugita H, Mizuno M. Influences of electrical conductivity of wall on magnetohydrodynamic control of aerodynamic heating. Journal of Spacecraft and Rockets. 2006;**43**:63-70

[20] Benard N, Moreau E. E H D force and electric wind produced by surface dielectric barrier discharge plasma actuators used for airflow control. In: AIAA 2012 3136. Reston: AIAA; 2012

[21] Hiraoka H, Lazare S. Surface modifications of kapton and cured polyimide films by Arf excimer laser—Applications to imagewise wetting and metallization. Applied Surface Science. 1990;**46**(1-4):264-271

[22] Sung K, Hwang SM, Lee CM, et al. Effects of chemical etching and functionalization times on the properties of Cu/polyimide films. Journal of the Korean Physical Society. 2010;**57**(6):1707-1712

[23] Noh BI, Yoon JW, Jung SB. Effect of laminating parameters on the adhesion property of flexible copper clad laminate with adhesive layer. International Journal of Adhesion and Adhesives. 2010;**30**(1):30-35

[24] Shin JW, Jeun JP, Kang PH. Surface modification and characterization of N+ ion implantation on polyimide film. Macromolecular Research. 2010;**18**(3):227-232

[25] Miron C, Sava I, Jepu I, et al. Surface modification of the polyimide films by electrical discharges in water. Plasma Processes and Polymers. 2013;**10**(9):798-807

[26] Chung TS, Shao L, Tin PS. Surface modification of polyimide membranes by diamines for H-2 and CO_2 separation. Macromolecular Rapid Communications. 2006;**27**(13):998-1003

[27] Park SC, Youn SW, Takagi H, et al. A study on surface modification of soluble block copolymer polyimide by UV irradiation and its application to electroless plating. Journal of Photopolymer Science and Technology. 2013;**26**(3):297-302

[28] Shao T, Zhang C, Long KH, et al. Surface modification of polyimide films using unipolar nanosecond-pulse DBD in atmospheric air. Applied Surface Science. 2010; **256**(12):3888-3894

[29] Fang Z, Xie Z, Li J, et al. Comparison of surface modification of polypropylene film by filamentary DBD at atmospheric pressure and homogeneous DBD at medium pressure in air. Journal of Physics D: Applied Physics. 2009;**42**:085204. 9 pp

[30] Park SC, Yoon SS, Nam JD. Surface characteristics and adhesive strengths of metal on O-2 ion beam treated polyimide substrate. Thin Solid Films. 2008;**516**(10):3028-3035

[31] Bulycheva NA, Kazaryana MA, Chaikov LL, et al. Nanoscale metal oxide particles produced in the plasma discharge in the liquid phase upon exposure to ultrasonic cavitation. 1. Method for producing particles. Bulletin of the Lebedev Physics Institute. 2014;**41**(9):264

[32] Hao Y, Wenchun W, Dezheng Y, et al. Atmospheric air dielectric barrier discharge excited by nanosecond pulse and AC used for improving the hydrophilicity of aramid fibers. Plasma Science and Technology. 2017;**19**:125401. 9 pp

[33] Zouaghi A, Mekhaldi A. Analysis of nanosecond pulsed and square AC dielectric barrier discharges in planar configuration: Application to electrostatic precipitation. IEEE Transactions on Dielectrics and Electrical Insulation. 2017;**24**(4)

[34] Athanasios CM, Igor VT, et al. Superposition of DC voltage and submicrosecond impulses for energization of electrostatic precipitators. IEEE Transactions on Plasma Science. 2012;**40**(10)

[35] Sato S, Kimura M, et al. A removal system of diesel particulate using electrostatic precipitator with discharge plasma. In: IAS. 2005. pp. 2203-2206

[36] Lee YH, Jung WS, Choi YR, et al. Application of pulsed corona induced plasma chemical process to an industrial incinerator. Environmental Science & Technology. 2003; **37**(11):2563

[37] Chang JS, Urashima K, Tong YX, et al. Simultaneous removal of NO_x and SO_2 from coal boiler flue gases by DC corona discharge ammonia radical shower systems: Pilot plant tests. Journal of Electrostatics. 2003;**57**(3-4):313

[38] Chang MB, Kushner MJ, Rood MJ. Removal of SO_2 and the simultaneous removal of SO_2 and NO from simulated flue gas streams using dielectric barrier discharge plasmas. Plasma Chemistry and Plasma Processing. 1992;**12**(4):565

[39] Obradović BM, Sretenović GB, Kuraica MM. A dual-use of DBD plasma for simultaneous NO_x and SO_2 removal from coal-combustion flue gas. Journal of Hazardous Materials. 2011;**185**(2-3):1280

[40] Masuda S, Hosokawa S, Tu X, et al. Novel plasma chemical technologies-PPCP and SPCP for control of gaseous pollutants and air toxics. Journal of Electrostatics. 1995;**34**(4):415

[41] Fan X, Zhu T, Sun Y, et al. The roles of various plasma species in the plasma and plasma-catalytic removal of low-concentration formaldehyde in air. Journal of Hazardous Materials. 2011;**196**:380

[42] Zhang X, Feng F, Li S, et al. Aerosol formation from styrene removal with an AC/DC streamer corona plasma system in air. Chemical Engineering Journal. 2013;**232**:527

[43] Harling A, Glover DJ, Whitehead JC, et al. Novel method for enhancing the destruction of environmental pollutants by the combination of multiple plasma discharges. Environmental Science & Technology. 2008;**42**(12):4546

[44] Zhu T, Wan Y, Li H, et al. VOCs decomposition via modified ferroelectric packed bed dielectric barrier discharge plasma. IEEE Transactions on Plasma Science. 2011;**39**(8):1695

[45] Dou B, Bin F, Wang C, et al. Discharge characteristics and abatement of volatile organic compounds using plasma reactor packed with ceramic Raschig rings. Journal of Electrostatics. 2013;**71**(5):939

[46] Zheng C, Zhu X, Gao X, et al. Experimental study of acetone removal by packed-bed dielectric barrier discharge reactor. Journal of Industrial and Engineering Chemistry. 2014;**20**(5):2761

[47] Zhang H, Li K, Shu C, et al. Enhancement of styrene removal using a novel double-tube dielectric barrier discharge (DDBD) reactor. Chemical Engineering Journal. 2014;**256**:107

[48] Jiang N, Lu N, Shang K, et al. Innovative approach for benzene degradation using hybrid surface/packed-bed discharge plasmas. Environmental Science & Technology. 2013;**47**(17):9898

[49] Conrads H, Schmidt M. Plasma generation and plasma sources. Plasma Sources Science and Technology. 2000;**9**:441-454

[50] Jingfeng T, Liqiu W, Nan L, et al. Repetitive nanosecond volume diffuse discharge under airflows. IEEE Transactions on Plasma Science. 2014;**42**(3):1-5

[51] Jingfeng T, Liqiu W, Yuxin H, et al. Effect of airflows on repetitive nanosecond volume discharges. IEEE Transactions on Plasma Science, Plasma Science and Technology. 2016;**18**(3):273-277

Chapter 14

Modeling of Novel Plasma-Optical Systems

Iryna Litovko and Alexey Goncharov

Additional information is available at the end of the chapter

http://dx.doi.org/10.5772/intechopen.77512

Abstract

This is the review of the current status an ongoing theoretical, simulations and some experimental researches of the novel plasma dynamical devices based on the axial-symmetric cylindrical electrostatic plasma lens (PL) configuration and the fundamental plasma-optical principles of magnetic electron isolation and equipotentialization magnetic field lines. The crossed electric and magnetic fields plasma lens configuration provides us with the attractive and suitable method for establishing stable plasma discharge at low-pressure. Using plasma lens configuration in this way some novel cost-effective, low-maintenance, high-reliability plasma devices using permanent magnets were developed. In part, it was proposed and created device for ion treatment and deposition of exotic coatings, device for filtering dense plasma flow from micro-droplets, electrostatic plasma lens for focusing and manipulating wide aperture, high-current, low-energy, heavy metal ion plasma flow, and the effective lens was first proposed for manipulating high-current beams of negatively charged particles (negative ions and electrons). Here we mainly describe models and results of theoretical consideration and simulation of the novel generation cylindrical plasma devices.

Keywords: plasma lens, plasma accelerator, space charge, ion beam, electron beam, beam transport

1. Introduction

The electrostatic plasma lens (PL) is a well-investigated tool for focusing and manipulating large area, high-current, moderate energy ion beams, where the concern of beam space charged compensation is critical. The fundamental concept of the PL was first described by Morozov [1]. It is based on application of the plasma-optical principles of magnetic insulation of electrons and equipotentialization of magnetic field lines for the control of electric fields introduced into the plasma medium [2].

These kinds of devices are part of a larger sort of plasma devices (plasma accelerators, ion magnetrons, thrusters, plasma lenses, etc.) that use a discharge in crossed electric and magnetic fields with closed electron drift for the generation, formation, and manipulation of intense ion beams and ion plasma flows. In accordance with the basic idea of plasma optics [1], spatial over thermal E-fields can be introduced in the plasma medium of an intense ion beam, which makes possible high-current ion beams manipulation and focusing including beams of heavy ions. An idea to use the space charge for that purpose appeared to be very fruitful and successful [2, 3]. A number of effective plasma lenses for positive ion beams focusing were made and tested. The robust construction, low-energy consumption, and high-cost-effectiveness make these tools attractive for practical applications.

Some new ideas for using these plasma-optical principles for creation axial symmetric high-current mass separation devices were described firstly in [4]. Note also that this approach is appropriate for the creation of linear or curved magneto-electrostatic plasma guiding ducts for use in vacuum-arc plasma filtering system. Following these plasma-optical principles, changing the magnetic field line configuration and the distribution of electric potential enables the formation and control of high-current ion beams while maintaining their quasi-neutrality. This makes the application of such devices attractive for the manipulation of high-current beams of heavy ions.

The plasma lens configuration of crossed electric and magnetic fields provides a suitable and attractive method for establishing a stable discharge at the low-pressure. Using plasma lens configuration in this way were elaborated, explored and developed some cost efficiency, low-maintenance plasma devices for ion treatment, and deposition of exotic coatings with given functional properties. These devices make using of permanent magnets and possess considerable flexibility with respect to spatial configuration. They can be operated as a stand-alone tool for ion treatment of substrates, or as part of integrated processing system together with cylindrical magnetron sputtering system, for coating deposition. The cylindrical plasma-optical magnetron sputtering device with virtual anodes and cylindrical plasma production device for the ion treatment of substrates with complicated cylindrical was proposed and created [5–7]. These devices can be applied both for fine ion cleaning and activation of substrates before deposition and for sputtering.

One particularly attractive result of this background work was observation of the essential positive potential at the floating substrate treated by cylindrical ion cleaning device. This suggested to us the possibility of an electrostatic plasma lens for focusing and manipulating high-current beams of negatively charged particles (electrons and negative ions) that is based on the use of the cloud of positive space charge in conditions of magnetic insulation electrons. The idea of the plasma lens based on electrostatic electron isolation for creation positive space charge was first proposed in [8]. Later, it was proposed to use magnetic electron insulation for creation of a stable positive space charge cloud [9].

Here we describe computer modeling for some of these novel plasma-optical systems.

2. Plasma lens with a positive space charge cloud for focusing intense negative charged particle beams

2.1. Model description

The plasma lens is a cylindrical plasma accelerator with an anode layer and used as a device with magnetic insulation of electrons for creation of the dynamic cloud of positive space charge. The scheme of the plasma lens with magnetic insulation used for creation of the dynamic cloud of positive space charge is shown in **Figure 1a**.

The lens has a system of permanent magnets that produce an axially symmetric magnetic field between the poles of a magnetic circuit serving as cathode. The magnetic field is controlled by varying the number of magnets. The magnetic field configuration is typical for the single magnetic lens configuration, because of the lens could focus the transported electron beams. When a positive potential is applied to the anode, a discharge in the axial magnetic, and radial electric crossed fields is ignited between the anode and the cathode. The electrons are magnetized in anode layer and drift along closed trajectories in the azimuthal direction, repeatedly ionize atoms of the working gas, and gradually diffuse to the anode. The ions thus formed are accelerated in the strong electric field created by the electron space charge and leave the ion source through a hole in the acceleration channel. The fast ions reach the system axis and accumulate in the region around it, as it schematically shown in **Figure 1b**. In this way, the axially converging ion beam creates a positive space charge. In the experiments, the energy of the argon ions converging beam could reach 2.5 kV. Maximum potential will be in the center on cylindrical axis. Ions are stored in the cylinder volume until their own space charge creates a critical electric field. This field forces ions to leave the volume, and the system comes to dynamic equilibrium after some relaxation time.

Electrons are magnetized in the anode layer, so their influence on ion dynamics can be neglected. The ion flow coming through the cylindrical surface will be equal to ions streaming down from the axis and leaving the cylindrical volume under action of the Coulomb force of their own space charge. Therefore, the set of equations describing this process in the cylindrical

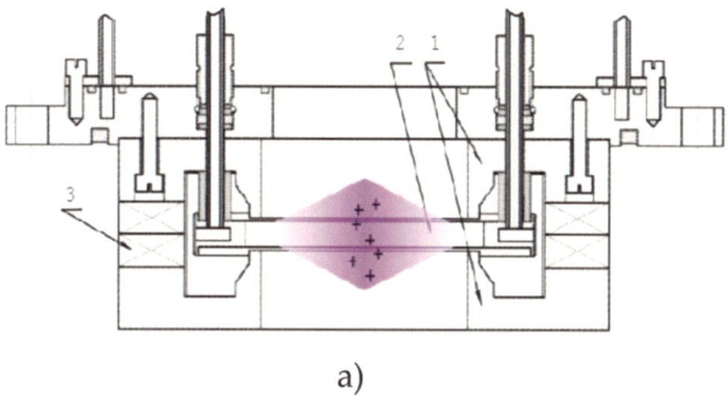

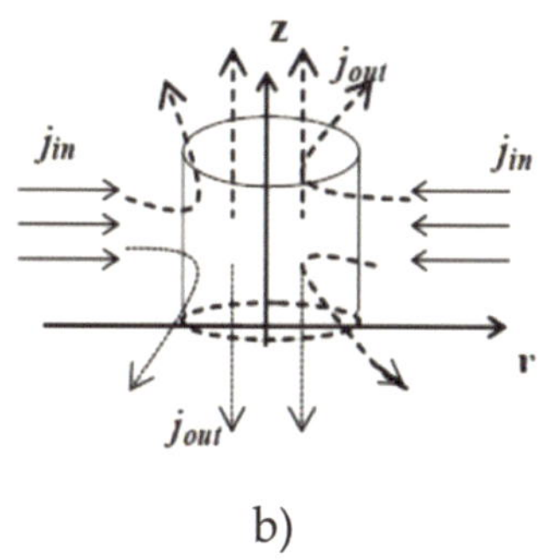

Figure 1. (a) Scheme of the plasma lens with magnetic electron insulation: 1 – cathode; 2 – anode; 3 – magnetic system based on permanent magnets; (b) Scheme the positive space charge cloud creation.

coordinate system can be written in the form that includes the Poisson, particles motion, and continuity law equations:

$$\frac{1}{r}\frac{\partial}{\partial r}\left(r\frac{\partial U}{\partial r}\right)+\frac{\partial^2 U}{\partial z^2}=-4\pi q_i n_i \tag{1}$$

$$M_i\frac{dv_i}{dt}=q_i E+\frac{1}{c}[v_i\times B] \tag{2}$$

$$V_i\cdot\left(\frac{\partial n_i}{\partial t}+div\left(j_{out}\right)\right)=S\cdot j_{in} \tag{3}$$

where M_i, q_i, v_i, n_i are ion mass, charge, velocity, and ion density, respectively, E – electric field: $E_r = -\partial U/\partial r$, $E_z = -\partial U/\partial z$, U – potential, B – magnetic field, V – cylindrical volume, j_{in} – current density at the boundary of current-collecting surface S of radius r and height h, and j_{out}– the ion current density leaving the cylindrical volume V. Knowing the space charge distribution, we can determine the expulsive force that acts on the particle on the boundary of space charge volume and calculate ions trajectories.

We can obtain estimations for electrical field and space charge in cloud in stationary case with simplifying assumptions. Let us suppose uniform ion distribution and that they enter to the cylinder perpendicular to its lateral surface and leave it along z-axis under action of electric field created by the ions own space charge. Thus considering of one-dimensional ions motion along the radius, we can write equation for potential in form:

$$\varphi_r''=\frac{4\pi j_{in}}{\sqrt{2q_i(\varphi_0-\varphi(r))/M_i}} \tag{4}$$

Solving it, we get solution similar to the low 3/2:

$$(\varphi_0-\varphi(r))^{3/2}\approx\frac{9\pi j_{in}}{\sqrt{2q_i/M_i}}r^2$$

and obtain estimation for electric field:

$$E_r=-\frac{\partial\varphi}{\partial r}\approx\frac{4}{3}\left(\frac{9\pi j_{in}}{\sqrt{2q_i/M_i}}\right)^{2/3}r^{1/3}.$$

Let us now consider a cylindrical layer of ion space charge with radius r and assume that the ions leave it along z-axis, then neglecting radial coordinate we can write Poisson equation in layer in form: $\varphi_z''=4\pi\rho$, where $j_{out}=\rho v$. In stationary case from (Eq.(3)), we obtain: $j_{in}=r\frac{\partial j_{out}}{2\partial z}$. Integrating this expression with using expression $v=\sqrt{2q_i(\varphi(z)-\varphi(z'))/M_i}$ for velocity of ions leaving the layer, we obtain expression for the distribution of ion space charge density in form: $\rho\approx\frac{2j_{in}}{r\sqrt{2q_i(\varphi(z)-\varphi(z'))/M_i}}z$. Substituting last expression in Poisson equation for the layer and solving it we get that potential and space charge density in layer is proportional to next expressions:

$$\varphi(r,z) \approx \left(j_{in}/r\right)^{2/3} z^2, \rho \approx 2\left(4\pi \frac{j_{in}}{r\sqrt{2q_i/M_i}}\right)^{2/3} \tag{5}$$

Substitute character system parameters we obtain estimations for ion density: n~10^{10} $cm^{-3,}$ and for electric field strength about 1000 V/cm, that is sufficient for high-current electron beam focusing.

Eqs. (1)-(3) were solved numerically by particle in cell (PIC)-method [10]. Every time interval Δt (that corresponded to the actual time interval approximately equal to $4{\cdot}10^{-8}$ sec) N new particles of charge q_i and mass M_i come to the volume considered. The magnitudes of N, Δt, q_i satisfy the relation: $\frac{Nq_i}{\Delta t} = j_i S$. Let us suppose that particles with energies from 0 to ε_{max} are distributed according to law:

$$N(\varepsilon) \approx \frac{1}{\sqrt{2\pi}\langle\varepsilon\rangle} \exp\left(-\frac{(\varepsilon - \varepsilon_0)^2}{2\langle\varepsilon\rangle}\right) \tag{6}$$

where $\varepsilon_0 = \varepsilon_{max}/2$ and $< \varepsilon >$ is average energy. They moved from cylinder surface to the system axis with the angular distribution according to cosine law: N(Θ)/N(0) ≈ cos (Θ), where N(Θ) are quantities of ions going out under angle Θ, N(0) is angular distribution amplitude. It should be noted that the distributions above are inherent to this kind of plasma accelerators with anode layer. As the first step the potential was specified at the anode, and the Laplace equation was solved. After that, the particles were launched and the equations of motion (Eq. (2)) for particles in the magnetic and calculated electric field were solved. The time step for motion equation solving was $\Delta\tau \ll \Delta t$ (about 10^{-11} sec). After time Δtby collecting of all particles with the use the "cloud in cell" method [10], the densities distributions of ions were calculated. The Poisson equation has been solved and electric field potential U(r, z) was calculated for this time moment. Electric field was calculated by the distribution of total space charge. After that in corrected electric field, the calculation of particles motion was resumed, and introducing the new portion of ions was performed. Equations of motion were solved both for "new" particles and for those that still left in the volume. **Figure 2** shows calculated

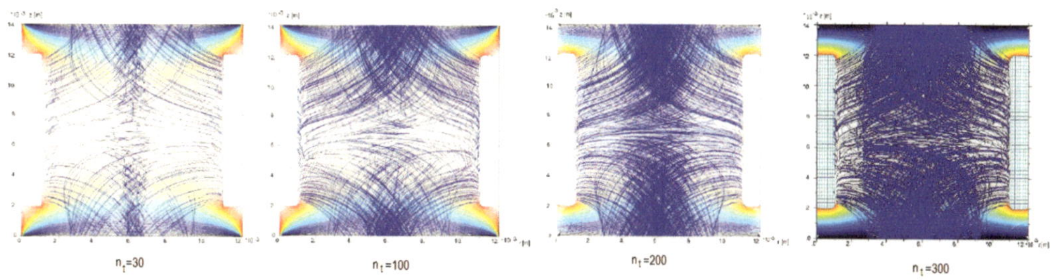

Figure 2. Positive space charge cloud formation and Ar+ ions trajectories for different time step.

Ar + ions trajectories in different time step. The calculation continued until reaching a self-consistent solution. The calculation time comprised 10^{-5} sec. For that time the stationary state of the lens operation was achieved.

2.2. Simulation results

The model was applied for calculating the lens volume based on the local area with diameter of 80 mm and a height of 50 mm. In our simulations, we considered Ar^+ and Xe^+ ion beams with maximal energy from 1 to 3 keV and total current of 20 mA that moved in the magnetic field. Magnetic field is similar to the experimental one and changes from 0.07 T near electrodes to 0.01 T on the system axis. The results of the calculations of the potential distribution when steady-state dynamical equilibrium is reached for Ar^+ ions with maximal energy 1.2–2.4 keV are shown in **Figure 3**. One can see that with ions energy increasing, the spatial distribution shape changes markedly. Whereas maximum of potential for ion beam's energy 0–1.5 keV (see left **Figure 3**) is double-humped situated in the coaxial region around the axis, the maximum for energy 0–2.4 keV (right) is single-humped onto the axis.

The same result we can see with ion mass increasing (see **Figure 4**). The maximum of potential for Ar + ion beam (**Figure 4** left) is in the coaxial region around the axis, but the maximum for

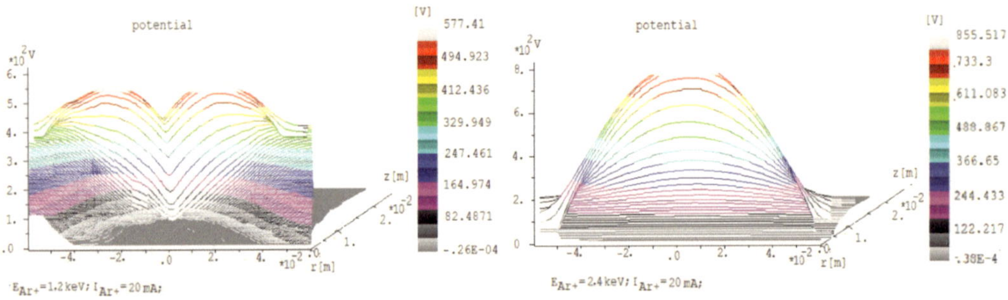

Figure 3. Potential distribution in plasma lens midplane for Ar + ion beam with total current 20 mA and E_{max} = 1.2 keV (left) and 2.4 keV(right).

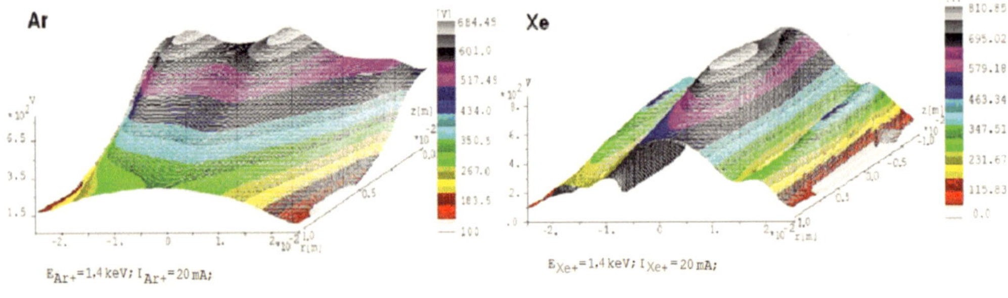

Figure 4. Potential distribution for Ar+ ion beam (left) and Xe+ ion beam (right) with total current 20 mA and E_{max} = 1.4 keV.

heavier Xe + ions (**Figure 4** right) is at the axis. This can be explained by a smaller influence of the momentum aberration on the converging Xe + ion beam dynamics.

The calculated ion density that could accumulate around system axis reaches 10^9–$2.7{\cdot}10^{10}$ $cm^{-3,}$ and electric field strength was up to 600 V/cm that is sufficiently for focusing intensive negative charged particle beams and correspond to our estimations. Thus the possibilities formation of the stable space charge cloud was demonstrated, and next task is to consider negative charge particle beam transport through space charge PL.

2.3. Negative charge particle beams focusing on space charge plasma lens

We investigated transport electron beam with energy from 5 to 20 keV through the plasma lens. As first step was solved equations for ion's part and as result – obtaining stable positive charge cloud inside plasma lens. Next step was launch e-beam through the lens with cloud. For correct description, we must solve equations for ions and electrons parts together, so we must include electron motion equations in our consideration and modify Poisson equation to form:

$$\frac{1}{r}\frac{d\varphi}{dr}+\frac{d^2\varphi}{dr^2}+\frac{d^2\varphi}{dz^2}=4\pi e(n_e+n_{eb}-n_i) \tag{7}$$

For simulation high-current electron beam transport need also taking into account the space charge of the particle and the magnetic self-field that may affect the dynamic beam particles in addition to the external fields. The possibility of ionization residual gas by electron beam is necessary taking into account also.

Equations of motion for electrons are solving by current tubes of variable width with central trajectory. A shape of trajectories in an electromagnetic field is calculated using Boris scheme [11]. A space charge beam density is calculated using equation of continuity: $div(\rho_e v_e) = 0$.A self-consistent solution can be found by repeated solving of Poisson equation, motion equations for all particles, and re-determination of the space charge distribution on every time step. An iteration method with relaxation was used for faster convergence.

Numerical simulations results show clearly that for electron beam current less than 1A the electrostatic beam focusing occurs [12, 13]. The results of simulation for space charge plasma lens and magnetic lens (ML) with the same magnetic field are shown in **Figure 5**. The comparison shows that beam compression is stronger, and beam divergence is less in PL case than in ML case. The experimental results [12, 13] confirm our simulations. The lens can be used even more effectively for negative ions beam focusing. The simulation results for H-beam with passing through plasma lens and magnetic lens with the same magnetic field are shown in **Figure 6**. One can see that lens effectively compresses H-ions beam, whereas the magnetic lens does not focus it at all.

However, it should be noted, some part of cloud ions can be captured by a beam and carried away from the cloud. It is positive for beam transport, but as a result of this, cloud potential decreases, and its focusing properties deteriorate. It is not critical for electron beams with current up to 1 A, because a new Ar + ions come to the cloud and renew its focusing properties.

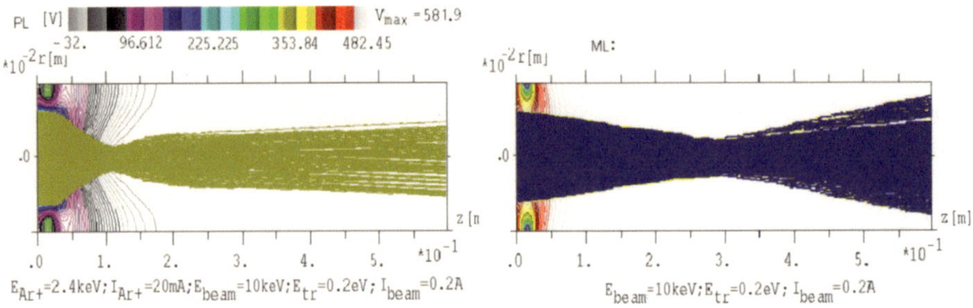

Figure 5. Electron beam (E_b = 10 keV, I_b = 0.2A) passing through the positive space charge plasma lens (left) and magnetic lens (right).

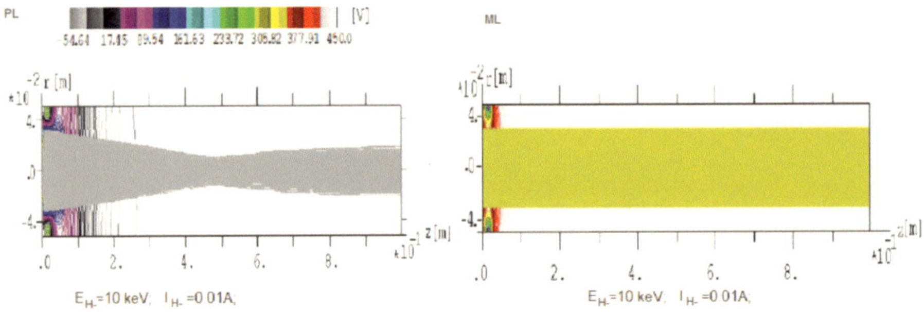

Figure 6. Trajectories of H-ions beam (E_b = 10 keV, I_b = 0.01A) passing through the positive space charge plasma lens (left) and magnetic lens (right).

However, for beam current of about or more 1 A the potential maximum in the positive space charge region decreases (from 580 to 210 V for I_{eb} = 1 A), the distribution is getting double-humped and electrostatic focusing destroyed (see **Figure 7** top).

It is due to that some part of ions comes out from cloud with the propagating electron beam and their number grows with beam current increasing [13]. A significant part of cloud particles carry out by e-beam along beam line, and ions are continuing to come in cloud from electrodes cannot support renewal processes. Thus cloud potential decrease and it distribution changes from one-hump to two-humps. Note that if it corresponded to case when beam space charge density a bit exceeds to space charge cloud density is possible to improve plasma lens electrostatic focusing property by increasing energy and current density Ar + ions beam that creates positive space charge cloud. **Figure 7** (down) shows potential distribution by electron beam propagating for increasing Ar + ions beam current from 20 up to 60 mA. One can see potential distribution come back to one-peak form and focusing properties of plasma lens recovered.

However, the positive space charge cloud quickly destroys with further increasing of electron beam current when beam space charge density significantly exceeds space charge cloud density, (see **Figure 8**) and it is not possible to renew electrostatic focusing properties anymore. In this case, for an electron beam with current on the order of tens of ampere for which the beam space charge density much more than space charge plasma lens the only the magnetic focusing

http://dx.doi.org/10.5772/intechopen.77512

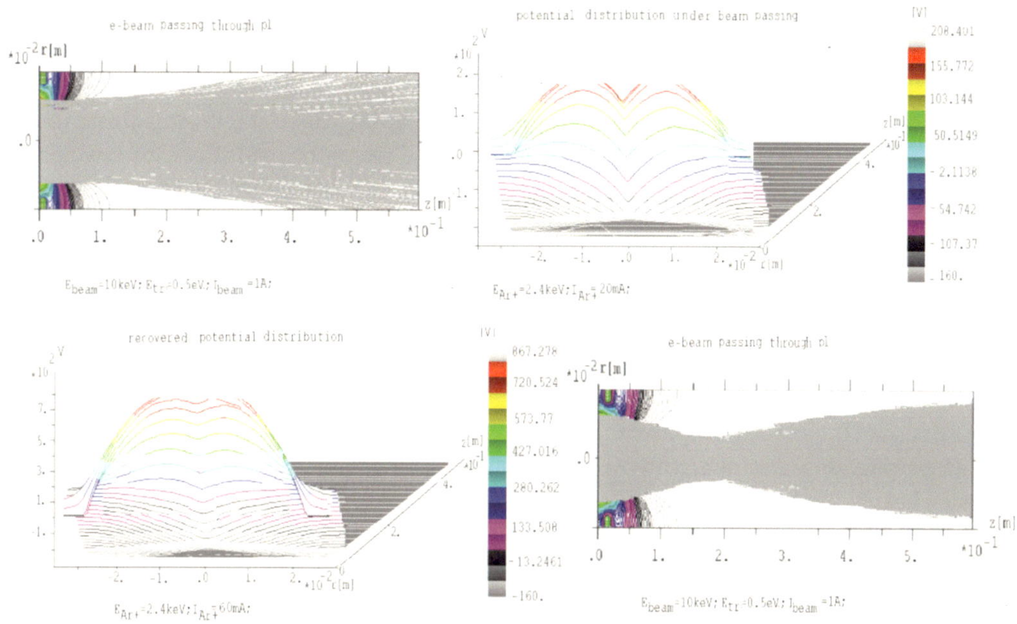

Figure 7. Electron beam trajectories (E_b = 10 keV, I_b = 1A) and positive charge cloud potential distribution created by Ar + ions beam with maximal energy 2.4 keV and current I = 20 mA(up) and I = 60 mA(down).

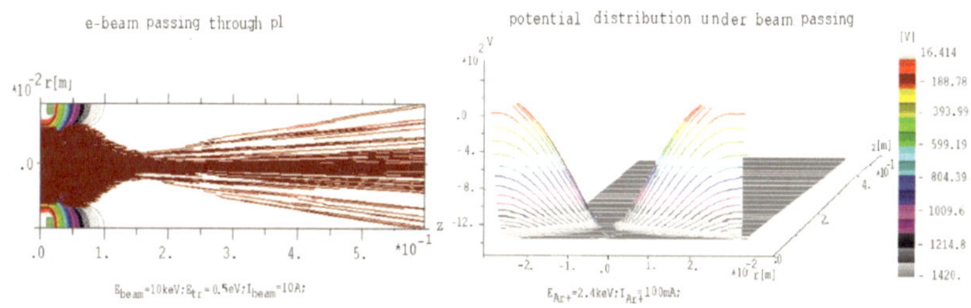

Figure 8. Electron beam trajectories (E_b = 10 keV, I_b = 10A) and positive charge cloud potential distribution created by Ar + ions beam with maximal energy 2.4 keV and current I = 100 mA. Magnetic field twice as much compared with previous case.

of the beam provides. However, even in this case use of PL is useful since it improves e-beam transport, providing additional compensation of beam space charge and decreasing beam divergence. Note that taking into consideration an ionization residual gas by electron beam do not lead to essential changing in simulations of final results.

2.4. Conclusion remarks

The research described has shown a principal possibility of creation of a positive space charge in the cylindrical anode layer accelerator due to a positive ion flow that converges onto its axis. This makes possible to develop a plasma lens with essentially under-compensated positive space charge. The lens will be used for focusing and manipulating beams of negatively charged particles. The value of positive charge potential formed at the axis and the steady state of the space charge depending on plasma dynamical parameters of the system are determined experimentally [12, 13]. Electric field value reaching 600–1000 V/cm realized under experimental condition is determined. Such electric field strength is sufficient for creation of short-focus elements to be used in systems for manipulating intense beams of negative ions and electrons. Experimental results [14] demonstrate an attractive possibilities application positive space charged plasma lens with magnetic electron insulation for focusing and manipulating wide-aperture high-current no relativistic electron beams. For relatively low-current mode for which electron beam space charged less than positive space charged plasma lens, it realizes electrostatic focusing is passing electron beam. In case of high-current mode, when electron beam space charge much more than space charge plasma lens the lens operates in plasma mode to create transparent plasma accelerating electrode and compensate space charge propagating electron beam. The lens magnetic field in this case uses for effective focusing beam.

In experiment was demonstrated a perspective applications of positive space charged PL with magnetic electron insulation for focusing and manipulating wide aperture high-current no relativistic electron beams (E_b = 16 keV; I_b = 100 A) [14]. Particularly, it was shown experimentally that under focusing these beams maximal compression factor was up to 30x, and beam current density at the focus was about 100 A/cm^2.

3. Plasma accelerator with open gas wall and closed electron drift

The accelerator with closed electron drift is one of the kinds of the electric rocket engines and devices for ion plasma treatment of the surface material. However, the accelerators with closed electron drift and open (gas) walls were not researched for now in contrast to the well-known and widely used plasma accelerators with anode layer and accelerators with closed electron drift and dielectric walls [15, 16]. Therefore, this type accelerator could be interested in manipulating high-current flow of charged particle as well as can be attractive for many different high-tech applications for potential devices of low-cost and compact thrusters. More than, it has some advantages since the wall absence leads to exclusion of the wall material inclusions into the ion beam and exclusion of the secondary electrons formation due to emission and thus to conservation of the plasma electrons dynamics.

The sample of plasma accelerator with closed electron drift and open walls is shown in **Figure 9** (left). This sample of cylindrical Hall-type plasma ion source that produced ion plasma flow converging toward the axis system was created for the properties exploration [17]. The discharge in the system burns due to ionization of the working gas by the electrons. Electrons are magnetized and formed stable negative space charge. The created ions accelerated from ionization zone to the cathode. As follow from discharge geometry an accumulation of ion space charge occurs as it is in the positive space charge lens (see above). The main part of generated ions leaves the system across radius, along with system axis, due to jets can appear

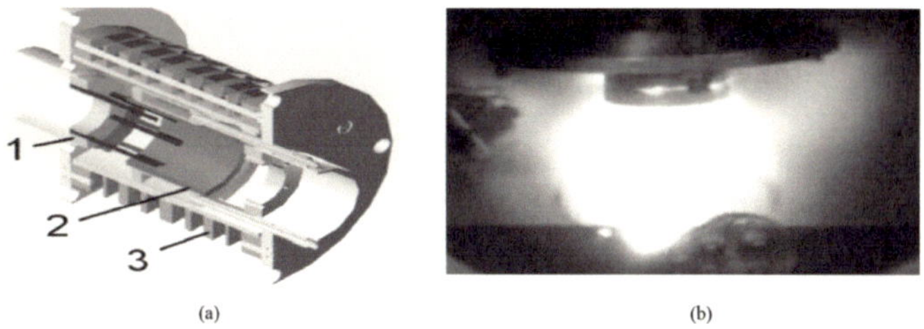

Figure 9. (a) Experimental sample: 1- cathode, 2- anode, and 3- permanent magnets system; (b)plasma jet in high-current operation mode.

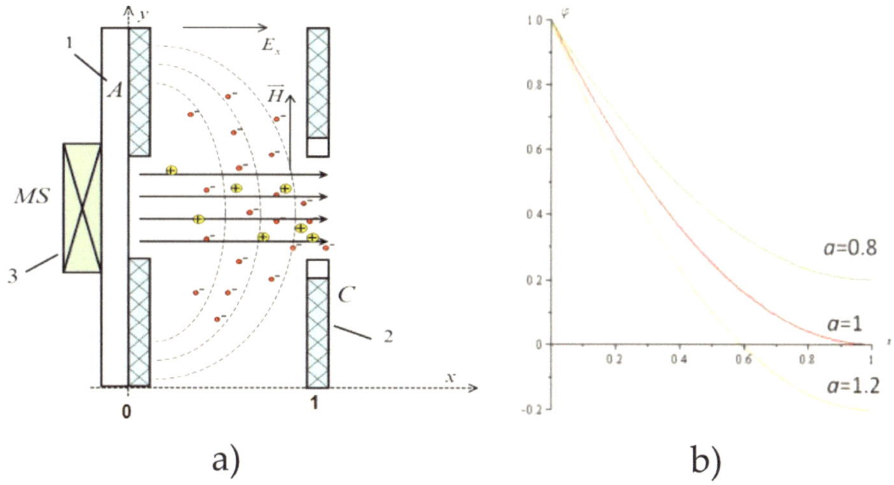

Figure 10. (a) Model of discharge gap: 1-anode, 2-cathode, 3-permanent magnets system; (b) potential distribution in the gap for different parameters *a* value.

on the edges of system. As follow from experiment [17, 18], the accelerator has two operating modes: low-current with narrow anode layer and clear-cut plasma flow and high-current when plasma fills the entire volume of the accelerator. The transfer to the high-current mode occurs under influence of two parameters: worked gas pressure and applied voltage. In high-current quasi-neutral plasma mode of accelerator operation, plasma jet is observed (see, **Figure 9** right). The preliminary results show that along the jet axis potential drop arise, which can be used for ion beam accelerating. The radial studies of plasma flow along system axis showed the significant increasing current density on the axis. It may indicate on plasma acceleration in that direction.

3.1. One-dimensional hydrodynamic and hybrid model

We will consider discharge gap, where ions production occurs due to ionization of the working gas by electrons (see **Figure 10a**). Electrons are magnetized, move along magnetic strength lines, and drift slowly to anode due to collisions. Ions are free and accelerated by electric field move to the system axis. Will assume that the discharge current density in gap volume is the sum of the ion and electron components:

$$j_e + j_i = j_d \tag{8}$$

where j_i, j_e are ion and electron current density consequently:

$$j_i(x) = e\nu_i \int_0^x n_e(x)dx \tag{9}$$

$$j_e(x) = e\mu_\perp \left(n_e E(x) - \frac{d}{dx}(n_e T_e) \right) \tag{10}$$

ν_i-is the ionization frequency, $\mu_\perp = \frac{e\nu_e}{m\omega_{eH}^2}$ electron transverse mobility, $E(x) = -\frac{d\varphi}{dx}$-electric field, φ - potential, ν_e is the frequency of elastic collisions with neutrals and ions, ω_{eH} is the electron cyclotron frequency, T_e – electron temperature that could write in form [15]:

$$T_e(x) = \frac{\beta}{j_e(x)} \int_0^x j_e \frac{d\varphi}{ds} ds \tag{11}$$

for ion density we could write:

$$n_i(x) = \sqrt{\frac{M}{2e}} \int_0^x \frac{n_e(s)\nu_i ds}{\sqrt{\varphi(x) - \varphi(s)}} \tag{12}$$

and Poisson equation closed this system of equations:

$$n_e - n_i = \frac{1}{4\pi e}\varphi'' \tag{13}$$

In some cases equations system Eqs.(9) and (13) allows exact analytical solutions. Dimensionless these equations and assume that position x = 0 correspond to anode and initial potential on this boundary is equal applied voltage, position x = 1 correspond to cathode, and ion current on the cathode is equal to discharge current. If we neglect diffusion for case $n_e >> n_i$, that corresponds to a low-current mode, we can get exact solution in form [18, 19]:

$$\varphi = a\left((x-1)^2 - 1\right) + 1, a = \frac{\nu_i d^2}{2\mu_\perp \varphi_a} \tag{14}$$

where d is gap length. One can see that at the parameter $a = 1$, the total applied potential falls inside the gap (see **Figure 1b**). For this case, gap length is equal: $d = \sqrt{\frac{2\mu_\perp \varphi_A}{\nu_i}}$. Under assumption that electrons originated from the gap only by impact ionization, and go to the anode due to classical transverse mobility, last expression can rewritten in the form:

$$\delta = \rho_e(\varphi_A)\sqrt{\frac{2\nu_e}{\nu_i}} \tag{15}$$

This expression coincides with one for classical anode layer [20] accurate within √2. Note in case when parameter $a < 1$ (the gap length less than δ) potential drop is not completed (see **Figure 1b**). For case $a > 1$, when the gap length d > δ potential drop exceed applied potential. This can be due to the electron space charge dominated at the accelerator exit.

Another case is quasi-neutral plasma: $n_i \approx n_e$, that correspond to high-current mode [19]. In this case from the solution follows one interesting conclusion: when electron density does not change along the gap we could get generalization condition of self-sustained discharge in crossed electrical and magnetic fields (EXH), taking into consideration both electron and ion dynamics peculiarity. More than in this case solution for high-current mode is reduced to form (Eq. (14)) also. Even if we consider more general model description, assuming that electron heating losing occurs mostly by different kind of collisions it can show that potential distribution along gap weakly depend on electron temperature saving close to parabolic distribution. And under assumption that time of energetic losses equal electron lifetime in the gap we again come back to potential distribution (14) [18, 19].

For a more detailed study of the influence of ion dynamics on the system process, we applied a one-dimension mode hybrid model used for calculations of a span mode with neutral-particle ionization. In this model dynamics of ions and neutrals is described by kinetic equations, wherein right parts take into account only single ionizations:

$$\frac{\partial f_o}{\partial t} + v_0\frac{\partial f_0}{\partial x} = -\langle\sigma_{ie}v_e\rangle n_e f_0, \ \frac{\partial f_i}{\partial t} + v_i\frac{\partial f_i}{\partial x} + \frac{e}{M}E\frac{\partial f_i}{\partial v} = \langle\sigma_{ie}v_e\rangle n_e f_0 \tag{16}$$

where f_0, f_i distribution function of neutrals and ions consequently that satisfies boundary conditions:

$$f_0(0,v,t) = \frac{1}{(2\pi MT)^{3/2}}\exp\left(-\frac{Mv^2}{2T}\right), \mathrm{v} > 0; f_0(0,\mathrm{v},\mathrm{t}) = 0, \mathrm{v} < 0; f_i(0,\mathrm{v},\mathrm{t}) = 0 \tag{17}$$

Right part (Eq. (16)) we can write:

$$\langle\sigma_{ie}v_e\rangle = \sigma_{\max}v_e(T_e)\exp\left(-U_i/T_e\right), \tag{18}$$

where σ max – maximal ionization cross-section, ve(Te) – average electron thermal velocity, Ui – potential of ionization.

For electron part we will use hydrodynamic description and solve (Eqs. (8), (10), and(13)), where for ion density and current density is valid:

$$n_i = \int f_i dv, j_i = \int v_i f_i dv \tag{19}$$

The result of simulation hybrid model after achieving dynamic equilibrium by system in comparisons with hydrodynamic model is shown in **Figure 11**. One can see that taking into

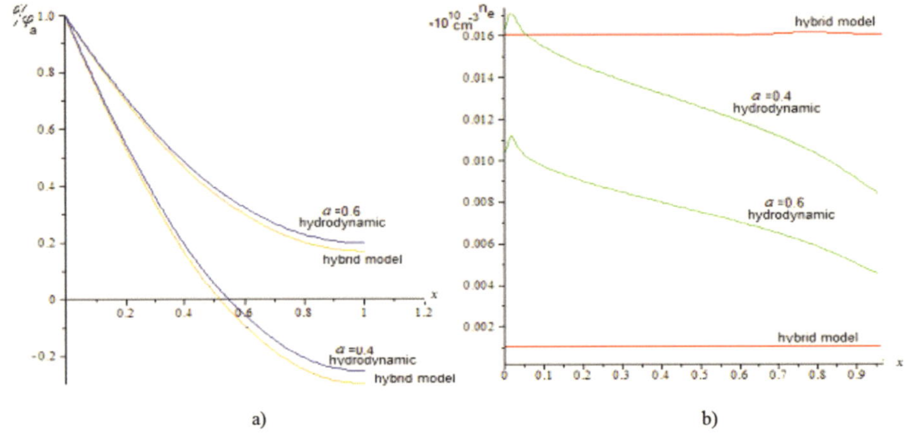

Figure 11. Comparison simulation results for hybrid and hydrodynamic models:(a) potential distribution in gap, (b) electron density.

account span mode with single ionization does not have a big impact on the result potential distribution (see **Figure 11a**). Electron density distribution (**Figure 11b**) looks more consistent along gap in hybrid model as compared to a hydrodynamic model. It is the result of influence on the system processes of neutrals dynamics and ionization, which are not taken into account in the hydrodynamic model.

Figure 12 shows ion density changing in the gap during the time in high-current mode (under applied anode potential equal 1200 V). It can be seen that at first ions number increases in near-anode region and remains almost constant in gap, then it increases almost linearly throughout the gap, and finally increases sharply at the cathode region.

Note that for correct description (especially high-current mode), it is necessary to model ionization and plasma creation, as well as motion of neutrals and formed ions in the whole volume of accelerator, thus need consider two-dimensional model.

3.2. Two-dimensional model

We will consider cylindrical geometry, where the anode is a cylinder with diameter 6.7 cm and applied potential about 1–2.5 kV, and cathode consist from two cylinders with diameter

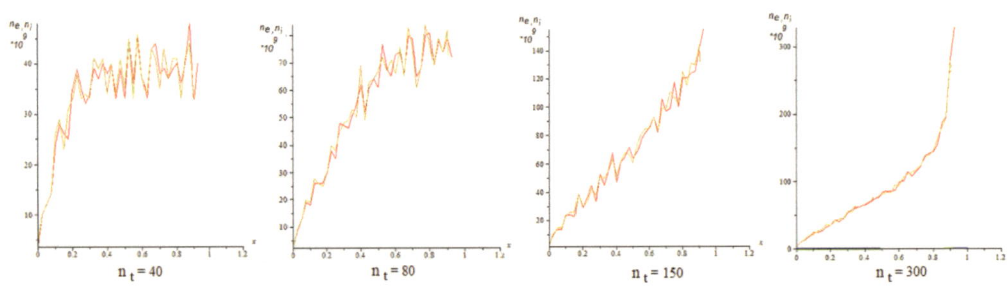

Figure 12. Ion density in the gap in high-current mode depending on time.

4.5 cm, spaced by some distance from each other. For ions and neutrals description we use Boltzmann kinetic equation:

$$\frac{\partial f_{i,n}}{\partial t} + \vec{v}_{i,n}\frac{\partial f_{i,n}}{\partial \vec{r}} + \frac{e}{M}\left(E + \frac{1}{c}[v \times B]\right)\frac{\partial f_i}{\partial v_i} = St\left\{f_{i,n}\right\} \tag{20}$$

We solved this equation by splitting on the Vlasov equation for finding ions and neutrals trajectories:

$$\frac{\partial f_{i,n}}{\partial t} + \vec{v}_{i,n}\frac{\partial f_{i,n}}{\partial \vec{r}} + \frac{e}{M}\left(E + \frac{1}{c}[v \times B]\right)\frac{\partial f_i}{\partial v_i} = 0 \tag{21}$$

and to correct the found trajectories taking into account the collision integral in which we took into account the processes of ionization and elastic and inelastic collisions:

$$\frac{Df_{i,n}}{Dt} = St\left\{f_{i,n}\right\} \tag{22}$$

The Vlasov equations were solved by the method of characteristics:

$$\frac{d\vec{v}_k}{dt} = \frac{q_k}{M}\left(\vec{E} + \frac{1}{c}[v_k \times B]\right], \frac{d\vec{r}_k}{dt} = \vec{v}_k \tag{23}$$

To solve these equations, the PIC method [10] with Boris scheme [11] was used to avoid singularities at the axis. For initial electric field distribution was taken electric field in the plasma absence: $E(r) = \frac{U_a}{r\ln(r_c/r_a)}$. The Monte-Carlo method was used for modeling of ionization in this field. The probability of a collision of a particle with energy ε_j during time Δt was found from expression [21]:

$$P_j = 1 - \exp\left(-v_j \Delta t \sigma(\varepsilon_j) n_j\left(\vec{r}_j\right)\right) \tag{24}$$

where $\sigma(\varepsilon)$ – collision cross-section (elastic, ionization or excitation), n_j – density of similar particles at the point r_j. To determine the probability of collision a random number β is chosen from interval [0,1] with the help of a random number generator. If $\beta < P_j$, then assumed that collision has occurred. It is determine the ratio of the cross-sections with the random number generator, which collision has occurred – elastic, excitation, or ionization. Independent of this particle parameters change or new ion adds in computational box. The emerging ions begin to move toward the system axis. The evolution of all particles that are in the modeling region is traced at each time step. For this motion, equations were solved, and new velocities and positions of the particles were found. Particles that move out the modeling box boundaries are excluded from consideration. After sufficiently long-time particle density distribution was found. The ion charge density and current density are calculated from coordinates and velocities of particles according to formulas:

$$\rho(r,t) = \frac{1}{V}\sum_j q_j R\left(\vec{r}, \vec{r}_j(t)\right), j(r,t) = \sum_j q_j v_j(t) R\left(\vec{r}, \vec{r}_j(t)\right) \tag{25}$$

where $R(r,r_j)$ – usual standard PIC – core, that characterizes particle size and shape and charges distribution in it. After that, the Poisson equation was solved, and new electric field distribution was found. Since electrons are magnetized we consider their movement in radial plane only, thus can solve for electrons one-dimensional hydrodynamic equations on each layer at z separately. Solve it we find electron density, calculate electric field on each layer and correct particle trajectories. After that, the procedure was repeated again. Modeling time is large enough for establish of ion multiplication process. The formation of the sufficient number of ions is possible due to magnetic field presence, which isolates anode from the cathode. Ions practically do not feel the magnetic field action and move from anode to the axis, where create a space charge, first in the center of the system. Electrons move along the magnetic field strength line, but due to collisions with neutrals, they start moving across the magnetic field. An internal electric field is formed, which slow down the ions and pushes out them from the volume along system axis. **Figure 13** shows results of modeling high-current mode ($U_a = 1.2$ kV, pressure 0.15 Pa, and magnetic field at the axis is 0.03 T). **Figure 13a** shows how the ions number to axis increases when ionization process is steady-state. One can see that number of ions increase not only to axis but also along axis from center to edge too. **Figure 13b** shows ion space charge distribution for different time step. One can see that ions create space charge in center of the system first, but then under electric field action they leave center and move along z-axis in both direction.

3.3. Conclusion remarks

First, the original approach to use plasma accelerators with closed electron drift and open walls for creation cost-effective low-maintenance plasma device for production converging toward axis accelerating ion beam was described. Based on the idea of continuity of current transferring in the system are found exact analytical solutions describing electric potential distribution along acceleration gap. It was shown that potential distribution is parabolic for different operation modes as in low-current mode as well as in high-current quasi-neutral plasma mode and cannot depend on electron temperature. It is found under conditions that everything electrons originated within the gap by impact ionization only, and go out at the anode due to mobility in transverse magnetic field, the condition full potential drop in the accelerating gap corresponds to equality gap length to the anode layer thickness. In case when the gap length less than anode layer thickness potential drop is not completed. For case when the gap length more than anode

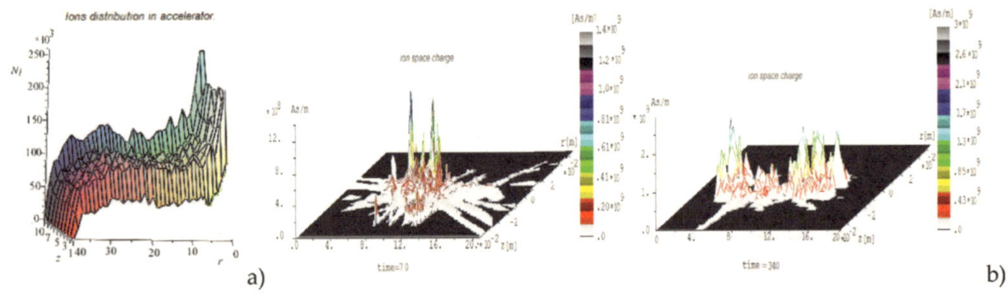

Figure 13. (a) Ions number dependence on r and z (r = 0,z = 0–center of the system); (b) ion space charge for time step 70 (left) and 340(right).

layer potential drop exceed applied potential. The generalization condition of self-sustained discharge in crossed ExH fields taking into consideration both electron and ion dynamic peculiarity were obtained. The performed modeling showed that in high-current mode the ions moving to the system center and then along the axis in both directions are able to create space charge. Experimental model of accelerator that formed ion flow converging toward the axis system was created [17, 18]. In high-current mode of accelerator operation is observed plasma jet. It is shown at the jet axis forms potential drop that could be used for ion beam accelerating. The experimental results are in good accordance with theory data.

Note also that the presented plasma device is attractive for many different high-tech practical applications, for example, like PL with positive space cloud for focusing negative, intense charge particles beams (electrons and negative ions), and for potential devices small rocket engines.

4. The model of the plasma-dynamical filter for micro-droplets eliminations

Now the vacuum-plasma technologies are widely used in erosion plasma sources, such as cathodic arc and laser-produced plasma, for thin films synthesis, and coatings with control properties. However, the micro-droplets present in the formed ion plasma flow restrict the applicability of this method of film synthesis. The micro-droplets component of erosion for the majority of metals is an essential part of general losses of the cathode material in a vacuum-arc, comparable with ion component. Therefore, for prevention of effect of micro-droplets on a substrate, it is necessary to eliminate droplets. The formation and propagation of micro-droplets, and also the mechanisms of decreasing of their effects on quality and rate deposition films and coatings were investigated in detail in works [22–26]. It is known from several approaches micro-droplets density reduction in ion plasma flows. Therefore, proposed methods do not provide the complete solution of the problem of micro-droplets elimination.

In paper Anders [26], the conclusion has been formulated that micro-droplets cannot be evaporated in the arc plasma flow without additional energy source. A new approach to the elimination or reduction of micro-droplets from the dense metal plasma flow based on the use of a cylindrical electrostatic PL configuration to generate an energetic radial electron beam within the low-energy ion plasma flow has been proposed and described in [27, 28]. The pumping of energy into arc plasma flow by the self-consistently formed radial beam of high-energy electrons for evaporation of micro-droplets could serve as additional energy source. The beam is formed by double layer, appeared in a cylindrical channel of the novel plasma-optical system in crossed radial electrical and longitudinal magnetic fields. Here we detailed describe the model of device for filtering dense plasma flow from micro-droplets.

4.1. Model approach

We will consider electrostatic PL configuration through which a low-energy arc ion plasma flow passes. **Figure 14** shows experimental device (left) and simplified model (right). A dense arc plasma flow with micro-droplets is propagated from the cathodic arc source and passes

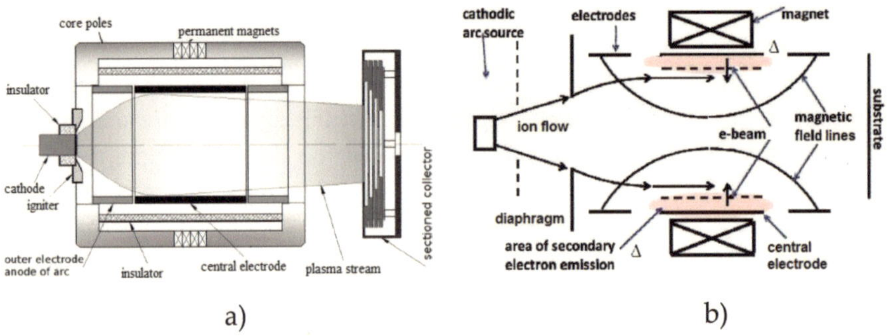

Figure 14. (a) Schema of the experimental set-up, (b) simplify model (Δ–spatial layer in which the strong radial electrical field is supported).

through diaphragm into the plasma-optical system. A part of the micro-droplets settles on the diaphragm, which is at the beginning of plasma-optical system, and remaining ones along with the flow propagate through the system.

The system consists of a cylinder (central electrode) of length L and diameter D on which the negative voltage U ~ 1–3 kV is applied, and a pair of external ring grounded electrodes arranged symmetrically to the central electrode. The system is in the magnetic field of a short coil or permanent magnets. When arc plasma flow reaches the zone of the magnetic field action, it penetrates into the flow. The magnetic field must satisfy to the condition: $\rho_{Le} \ll D \ll \rho_{Li}$, where ρ_{Le}, ρ_{Li} – are the electron and ion Larmour radiuses. In such a magnetic field the electrons are magnetized, and ions are not magnetized. A dense plasma flow propagating inside the cylinder creates a thin wall layer Δ with a large radial electric field E_r. The magnetic field lines are equipotential inside flow, and electrical field, which created is similar to magnetic field lines. Because of the electrons of the flow are magnetized, they in the magnetic field are displaced to axis and prevent radial expansion of the flow. The density of flow increases with increasing of the magnetic field near axis, so control by magnetic field we can control by flow, keeping it from expansion.

High-energy electrons appear near the inner cylindrical surface by secondary ion-electron emission at this surface bombardment by peripheral flow ions. In result, the fast electrons beam created with current density $j_{eb} = \gamma j_{is}$ in near-wall layer (γ – secondary ion-electron emission coefficient).This electron beam injected from near-wall layer to plasma flow axis can serve as additional effective energy source for micro-droplet elimination.

4.2. Modeling and results

The modeling of high-energy radial electron beam formation was the first task of our consideration. For simulation, we used PIC-method with Boris scheme, analogically to previous section. The copper plasma flows transport through the plasma lens with applied potential −3 kV on the central electrode and magnetic field about 300–400 E was considered. **Figure 15a** shows initial plasma flow. The peripheral ions of flow collide with the inner wall of the central electrode and emitted electrons. The Monte-Carlo method was used for modeling emission

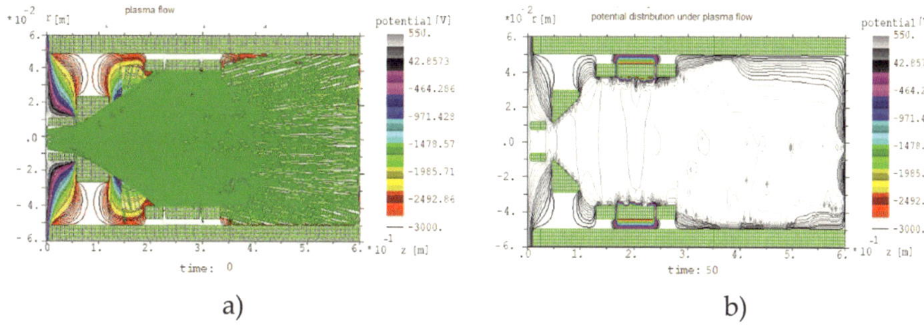

Figure 15. (a) Initial copper plasma flows, (b) potential distribution for time step 50.

process and electrons multiplication. The potential distribution is found by solving Poisson equation (see Eq.(7)). Ion density we find from the continuity equation: $div(j_i) = 0$. For electron density we can use expression: $n_e = n_{e0} \cdot \exp.(e\varphi/T_e)$, where n_{e0} – electron density of quasi-neutral plasma. For simplicity, we did not take into account the plasma heating and used finite T_e and T_i at the first stage. For all particles in the calculation box, the motion equations solve and find new positions and velocities. Part of the ions reaches the cathode and knocks out γ-electrons from it. We took them into account for Poisson equation solving too. We also take into account their ability to ionize when acquiring energy exceed the ionization energy. Very soon (in real time it corresponds to 1.6–2.5 μs) the potential jump near cathode is formed (see **Figure 15b**).

One can see that size of the potential jump layer is very small (about $5 \cdot 10^{-7}$ cm, so it is $\ll \rho_{Le}$), thus electric field formed in the layer is large enough to accelerate the electrons to high-energy in the layer and force them to move toward the axis. We assumed the initial energy of emitted electron is 5 eV. Initial electrons move along magnetic field line as it shown in **Figure 16a**. However, due to appearance strong electric field in the layer, the average electron energy increases. It begins to exceed 10 eV at field strength is about 100 V/cm, and when electric field strength reaches about 500 V/cm it exceeds 100 eV. At such energies, the secondary electrons acquire the ability to ionize neutrals with the new secondary electrons appearance. As results, an avalanche-like ionization begins. Electrons move toward the axis and can accumulate there (see **Figure 16b**)). Note that application of the plasma lens to the transport of low-energy

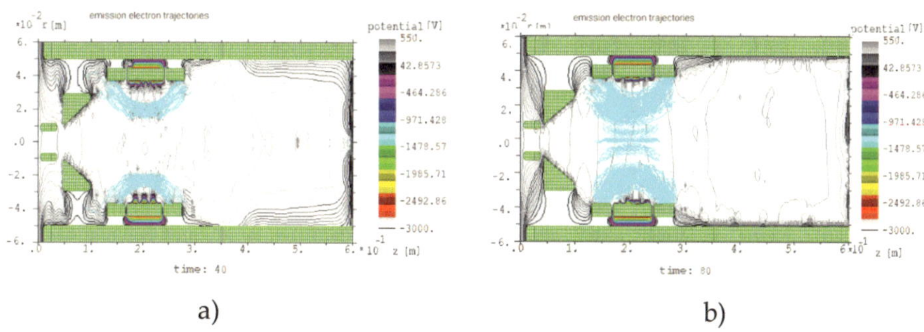

Figure 16. Trajectories of the emission electrons depends on time (a) $N_t = 40$, (b) $N_t = 80$.

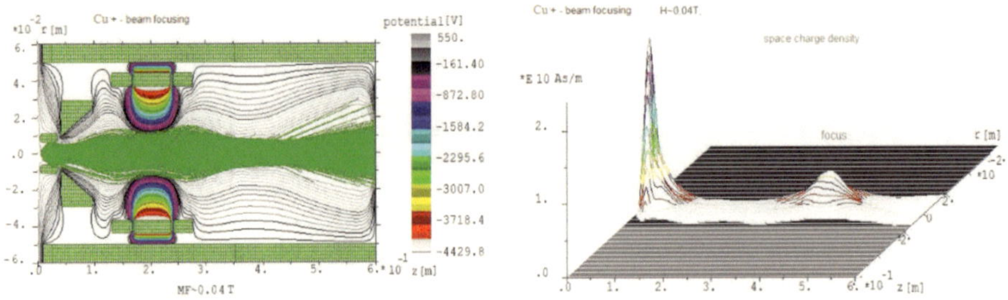

a) b)

Figure 17. Influence radial electron beam on beam focusing: (a) trajectories of cu + beam; (b) space charge density in the beam.

high-current ion beam can improve the delivery of plasma flow to a substrate, as well as providing micro-droplet removal via the fast electrons within the lens region.

We considered the transport aspect and effect of fast electrons on transport characteristics low-energy ion plasma beam. **Figure 17** shows results of modeling transport copper ion beam with current 100A through plasma lens with presence of fast electrons in the volume. One can see that accumulation high-energy electrons on the axis and their space charge can provide additional compensation and focusing of the high-current ion beam.

4.3. Conclusion remarks

A new approach was proposed for the elimination of micro-droplets from the dense metal plasma, based on evaporation and thus removal of micro-droplets from the arc plasma by energetic electrons within the electrostatic PL. These electrons are generated self-consistently by secondary emission in the near-wall plasma layer from the internal surface of the lens central electrode and serves to evaporate, and thus remove micro-droplets from the plasma flow. The experiments [29] demonstrate the effectiveness of the electrostatic PL for focusing and manipulating wide-aperture, high-current, low-energy, streaming metal ion plasma flows. In these experiments, the self-sustained focusing of high-density, wide-aperture, low-energy ion plasma flow was observed. It has been shown that the presence of fast electrons in the volume of the plasma lens both improves the propagating ion plasma flow toward the substrate and introduces additional energy for effective evaporation and elimination of micro-droplets from the plasma flow.

5. Conclusions

We review some new results of development novel generation of cylindrical plasma devices based on the concept of magnetic insulation of electrons and equipotentialization magnetic field lines and an electrostatic PL configuration. The PL configuration of crossed electric and magnetic fields provides an attractive and suitable method for establishing stable plasma discharge at low-pressure. Use of plasma lens configuration in this way was elaborated, explored and developed some cost efficiency, low-maintenance plasma devices for ion treatment, and deposition of exotic coatings with given functional properties. These devices make using permanent magnets and possess considerable flexibility with respect to spatial configuration. They can be operated as a stand-alone tool for ion treatment of substrates, or as part of integrated processing system together with cylindrical magnetron sputtering system, for coating deposition.

The research described has shown a principal possibility of a positive space charge cloud creation for negative charged particle beam focusing. It was demonstrated that possible create a positive space charge due to the magnetic electron insulation and reach strong focusing electric field, which sufficient for creation of short-focus elements to be used in systems for manipulating intense beams of negative ions and electrons. The further development of focusing properties of these devices demands study of intense negatively charged particle beams passing through the systems.

The simulations and experimental results demonstrate the high-efficiency of the electrostatic PL for focusing and manipulating wide aperture, high-current, low-energy, heavy metal ion plasma flow. These results open up new attractive way for further development and application erosion plasma sources for synthesis of exotic films and coatings with given functional properties. Some preliminary theoretical and experimental studies [27–29] have been carried out, providing confidence and optimism that the proposed idea for micro-droplet elimination has good potential for success.

Acknowledgements

This work was support by Grant NASU No PL-18 and No P-13_18.

Author details

Iryna Litovko[1]* and Alexey Goncharov[2]

*Address all correspondence to: ilitovko@ukr.net

1 Institute for Nuclear Research NAS of Ukraine, Kiev, Ukraine

2 Institute of Physics NAS of Ukraine, Kiev, Ukraine

References

[1] Morozov A. Focusing of cold quasi-neutral beams in electromagnetic fields. Doklady of the Academy of Sciences of the U.S.S.R. 1965;**163**(6):1363

[2] Morozov, Lebedev S. Plasmaoptics. In: Leontovich M, editor. Reviews of Plasma Physics. New York: Consultants Bureau; 1975

[3] Goncharov A, Dobrovolsky A, Zatuagan A, Protsenko I. High-current plasma lens. IEEE Transactions on Plasma Science. 1993;**21**(5):573

[4] Morozov AI, Semashko NN. On the mass separation of quasineutral beams. Technical Physics Letters. 2002;**28**(12):1052-1053

[5] Goncharov A, Brown I. High-current heavy ion beams in the electrostatic plasma lens. IEEE Transactions on Plasma Science. 2004;**32**(1):80

[6] Goncharov AA, Brown IG. Plasma devices based on the plasma lens-Areview of results and applications. IEEE Transactions on Plasma Science. 2007;**35**(4):986-991

[7] Goncharov A. The electrostatic plasma lens. Review of Scientific Instruments. 2013;**84**: 021101

[8] Goretskii V, Soloshenko I, Shchedrin A. Space charge lens for focusing negative-ion beams. Plasma Physics Reports. 2001;**27**(4):335-339

[9] Goncharov A, Evsyukov A, Dobrovol'skii A, Litovko I. Computer model for plasma devices based on the plasma lens configuration. Advances in Applied Plasma Science. 2007;**6**:5-8

[10] Potter D. Methods of Calculations in Physics. Moscow: Mir; 1975

[11] Boris JP, Lee R. Optimization of particle calculations in 2 and 3 dimensions, Communications in Mathematical Physics. 1969. 12. p. 131

[12] Goncharov A, Dobrovolskiy A, Dunets S, Evsyukov A, Litovko I, Gushenets V, Oks E. Positive-space-charge lens for focusing and manipulating high-current beams of negatively charged particles. IEEE Transactions on Plasma Science. 2011;**39**(6):1408-141113

[13] Goncharov AA, Dobrovolskiy AM, Dunets SP, Litovko IV, Gushenets VI, Oks EM. Electrostatic plasma lens for focusing negatively charged particle beams. The Review of Scientific Instruments. 2012;**83**(02B)

[14] Gushenets V, Goncharov A, Dobrovolskiy A, Dunets S, Litovko I, Oks E, Bugaev A. Electrostatic plasma lens focusing of an intense electron beam in an electron source with a vacuum arc plasma cathode. IEEE Transactions on Plasma Science. 2013;**41**(4, Part 3):2171-2174

[15] Morozov AI. Introduction to Plasmadynamics (Fismatlit, Moscow);2008. p. 572

[16] Goncharov AA, Brown IG. Physics and application of plasma accelerators. In: Morozov AI, editor. Minsk: Science and Technology; 1974

[17] Goncharov AA, Dobrovolsky AN, Naiko IV, Naiko LV, Litovko IV. Modes of plasmadynamical system with closed electron drift and open walls. 2017 IEEE International Young Scientists Forum on Applied Physics and Engineering (YSF-2017). Book of papers YSF-2017, p. 267-270. DOI: 10.1109/YSF.2017.8126633

[18] Goncharov AA, Litovko IV, Dobrovolsky AN, Najko LV, Najko IV. Novel modification of hall-type ion source (study and the first results). Review of Scientific Instruments. 2016;**87**: 02A501

[19] Litovko IV, Goncharov AA, Dobrovolskiy AN, Naiko LV, Naiko IV. Modelling of new generation plasma optical devices. Nukleoika. 2016;**61**(2):207-212

[20] Grishin D, Leskov L, Kozlov N. Plasma Accelerators, Mashinostroenie, Moscow-231 c1983

[21] Vahedi V, Surendra M. A Monte Carlo collision model for the particle-in-cell method application for argon and oxigen discharges. Computer Physics Communications. 1995; **87**:179-198

[22] Anders A. Approaches to rid cathodic arc plasmas of macro- and nanoparticles: A review. Surface and Coatings Technology. 1999;**120–121**:319

[23] Boxman RL, Goldsmith S. Macroparticle contamination in cathodic arc coatings: Generation, transport and control. Surface and Coatings Technology. 1992;**52**:39

[24] Schülke T, Anders A. Velocity distribution of carbon macroparticles generated by pulsed vacuum arcs. Plasma Sources Science and Technology. 1999;**8**:567

[25] Aksenov II. The vacuum arc in erosion plasma sources. Kharkov: NSC KIPT. 2005:212

[26] Anders A. Growth and decay of macroparticles: A feasible approach to clean vacuum arc plasmas? Journal of Applied Physics. 1997;**82**:3679-3688

[27] Goncharov AA, Maslov VI, Fisk A. Novel plasma-optical device for the elimination of droplets in cathodic arc plasma coating. In: 55th Annual Technical Conference Proceedings of the Society of Vacuum Coaters (SVC); April 28-May 3, 2012; Santa Clara, California, USA. pp. 441–444

[28] Goncharov AA. Recent development of plasma optical systems (invited). Review of Scientific Instruments. 2016;**87**:02B901. DOI: 10.1063/1.4931718

[29] Bugaev A, Dobrovolskiy A, Goncharov A, Gushentets V, Litovko I, Naiko I, Oks E. Self-sustained focusing high-current heavy metal ion plasma flow produced by vacuum arc. Journal of Applied Physics. 2017;**121**(4):043301

Chapter 15

Plasma Damage on Low-*k* Dielectric Materials

Yi-Lung Cheng, Chih-Yen Lee and Chiao-Wei Haung

Additional information is available at the end of the chapter

http://dx.doi.org/10.5772/intechopen.79494

Abstract

Low dielectric constant (low-*k*) materials as an interconnecting insulator in integrated circuits are essential for resistance-capacitance (RC) time delay reduction. Plasma technology is widely used for the fabrication of the interconnects, such as dielectric etching, resisting ashing or stripping, barrier metal deposition, and surface treatment. During these processes, low-*k* dielectric materials may be exposed to the plasma environments. The generated reactive species from the plasma react with the low-*k* dielectric materials. The reaction involves physical and chemical effects, causing degradations for low-*k* dielectric materials. This is called "plasma damage" on low-*k* dielectric materials. Therefore, this chapter is an attempt to provide an overview of plasma damage on the low-*k* dielectric materials.

Keywords: plasma, low dielectric constant, porosity, plasma damage, Cu interconnects, back end of line, damascene

1. Introduction

To improve the performance of integrated circuits (ICs), the device dimensions are continuous scaling down. However, as the technology node of ICs is advanced to 0.25 μm, the interconnect-induced delay outpaces the gate delay, becoming the main obstacles for the downscaling [1–3]. This interconnect-induced delay is so-called resistance-capacitance (*RC*) delay, which is produced by the conductors and insulators in the interconnects [4–6]. With decreasing the device dimensions, both the resistance and the interline capacitance increase due to the decrease of the conductor cross section, the increase of the wire length, and the reduction of interline spacing. Hence, the *RC* delay is significantly increased with the advance of the technology node.

In order to slow down the increase of *RC* delay, the introduction of new materials to the back-end-of-line (BEOL) interconnects is needed. Aluminum (Al) had been replaced by copper (Cu) as a conductor dielectric because Cu can provide a lower resistivity (ρ) [7]. In the case of the interconnecting insulator, the traditional SiO_2 dielectric had been replaced by the low-*k* materials with the relative dielectric constant (*k*) lower than 4.0 (SiO_2 *k* value) [8–10].

Additionally, the integration method for Cu/low-*k* interconnects must be changed because Cu etching is very challenging due to nonvolatile by-products. Traditional metal etching approach had been replaced by a damascene process [11]. In a damascene process, plasma technology is widely used because it can provide an isotropic process and a fast rate. Thus, these changes make the low-*k* materials to direct contact with the plasma, such as dielectric etching, photo strip, barrier metal deposition, and surface treatment. Under the plasma irradiation, low-*k* materials are sensitive to chemical modification, resulting in an increased *k* value. This is so-called plasma damage [12–15], becoming the main impediment to a successful integration of low-*k* materials into ICs.

In this connection, this chapter is an attempt to provide an overview of plasma damage on the low-*k* materials. This chapter is organized as follows: in Section 3, we introduce the low-*k* materials and plasma. Next, in Section 5, the processing with plasma damage on the low-*k* materials during interconnects fabrication is identified. Then, in Section 4, the results of plasma damage on the low-*k* materials based on our group's investigation are summarized. Finally, short conclusion is provided in Section 5.

2. Low-*k* materials and plasma

2.1. Low-*k* materials

The dielectric constant of materials can be typically described by Clausius-Mossotti Equation [16]:

$$\frac{k-1}{k+2} = \frac{4\pi N}{3}\alpha \tag{1}$$

where $k = \varepsilon/\varepsilon_0$, ε, and ε_0 are the dielectric constants of the material and vacuum, N is the number of molecules per unit volume (density), and α is the total polarizability, including electronic (α_e), distortion (α_d), and orientation (α_o) polarizabilities. According to Eq. (1), the dielectric constant of materials can be reduced by two strategies: decreasing the total polarizability (α) and density (N). Reducing the polarizability can be achieved by the use of low polar bonds (like C–C, C–H, Si–CH_3, etc.), and reducing the film's density can be obtained by means of the introduction of porosity. **Table 1** summarizes the classification of low-*k* materials and their corresponding dielectric constants.

Low-*k* materials can be divided into several categories: silica-based, silsesquioxane (SSQ)-based, organic polymers, and amorphous carbon low-*k* materials [17–20]. The last three categories have integration issue due to the weak mechanical strength; therefore, they are not officially production in the semiconductor industry.

Classification	Material	Dielectric constant (k)
Silica based	SiOF (FSG)	~3.5
	SiCOH	~2.8
Silsesquioxane (SSQ) based	HSQ	~3.0
	MSQ	~2.5
Polymer	Ploy(arylene ether) (PAE)	~2.6
	Polyimide	~2.3
	Parylene-N(-F)	~2.4
	Teflon (PTFE)	~2.0
Amorphous Carbon	C-C (-F)	~2.0
Porous	Porous SiCOH, MSQ, PAE	< 2.0
Air gaps	Air	~1.0

Table 1. Low-*k* dielectric classification.

The silica-based low-*k* materials have been successfully integrated in a microprocessor due to high chemical and thermal stability. The silica-based materials have the tetrahedral basic structure of SiO_2. Silica has a molecular structure, in which each Si atom is bonded to four oxygen atoms and each oxygen atom is bonded to two silicon atoms ($SiO_{4/2}$). Each silicon atom is at the center of a regular tetrahedron of oxygen atoms.

The first-generation low-*k* material in semiconductor production line was fluorinated silicon glass (FSG or SiOF), in which the Si–O bond is replaced by the less polarizable Si–F bond. FSG materials were used at the 0.18 μm technology node with the dielectric constant from 3.5 to 3.8, depending on the concentration of Si–F bond [21, 22].

Next, the second-generation low-*k* material was the organosilicate glass (SiCOH), in which the Si–O bond is replaced by the less polarizable Si–CH_3 bond. The *k* value of the SiCOH material is in the range of 2.6–3.0, depending on the number of CH_3 groups built into the structure. So, SiCOH materials were successfully integrated in some 130 nm and 90 nm products [23, 24]. Generally, FSG and SiCOH materials were deposited by plasma-enhanced chemical vapor deposition (PECVD). Moreover, both fluorine and carbon increase the interatomic distances or "free volume" of silica. This provides an additional decrease of dielectric constant but decreases the film density. Since the CH_3 group has a larger volume and is hydrophilic, SiCOH materials have a lower density (~1.2–0.4 g/cm^3) and tend to be hydrophilic.

The limitation of *k* value for SiCOH materials is ~2.6. To prevent or limit an increase in the BEOL capacitance in the advanced technology nodes (65 nm or below), it requires a new low-*k* material with a further lower-*k* value (< 2.5). To meet this goal, the introduction of porosity in the low-*k* SiCOH materials is required because air can provide the minimum *k* value of ~1.0. The produced low-*k* material is a so-called porous SiCOH dielectric, which can be fabricated either by the structural or the subtractive method [25–27]. The latter method is widely accepted because the produced film is more thermally stable and can provide a lower-*k* value. In the subtractive method, the films are deposited as a dual-phase material, using a mixture of a SiCOH skeleton precursor with an organic porogen precursor. The popularly used skeleton precursor is diethoxymethylsilane (DEMS). The used organic porogen precursor must have sufficient volatility for easy removal. The used molecules are alpha-terpinene (ATRP), bicycloheptadiene (BCHD), or cyclooctane (C_8H_{16}). Hence, in order to remove the labile organic fraction in the as-deposited films, curing process has to be done after the deposition [8, 10, 27]. By this way, a porous film can be formed. Thermal curing, electron beam, or ultraviolet (UV) irradiation can be used to achieve this work. Generally, UV-assisted curing for the fabrication of porous SiCOH dielectrics is widely adopted by the semiconductor industry because it can also rearrange the film's structure and enhance the cross-linking of the skeleton. This provides a big help to improve the mechanical strength for porous SiCOH dielectrics.

The *k* value of porous SiCOH dielectrics can be scaling down by increasing the porosity and pore size simultaneously. However, this makes materials to become softer. Moreover, both the dielectric breakdown field and leakage current are degraded. Furthermore, as the porosity or pore size increases to a critical value, the pores can be connected each other to form so-called open pores. The open pores can be served as the easier penetration path into the bulk of the low-*k* material for active reactants [28]. Thus, more challenges will be addressed as porous SiCOH dielectrics are integrated in the advanced technology nodes.

2.2. Plasma

In a vacuum system, plasma can be produced by introducing the process gas and applying the power. The process gas can be underwent ionization, excitation/relaxation, and dissociation under the power. Therefore, energetic ions, electrons, light (from deep vacuum ultraviolet (VUV) to infrared (IR)), and highly reactive radicals are produced in the plasma [29, 30]. In semiconductor processing, plasma technology can be used for ion implantation, etching, and deposition. The ion implantation processing is achieved by the energetic ions. The etching processing involves both physical and chemical reactions, which are related to the energetic ions and the highly reactive radicals, respectively. The deposition processing only relies on the highly reactive radicals for chemical reaction.

To produce the plasma, three main reactors are used: capacitively coupled plasma (CCP), inductively coupled plasma (ICP), and remote or downstream plasma (RP or DSP) [31]. The energy transfers are through capacitive coupling by parallel electrodes, inductive coupling by a coil, and microwaves for CCP, ICP, and RP systems, respectively. In the CCP and ICP systems, light from VUV to IR, energetic ions, electrons, and highly reactive radicals are presented. In the RP reactors, however, the plasma generation region is usually separated from the processing region. Additionally, a grid between the plasma and the substrate is used for charge neutralization, and a special measure is designed to minimize the photon flux. As a result, only reactive radicals or dissociated molecules or atoms can reach the surface of the wafer. This minimizes the damage from light and/or high-energy species. Due to the absence of ions, the RP reactors cannot provide patterning etching.

In the ICP systems, there are two applied RF power: one is source power (top power), and the other is bias power (bottom power). Therefore, plasma density and ion energy can be controlled separately. Additionally, the ICP system has the highest plasma density with 10^{11}–10^{12} electrons/cm^3 [31]. The plasma density of CCP system is 10^9–10^{10} electrons/cm^3. The RP system has the lowest plasma density. Due to anisotropic etching property provided by ion bombardment, ICP and CCP systems are usually used for pattern etching. Since dielectric films are very sensitive to ion bombardment and ICP reactors lack passivating species required by typical dielectric etching, CCP reactors are mostly used for dielectric patterning etching. On the other hand, ICP reactors are often used for conductor patterning etching due to the etching rate consideration. To avoid damage by ion bombardment and UV light irradiation or no need anisotropic etching in the plasma process, RP reactors are the best choice. So, cleaning and resist stripping processes during semiconductor fabrication can be done by RP reactors.

2.3. Plasma damage mechanism

The plasma-induced damage on the low-*k* dielectrics is a complex phenomenon involving both physical and chemical effects. Ion bombardment on the low-*k* dielectrics represents the physical effect. This effect depends on the energy distribution and flux for each ionic species. The chemical effect involves photochemistry induced by the UV radiation and chemical reaction between the radicals and low-*k* constituents. Under physical and chemical reactions in the plasma, the surface of low-*k* dielectrics is modified. The modification depth is related to the ion energy, diffusion of active radicals (O, H, F, etc.), and porosity and constituents in the low-*k* material [32, 33].

The plasma damage on low-*k* dielectrics makes the increase of the dielectric constant, the changes in bonding configuration, the formation of carbon-depleted layer, film shrinkage, and surface densification.

The depletion of carbon is mainly caused by active radicals through chemical reactions. Due to the loss of hydrophobic CH_3 groups, the surface of low-*k* dielectrics becomes hydrophilic and adsorbs moisture. Therefore, a drastically increase in the *k* value and leakage current and a degradation in the dielectric breakdown were detected for plasma-treated low-*k* dielectrics.

3. Low-*k* plasma damage during interconnects fabrication

As Al/SiO_2 interconnects had been transferred to Cu/low-*k* interconnects, the fabrication method was also changed. "Damascene" process has been used to fabricate Cu/low-*k* interconnects because Cu cannot be easily patterned by reactive ion etching (RIE) due to the low volatility of Cu etching by-products, such as Cu chlorides and Cu fluorides [34]. Generally, "dual-damascene" process, in which both via and trench are patterned simultaneously, is widely used. The sequence of via and trench patterning can be changed. Via-first dual-damascene process, in which via is first patterned, is preferred [35]. The process flow of via-first dual damascene is plotted step by step, as shown in **Figure 1**.

The induced plasma damage on low-*k* dielectrics during the fabrication of Cu/low-*k* interconnects by the use of via-first dual-damascene process is described below:

After processing of Metal-1 (M-1), the etching stop layer (Cu barrier dielectric layer) is firstly deposited by PECVD method. The used material can be SiN, SiC, or SiCN. Before deposition, NH_3 or H_2 plasma clean is performed to remove copper oxide (CuO_x) for adhesion improvement [36, 37]. Both these two steps would damage the underlying low-*k* dielectric. Then, a PECVD SiCOH low-*k* dielectric film is deposited for the Via-1 (V-1)/Metal-2 (M-2) patterning. Due to the presence of the etching stop layer, the plasma damage is seldom occurred in this step. Next, Via-1 and Metal-2 trench are subsequently patterned. Via-1 patterning is stopping on the etching layer. Then, before Metal-2 trench patterning, the plug is filled into the Via-1 to avoid etching during Metal-2 trench etching. Finally, resist removal and etching stop layer opening are subsequently performed to complete the dual-damascene patterning.

In the Via-1 and Metal-2 trench patterning, the etching process induces plasma damage not only on the horizontal surfaces but also on the vertical surfaces (sidewall).

The damaged layer on the horizontal surfaces can be removed as the etching proceeds. Therefore, the resulted damage on the low-*k* dielectric is the result of a competition between the etching rate and the diffusion rate of active species causing the damage. The non-damaging process can be achieved by using higher etching rate process. However, for vertical surfaces, the damage is still remained after etch. The damage is more minor due to the absence of ion

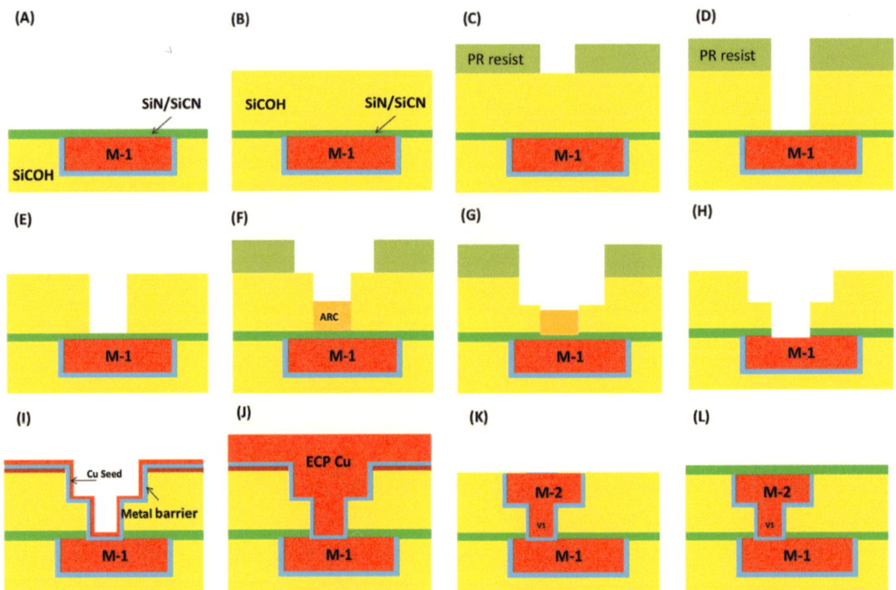

Figure 1. Via-first dual-damascene process flow for Cu/low-*k* interconnects. (A) Etching stopping layer deposition (SiN, SiC, SiCN, SiCOH, SiO_2). (B) Low-*k* dielectric (SiCOH) deposition. (C) Via-1 lithography. (D) Via-1 etching. (E) Photoresist ashing or stripping. (F) Via-1 ARC plug and M-2 trench lithography. (G) M-2 trench etching. (H) ARC plug removal, photoresist ashing, and stripping etching stop layer opening. (I) Cu metal barrier and seed layer deposition. (J) Cu ECP deposition. (K) Cu CMP. (L) Etching stopping layer deposition (repeated (A)).

bombardment. The detrimental effect on the low-*k* dielectrics is caused by photoresist process because oxygen (O_2) is widely used as plasma gas due to high reactivity of O radicals [38, 39]. To minimize the plasma damage during the photoresist process, H_2-based plasma in the RP reactor is a viable alternative. To facilitate the removal rate of photoresist, the operation temperature can be elevated [40, 41].

Then, metallization process is preceded in the dual-damascene structure. Cu barrier layer, Cu seed layer, and bulk Cu layer are subsequently deposited. Finally, Cu chemical-mechanical polishing (CMP) process is used to remove the excess metal over the field regions. Thus, a layer of Cu dual-damascene structure (via and trench) is finished. In these steps, Cu barrier layer and seed layer are performed by PVD sputtering with using plasma. The former step would cause damage on the low-*k* dielectrics due to the direct contact with the dielectric. The purpose of this step is to prevent Cu from diffusing into the dielectric, and the typically used material is a TaN/Ta barrier layer. It should be mentioned that plasma cleaning before Cu barrier layer deposition is necessary because the underlying Cu film is opening. This plasma cleaning can be done either by Ar physical bombardment or H_2 chemical reaction. However, low-*k* dielectrics are damaged under such plasma cleaning.

After completing the Cu metallization fabrication, the above steps are repeated for each metal level. After the last metal layer is fabricated, thick dielectric passivation layer (e.g., SiO_2/SiN bilayer) is deposited, and via is opened to the bond pads.

4. Low-*k* plasma damage

4.1. Plasma damage characterization

To characterize the plasma damage on the low-*k* dielectrics, several methodologies can be used to detect the physical and chemical changes of low-*k* dielectrics after irradiation of plasma. The plasma induces a dense, hydrophilic, SiO_2-like layer at the top surface of the low-*k* dielectric. The thickness of this layer can be measured using spectral reflectivity or ellipsometry with bilayer model, scanning electron microscope (SEM), or transition emitting microscopy (TEM). **Figure 2** displays TEM image of the porous low-*k* dielectrics after O_2 plasma treatment. A distinct layer is formed at the top surface of the film.

X-ray reflectivity (XRR) is another method to determine the density, thickness, and roughness of both pristine and damaged low-*k* layers through software data fitting [42]. **Figure 3** shows the XRR density profile of the low-*k* film after He plasma. The result demonstrates that He plasma creates a thin densification layer in the top part of the low-*k* film. The thickness of this densification layer is close to 17 nm. The density of the bulk layer in the pristine material density is constant and remained unchanged. However, the top of the densification layer has a higher density [43].

"HF decoration" method [44] can be used to detect the modification layer induced by plasma. This method is based on the fact that a pristine low-*k* dielectric is usually not dissolved or

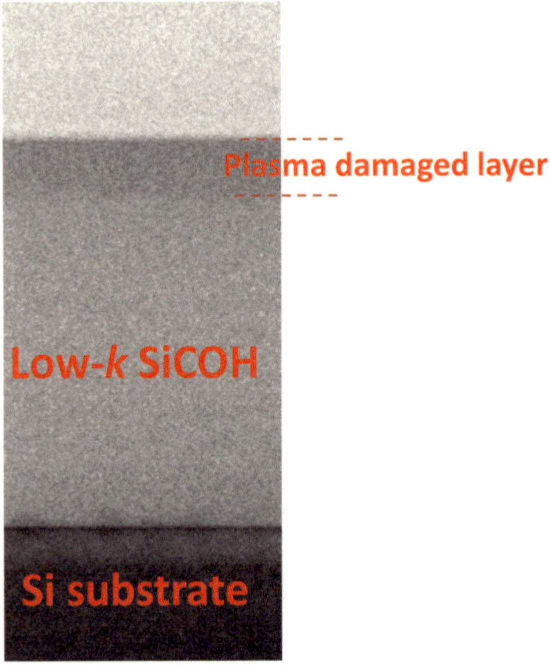

Figure 2. TEM image of porous low-*k* dielectric after O_2 plasma treatment.

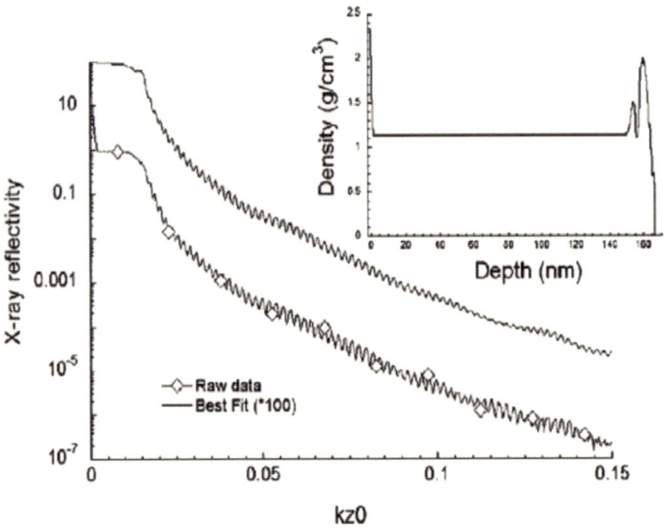

Figure 3. XRR spectrum of low-*k* dielectrics after He plasma treatment [43].

slowly dissolved in the diluted HF solution. In contrast, a plasma-induced damaged layer is attacked by HF very quickly. Therefore, following the HF decoration, the thickness loss is equal to the thickness of the damaged layer.

A given system of solid and liquid (or vapor) at a given temperature and pressure has a unique equilibrium contact angle. The measured angle is water contact angle (WCA). It can be used to quantify the wettability of a solid surface by a liquid via the Young equation. If the used liquid molecules are strongly attracted to the solid molecules, the liquid drop then will completely spread out on the solid surface, corresponding to a WCA of 0°. This case can be occurred at bare metallic or ceramic surfaces for water liquid. As an oxide layer or contaminant is on the solid surface, WCA value significantly increases. Generally, the solid surface tends to be hydrophilic if WCA value is smaller than 90°, while if WCA value is larger than 90°, the solid surface is considered to be hydrophobic. For low-*k* dielectrics, WCA measurement is a power method to determine the films' hydrophobicity. If the used low-*k* dielectrics are hydrophilic, they tend to absorb moisture in the air, increasing the dielectric constant. Moreover, as the plasma is treated on low-*k* dielectrics, Si–OH/H–OH bonds can be formed because the plasma-generated dangle bonds absorb moisture. **Figure 4** compares the WCA values and images of the pristine and plasma-treated SiCOH low-*k* dielectrics. The WCA value of the as-deposited SiCOH low-*k* dielectrics is larger than 85° due to the presence of hydrophobic Si–CH_3 groups. After plasma irradiation, the loss of Si–CH_3 groups and the formation of Si–OH/H–OH bonds result in a decreasing WCA value, making the low-*k* dielectric to be more hydrophilic.

Fourier transform infrared (FT-IR) spectroscopy is a common technique to characterize the structure of SiCOH low-*k* dielectrics [45, 46]. **Figure 5** compares the FTIR spectrum of the pristine and plasma-treated low-*k* dielectrics. Absorption bands located at ~950–1250 and ~1273 cm^{-1} correspond to Si–O–Si and Si–CH_3 groups, respectively, which are the main representative. Additionally, Si–H bending and C–H_x stretching located at 2200–2250 and 2850–3100 cm^{-1} are detected. The appearance of the absorbance of the Si–OH and H_2O groups at 3200–3500 cm^{-1} depends on hydrophobic properties of the film. For the pristine SiCOH low-*k* dielectrics, no peak at 3200–3500 cm^{-1} is detected, representing that no moisture is present in the film, which is consistent with WCA result. After NH_3 plasma irradiation, the intensities

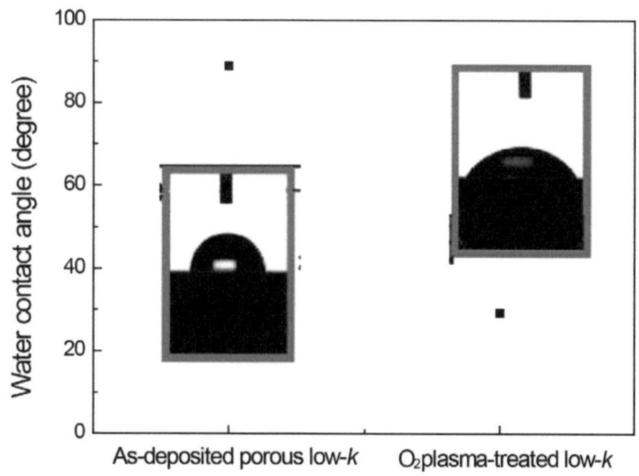

Figure 4. WCA values and images of porous low-*k* dielectrics after O_2 plasma treatment.

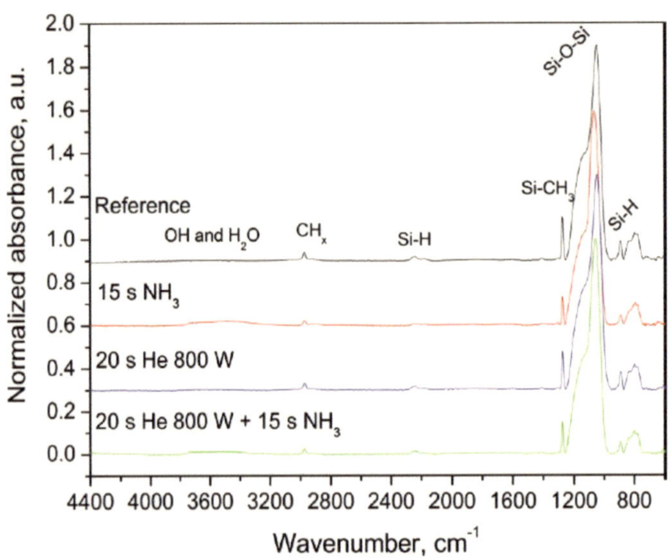

Figure 5. FT-IR absorption spectra of low-*k* dielectrics before and after plasma treatment in the range of 4400–400 cm^{-1} [47].

of the Si–CH_3 and Si–H absorbances are decreased, while the absorbances of the Si–OH/H_2O groups are increased. However, the low-*k* dielectric is pretreated with He plasma, and it can suppress the formation of the Si–OH/H_2O groups [47].

The Si–O–Si bridging in the 900–1250 cm^{-1} can be deconvoluted into three peaks centered at 1129, 1063, and 1023 cm^{-1}, corresponding to the Si–O–Si cage-like structure with a bond angle of approximately 150°, Si–O–Si network with a bond angle of 140°, and Si–O–Si suboxide structure with a bond angle of less than 140°, respectively. Other contributions from C–O–C and Si–O–Si asymmetric stretching will also be overlapped with the Si–O–Si asymmetric stretching in the broadband at 1000–1200 cm^{-1} [45]. The intensity of Si–O–Si bonds slightly increases, and this peak shifts to a higher wavelength after O_2 plasma treatment.

X-ray photoelectron spectroscopy (XPS) is a surface-sensitive spectroscopic technique to quantitatively measure a material's elemental composition. XPS can also be operated in a "depth-profiling mode" to analyze the elemental composition throughout the film by using ion etching/sputtering technology. For SiCOH low-*k* dielectrics, C, O, and Si elements can be detected, while H element cannot be detected.

In the pristine SiCOH low-*k* dielectrics, a homogeneous chemical composition was expected, but the ratio of these elements depends on the used materials. For the plasma-treated sample, the top surface exhibits a high initial oxygen concentration coupled to a very low carbon concentration. A gradual increase in carbon content and a concomitant decrease in oxygen concentration were observed with the film depth, as shown in **Figure 6**. As the atomic concentrations are back to a level same with the pristine low-*k* dielectric, the depth is corresponding to the plasma-damaged layer. Moreover, this plasma-damaged layer is not a homogeneous layer.

The dielectric constant (*k*), the leakage current, the breakdown voltage (or field), and the breakdown time of low-*k* dielectrics are measured using metal-insulator-semiconductor (MIS) capacitor structures, which can be fabricated by evaporation of aluminum through a metal shadow mask to form Al dots on the film. Before measurements, the samples are required to remove the physically absorbed water by annealing at 100–150°C. The *k* value of the low-*k* dielectric is determined from the measured capacitance by capacitance-voltage (*C-V*) measurements at a frequency of 10 kHz. The film thickness and the dot area must be precisely measured in order to obtain the reliable *k* value. The leakage current and the breakdown voltage (or field) are determined by current-voltage (*I-V*) measurements. The breakdown field is calculated by the measured breakdown voltage divided by film thickness. The leakage current *J* (A/cm^2) is recorded as a function of field strength until the breakdown field is reached. It is usually reported at a low field of 1–2 MV/cm. The breakdown time is measured by using time-dependent dielectric breakdown (TDDB) tests. In a TDDB test, a constant voltage (field) is applied to the MIS capacitor structure with a low-*k* dielectric, and the leakage current is monitored with stress time. The dielectric breakdown time is recorded as the stress time at a sudden rise of the leakage current density. The applied voltage (field) must be lower than the measured breakdown voltage (field). In a real Cu/low-*k* interconnects, comb/serpentine (also called meander fork) or comb/comb (fork/fork) patterns are typically used to measure the interline capacitance, the leakage current, the breakdown field, and the dielectric breakdown time.

4.2. Plasma damage on the electrical characterization and reliability

In this section, the results of plasma damage on the low-*k* dielectrics from our group's investigation are reported in terms of the effects on the electrical characterization and reliability. The experimental detail deposition can be found elsewhere [39, 48–50].

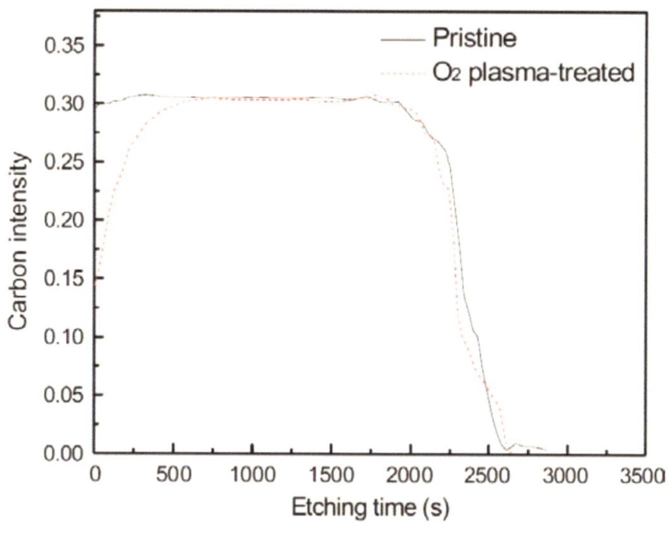

Figure 6. Carbon concentration of XPS depth-profiling for pristine and O_2 plasma-treated low-*k* dielectrics.

4.2.1. O_2 plasma damage

4.2.1.1. Plasma process dependence

Figure 7 shows the variation in the *k* value of a porous low-*k* dielectric after O_2 plasma treatments with various plasma conditions (power, treatment time, and O_2 flow rate). The *k* value of the pristine porous low-*k* dielectric was 2.56. The *k* values of porous low-*k* dielectrics increased after plasma treatment. The increasing magnitude increased with increasing the RF power and the treatment time but slightly deceased with increasing the O_2 flow rate. More reactive oxygen species (ions and radicals) formation and a deeper penetration depth for a higher RF power and a longer treatment time in an O_2 plasma treatment can be responsible for a larger change in the *k* value. The negative dependence on the O_2 flow rate can be attributed to the decreased dissociation rate of O_2 gas due to a fixed RF power, leading to a decreased reactive oxygen species (ions and radicals) in the plasma.

Figure 8(a) compares the leakage current density and the stress electric field for the pristine and O_2 plasma-treated low-*k* films. At a lower electric field, the leakage current density increases with increasing the electric field (region I). Then, the leakage current density reaches a plateau without significant variation (region II). Finally, the leakage current density suddenly jumps, whose value is over 10^{-2} A/cm^2. This electric field is defined as the dielectric breakdown electric field (region III). For O_2 plasma-treated low-*k* dielectrics, a higher leakage current in region I, a longer duration in region II, and a lower breakdown electric field in region III were detected.

Figure 8(b) compares the leakage current densities at 1 MV/cm and 2 MV/cm and the breakdown electric field of the O_2 plasma-treated low-*k* dielectrics with various treatment conditions. Similar to the result of *k* value change, a higher RF power, a longer treatment time, and a lower O_2 flow rate can result in the largest increase of the leakage current density for

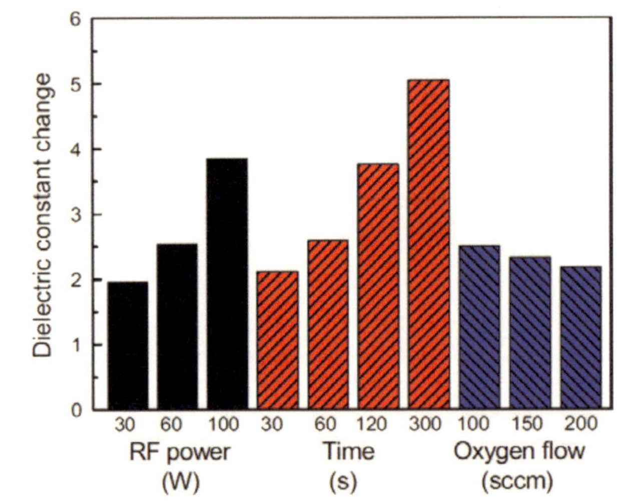

Figure 7. Dielectric constant change of porous low-*k* dielectrics under various O_2 plasma treatment conditions (standard condition: RF power = 60 W; time = 60 s; oxygen flow = 100 sccm).

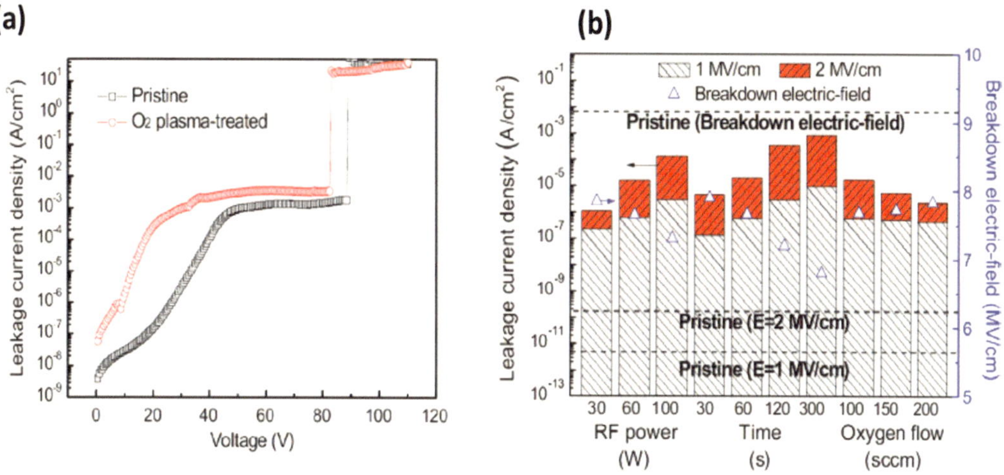

Figure 8. (a) Leakage current density versus applied voltage for pristine and O_2 plasma-treated low-*k* dielectrics. (b) Leakage current densities at 1 and 2 MV/cm and dielectric breakdown field of porous low-*k* dielectrics under various O_2 plasma treatment conditions.

O_2 plasma-treated low-*k* dielectrics. A higher leakage current density also leads to a lower breakdown electric field for the O_2 plasma-treated low-*k* dielectrics. The degrading electrical performance for O_2 plasma-treated low-*k* dielectrics can be attributable to more absorption of moisture, which provides ionic conduction pathways by releasing mobile ions (H^+, OH^-) [51].

Figure 9 compares the characteristic dielectric breakdown times ($T_{63.2\%}$) for O_2 plasma-treated low-*k* dielectrics with various treatment conditions. The characteristic dielectric breakdown time was determined from Weibull distribution, representing the time as 63.2% of the sample failed [52]. The degradation in the characteristic dielectric breakdown time becomes serious with an increase of RF power and treatment time or a reduction of O_2 flow rate. All process parameters in the O_2 plasma process would degrade low-*k* dielectric properties. The RF power and the treatment time of O_2 plasma treatment cause a more significant degradation.

4.2.1.2. Low-k dielectric dependence

Four kinds of low-*k* dielectrics were treated by O_2 plasma with various RF powers. The used low-*k* dielectrics were dense low-*k* (*k* = 3.02; called low-*k*_1), porogen low-*k* without UV (*k* = 2.92; called low-*k*_2) and with UV curing (2.56; called low-*k*_3), and porogen low-*k* with UV curing and followed by a RP H_2/He plasma treatment (*k* = 2.48; called low-*k*_4).

Figure 10 compares the change percentage of the *k* value as a function of RF power in O_2 plasma process for four different low-*k* films. The increasing magnitude is enlarged with an increase of RF power for all low-*k* dielectrics. The porogen-containing low-*k* dielectrics (low-*k*_2 and low-*k*_3) have a higher increase in the *k* value as compared to low-*k*_1. This suggests that porogen plays an important role for the low-*k* dielectrics under plasma irradiation. Furthermore, the highest increase in the *k* value is occurred in the low-*k*_2 (porogen-containing low-*k* film without UV curing). The UV irradiation on the low-*k* dielectrics not only removes the porogen to form pores but also strengthens the bonding strength of the low-*k* dielectrics

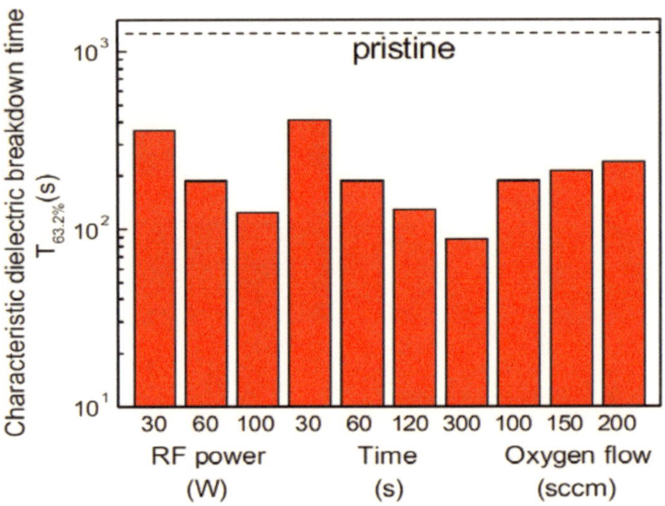

Figure 9. Characteristic dielectric breakdown times of porous low-*k* dielectrics under various O_2 plasma treatment conditions (standard condition: RF power = 60 W; time = 60 s; oxygen flow = 100 sccm).

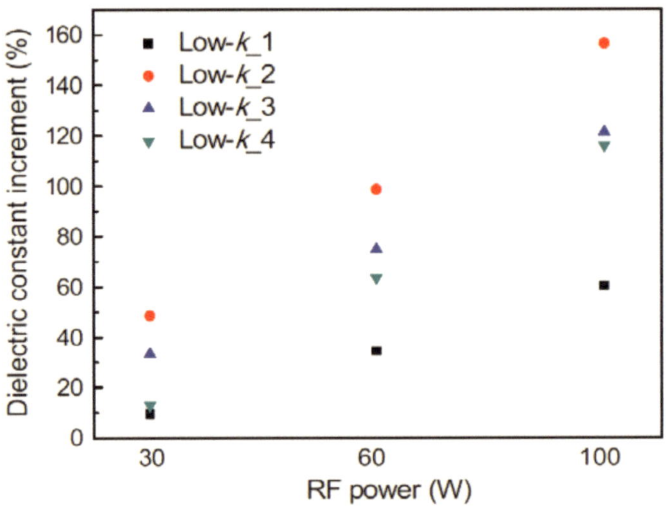

Figure 10. Change percentage in dielectric constant of different low-*k* dielectrics as a function of RF power in O_2 plasma treatment.

[53]. Therefore, the resistance to O_2 plasma damage can be reinforced. To enhance plasma resistance for porous low-*k* dielectrics, a RP H_2/He plasma treatment seems to be a possible method to alleviate the increase in the *k* value upon O_2 plasma process. The RP H_2/He plasma treatment can form a densification layer on the low-*k* dielectric's surface without damaging film's properties. This formation densification layer can effectively resist O_2 plasma damage and prevent active oxygen species to penetrate into the film. However, as the RF power further increases in O_2 plasma treatment process, the difference in the *k* value in low-*k*_3 and low-*k*_4

becomes smaller. This implies that the post-remote H_2/He plasma treatment on the porous low-*k* dielectric is becoming ineffective in preventing O_2 plasma damage as a higher RF power. As the RF power is increased in O_2 plasma treatment process, the more active oxygen species are produced, and these active species get more energy so as to penetrate the densification layer induced by remote H_2/He plasma treatment into a deeper region within the low-*k* dielectric, causing the bonding breakage and reaction with moisture.

Figure 11 compares the degradation in the characteristic dielectric breakdown times relative to those of the pristine low-*k* dielectrics as a function of RF power. The stress electric field was 6.8 MV/cm for all low-*k* dielectrics. The reliability performance continuously degrades with RF power. Additionally, for the same RF power, the degradation order is low-*k*_2 > low-*k*_3 > low-*k*_4 > low-*k*_1. This means that porogen, rather than pore, within a low-*k* dielectric is a key issue to cause the reliability degradation under O_2 plasma treatment. Since two phases (matrix and porogen) coexist in the low-*k* dielectrics, a weaker bonding strength can be deduced, resulting in a weaker resistance against O_2 plasma damage. By means of UV irradiation or the post-deposition plasma treatment, using remote H_2/He plasma to form a surface densification layer can alleviate the reliability degradation for the porous low-*k* dielectrics under O_2 plasma treatment.

4.2.1.3. Plasma component dependence

A "roof" structure, consisting of a top optical mask, is designed to isolate the ions, photons, and radicals to reach the porous low-*k* dielectrics [54]. Si, MgF_2, or CaF_2 was used as a mask in this study. The height of the gap was fixed at 1 cm. Under O_2 plasma treatment with different masks, the plasma species penetrating into the porous low-*k* dielectric through the gap is summarized in **Table 2**. For the porous low-*k* dielectric under plasma using various masks,

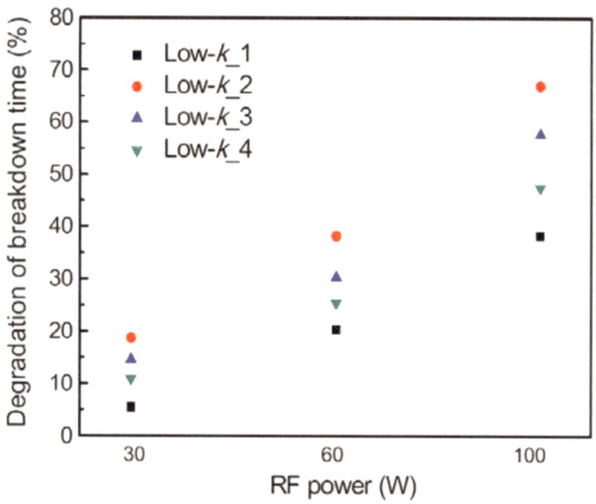

Figure 11. Degradation in characteristic dielectric breakdown time of different low-*k* dielectrics as a function of RF power in O_2 plasma treatment.

the thickness reduction, the Si–CH_3 group extraction, the Si–OH/H–OH bond formation, the top modification layer formation, the WCA value declination, the dielectric constant increment, and the dielectric breakdown field degradation were detected. The results indicate that all ions, photons, and radicals in the plasma cause negative impact on the porous low-*k* dielectrics, but they have different contributions. The maximum change is in the case of without mask. In Si mask case, the photons and the ions are blocked so that only oxygen radicals can react with the porous low-*k* dielectrics. Its plasma damage is less minor. In MgF_2 mask or CaF_2 mask, photons can penetrate but depends on the wavelength. With an addition of photon effect, the changes in the physical and electrical properties for the porous low-*k* dielectrics slightly increase. Furthermore, as ions are added in the plasma to react with the porous low-*k* dielectrics (without mask case), the changes become significant. This implies that the synergy between the radicals, the photons, and the ions in the plasma induces the highest degradation in the porous low-*k* dielectrics.

The plasma-induced damage mechanism is that the bonds inside the porous low-*k* dielectrics are broken by ion bombardment and then easily react with radicals to form the new bonds or Si–OH/H–OH bonds with a higher *k* value (~80). As for the contribution of photons, photons can weaken or broken the low-*k* dielectric's bonds, assisting the chemical reaction of radicals. The photons with a higher energy cause more bonding breakage, inducing a more degradation. Therefore, a higher degradation in the porous low-*k* dielectric underneath MgF_2 mask during O_2 plasma treatment was detected due to extra photon transmission with 120–250 nm wavelength.

Figure 12 plots the characteristic dielectric breakdown times ($T_{63.2\%}$) versus the applied electric field for O_2 plasma-treated low-*k* dielectrics with various masks. In a fixed electric field, the order of $T_{63.2\%}$ is pristine > Si mask > MgF_2 mask > CaF_2 mask > without mask, indicating that all ions, photons, and radicals in the plasma cause the dielectric reliability degradation. For example, in an electric field of 6.8 MV/cm, the dielectric lifetime degradation ratios are 43.17, 66.41, and 82.18% for Si mask, MgF_2 mask, and without mask cases, respectively, corresponding to radical, radicals + photon, and ions + photons + radical effects. By simple calculation, the contributions of radicals, photons, and ions were 43.17, 23.24, and 15.77%, respectively,

Optical mask	Plasma components on low-*k* film	Thickness reduction (%)	Top-modified layer (nm)	Etching rate of top-modified layer (nm/min)	Water contact angle (°C)	Methyl depletion (ΔSi – CH_3/Si – O) (%)	Moisture uptake (ΔSi – OH/Si – O) (%)	Dielectric constant increase (Δ*k*)	Leakage current density at 2 MV/cm (A/cm²)	Breakdown field (MV/cm)
Without	Ions, radicals, photons (VUV, UV, Vis, IR)	8.36	58.63	80.295	27.03	2.91	8.7	3.52	1.25×10^{-5}	8.42
MgF_2	Radicals, photons (VUV[a], UV, Vis, IR)	7.31	40.01	54.3	35.22	2.42	6.4	1.5	5.37×10^{-7}	8.98
CaF_2	Radicals, photons (VUV[b], UV, Vis, IR)	6.92	38.27	51.82	36.4	2.28	6.1	1.23	3.46×10^{-7}	9.03
Si	Radicals, photons (IR)	6.12	33.35	45.68	42.26	1.85	5.3	0.73	1.76×10^{-7}	9.23
Remark:	Pristine low-*k*:	-	-	2.86	88.96	-	-	2.56[c]	9.39×10^{-9}	9.82

[a] 120–390 nm.
[b] 250–350 nm.
[c] The dielectric constant of the studied low-*k* film.

Table 2. Change of physical and electrical characteristics of porous low-*k* dielectrics using different masks under O_2 plasma treatment.

http://dx.doi.org/10.5772/intechopen.79494

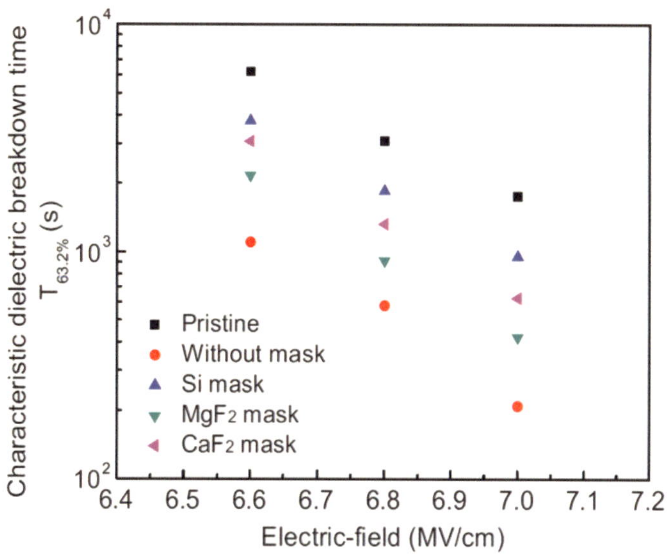

Figure 12. Characteristic dielectric breakdown time versus electric field for porous low-*k* dielectrics under O_2 plasma treatment with various masks.

indicating that radicals cause the greater degradation in the dielectric reliability than the other two plasma components. However, this finding is still needed to be demonstrated by more experiments which can be treated under the individual plasma component. The synergy between radicals, photons, and ions causes a considerable degradation. Getting rid of one or two components from the plasma environment is a workable strategy for the low-*k* dielectric's reliability improvement.

4.2.2. Hydrogen (H_2)/helium (He) plasma damage

H_2/He plasma treatments on the porous low-*k* dielectrics ($k = 2.56$) using CCP and RP systems at various operation temperatures (25–350°C) were investigated. The *k* value of porous low-*k* films after H_2/He plasma treatment in CCP and RP systems as a function of the operation temperature is presented in **Figure 13**. An increased *k* was detected after H_2/He plasma treatment in CCP system, and the increasing magnitude increased with increasing the operation temperature. However, as the operation temperature is raised above 250°C, the increasing rate of the *k* value tends to alleviate. This phenomenon can be explained by transforming Si–OH bonds to Si–O–Si bonds at an elevated temperature above 200°C. For porous low-*k* dielectrics treated by H_2/He plasma in RP system, the increase in the *k* value was lower owing to a relatively small Si–CH_3 depletion and Si–OH formation. Additionally, the *k* value was slightly reduced with an increase of the operation temperature. Furthermore, as the operation temperature is elevated to 350°C, the *k* value was reduced to be lower than 2.56. The result suggests that the effect of H_2/He plasma treatment in RP system on porous low-*k* dielectrics turns to be positive by raising the operation temperature to 350°C. As a consequence, a "damage-free" resist strip processing can be obtained by using H_2/He plasma treatment in RP chamber at elevated temperatures.

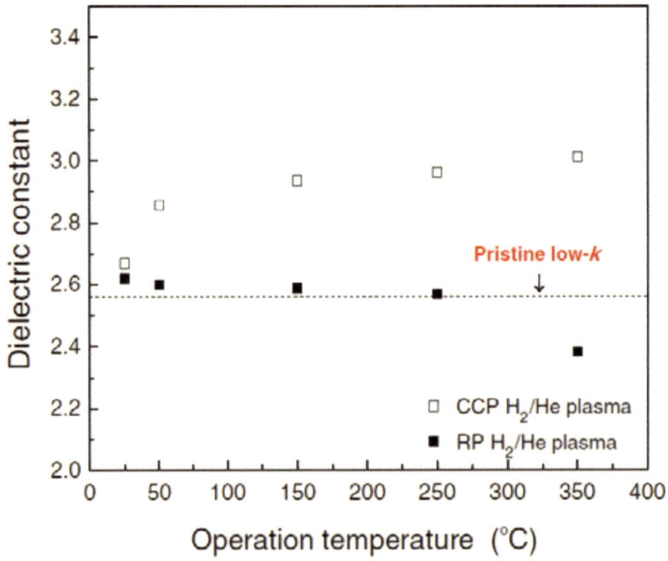

Figure 13. Dielectric constant of H_2/He plasma-treated low-*k* dielectrics operated in CCP and RP systems as a function of operation temperature.

Figure 14(a) and **(b)** shows the breakdown field and the dielectric breakdown time, respectively, of porous low-*k* dielectrics under H_2/He plasma treatment in the CCP or RP systems at various operation temperatures. Both results indicate that H_2/He plasma-treated low-*k* dielectrics in RP system exhibited a higher breakdown field and a longer breakdown time as compared to those in CCP system, indicating that deep UV light radiation and ion bombardment induced from H_2/He plasma treatment in CCP system on the low-*k* dielectric can accelerate the degradation of reliability. The trends of temperature dependence of reliability characteristics were different for H_2/He plasma treatments in the CCP and RP systems. The breakdown field and the breakdown time of H_2/He plasma-treated low-*k* dielectrics in CCP system were decreased, while those in CCP system were improved as the operation temperature is raised. Furthermore, H_2/He plasma-treated low-*k* dielectrics operated in CCP system displayed a strong temperature dependence of reliability, implying that the reaction induced by radicals is not enhanced by increasing the temperature. However, with the assistance of deep UV light radiation and ion bombardment, the reaction becomes stronger at a higher operation temperature. As the operation temperature of H_2/He plasma treatment in RP system was raised to 350°C, the reliability performance of the plasma-treated low-*k* dielectrics exceeded that of the pristine samples. A better reliability for H_2/He plasma-treated low-*k* films operated in RP system at 350°C can be attributable to another mechanism because the scission of Si–CH_3 bonds was still detected although the decreasing ratio was reduced. H_2/He plasma treatment on porous low-*k* dielectrics in RP system at evaluated temperatures reportedly removes carbon-based porogen residues, which are formed inside the porous low-*k* structure due to non-optimized incorporation of porogen molecules and non-optimized UV curing [55]. The removal of porogen residues from porous low-*k* dielectrics has also been demonstrated to promote reliability for low-*k* dielectrics. Therefore, H_2/He plasma treatment

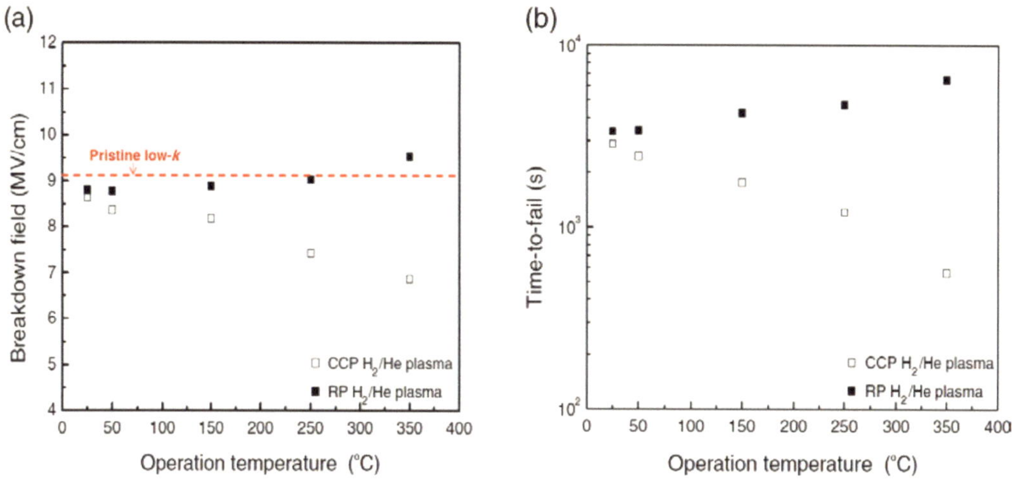

Figure 14. (a) Breakdown field. (b) Dielectric breakdown time of H_2/He plasma-treated low-*k* dielectrics operated in CCP and RP systems as a function of operation temperature.

on porous low-*k* dielectrics in RP system at 350°C efficiently removes porogen residues from porous low-*k* dielectrics, resulting in a better reliability.

The mechanism about the reaction between H_2 reactive plasma species and porous low-*k* dielectrics can be described as [56]:

$$\equiv\text{Si–}CH_3 + 2H \rightarrow \equiv\text{Si–H} + CH_4 \quad \Delta H_r = -411\ \text{KJ/mole} \tag{2}$$

$$\equiv\text{Si–O–Si}\equiv + 2H \rightarrow \equiv\text{Si–H} + \equiv\text{Si–OH} \quad \Delta H_r = -325\ \text{KJ/mole} \tag{3}$$

where ΔH_r is the estimated enthalpy. The reaction temperature of these two reactions is assumed to be 25°C. The negative enthalpies of reactions (2) and (3) represent that the reactions are exothermic and presumably occurred at room temperature [57]. Assuming that the amount of H atoms remains unchanged at an elevated temperature, these two reactions would become less favored with an increase of the reaction temperature according to Chatelier's principle. Therefore, the scission of Si–CH_3 and Si–O–Si groups should be stronger at a lower temperature. However, FT-IR analysis revealed that the losses of CH_3 and Si–O–Si groups were higher for porous low-*k* dielectrics treated by H_2/He plasma in CCP system at a higher temperature. The discrepancy can be explained by the fact that only H radical is considered to react with the low-*k* dielectric for reactions (2) and (3). However, H ions and VUV photons can be produced in H_2/He plasma operated in CCP system. Due to the presence of H ions and VUV photons, the above two reactions become possible because the bonding energies of Si–CH_3 and Si–O–Si bonds are weaken. Furthermore, at an elevated temperature, ions and photons can gain more energy and easily break Si–CH_3 and Si–O–Si bonds, causing a violent response for reactions (2) and (3). In the case of H_2/He plasma treatment in RP system, only H radicals can react with the porous low-*k* dielectric. According to FT-IR result, only Si–CH_3

group was found to reduce, and the concentration of Si–O–Si bond almost kept unchanged for H_2/He plasma-treated low-*k* dielectrics, implying that reaction (2) is favored over reaction (3) because of a lower dissociation energy of Si–CH_3 bond. Additionally, the reduction amount of Si–CH_3 bond is relatively small and no temperature dependence effect, indicating that reaction (2) is relatively weak even at a higher temperature for H_2/He plasma treatment in RP system.

In addition to the above reactions (2) and (3), H_2 plasma can break Si–CH_3 and Si–O–Si bonds to create Si dangling bonds. The subsequent air exposure makes these Si dangling bonds transform to Si–OH bonds. If OH- bonds are weak or physically bonded, dehydroxylation of Si–OH bonds can occur to form Si–O–Si bonds at a higher temperature [58]. This can be explained by the reduction of Si–OH bonds for H_2/He plasma-treated low-*k* films operated at temperatures above 250°C.

4.2.3. Ammonia (NH_3)/nitrogen (N_2) plasma damage

The effect of the NH_3/N_2 ratio in plasma treatment on the porous low-*k* dielectrics (k = 2.56) was investigated. The reaction mechanism between the porous low-*k* dielectric and NH_3/N_2 plasma can be described as follows: in pure N_2 gas plasma, only N, N_2, and N_2* active species are generated, and no hydrogen species is produced. Physical bombardment by N radicals is favorable, roughing the film's surface. Moreover, the weak bonds in the low-*k* dielectric, such as Si–H, Si–CH_3, and C–H_x bonds, can be broken by these active species in the plasma, forming Si–N and C–N bonds. As NH_3 gas was added into the plasma, other active species in addition to the N, N_2, and N_2* active species, such as H, NH_2, NH_4, and N_2H, may be generated. The Si–CH_3 group in the low-*k* dielectric is broken to form Si dangling bonds. This dangling bond easily absorbs H or NH_2 species to form Si–H or Si–NH_2 bonds due to a lower reaction energy, which is thermodynamically favorable [59–62]. The Si–H and Si–NH_2 bonds are not stable in air and easily react with ambient air to form Si–OH, which is more hydrophobic and has a higher *k* value. As the portion of NH_3 in the plasma increases, the number of H and NH_2 active species increases accordingly. At the same time, the amount of the generated N, N_2, and N_2* active species is limited because more energy is required to generate these active species. These changes in the plasma result in the significant replacement of –CH_3 groups by H and NH_2 active species, the formation of more Si–OH bonds, and the reduction of Si–N and C–N bonds.

Figure 15 shows the changes in the *k* value of NH_3/N_2 plasma-treated low-*k* dielectrics upon O_2 plasma treatment. After NH_3/N_2 plasma treatment, the *k* value of the plasma-treated low-*k* dielectrics increases. Under pure NH_3 or pure N_2 gas plasma treatment conditions, the increase is larger. This can be attributed to more formation of Si–OH bonds or Si–N/C–N bonds on the surface layer for pure NH_3 or pure N_2 gas plasma treatment, respectively. Treatment with O_2 plasma increases the *k* values of all NH_3/N_2 plasma-treated low-*k* dielectrics by the replacement of Si–CH_3 and Si–H bonds with Si–O bonds [63]. The increase in the *k* value becomes larger with the NH_3/N_2 gas ratio. The pure N_2 gas plasma-treated sample exhibits a smaller increase in the *k* value owing to the formation of protective Si–N/C–N layer. This layer suppresses the penetration of oxygen radical into the low-*k* dielectric.

Figure 16(a) plots the measured dielectric breakdown fields of NH_3/N_2 plasma-treated low-*k* dielectrics. All NH_3/N_2 plasma-treated samples had a poorer dielectric breakdown performance than the pristine low-*k* dielectric. Moreover, the breakdown field of the NH_3/N_2 plasma-treated low-*k* dielectrics decreases as the $NH_3/(N_2 + NH_3)$ ratio increases. The low-*k* dielectric that was plasma treated with pure NH_3 gas has the lowest breakdown field and the highest leakage current.

Figure 16(b) compares $T_{63.2\%}$ values as a function of applied electric field for the pristine and plasma-treated low-*k* dielectrics. All plasma-treated samples had shorter dielectric breakdown times and a wider distribution as compared to the pristine low-*k* dielectric. The reduction of the dielectric breakdown time may be caused by an accumulation of defects owing to

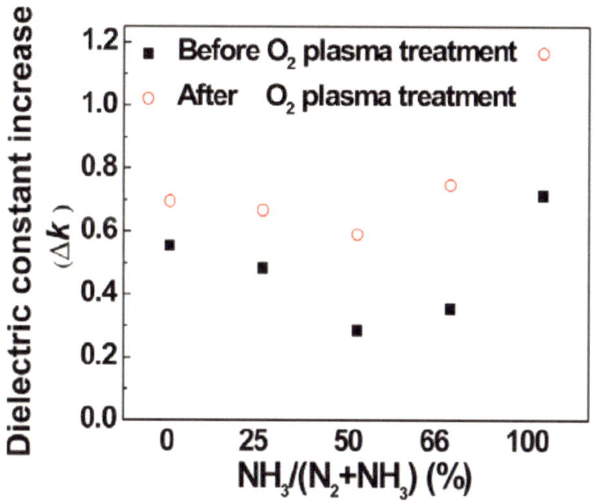

Figure 15. Change in dielectric constant of porous low-*k* dielectrics after NH_3/N_2 and O_2 plasma treatments.

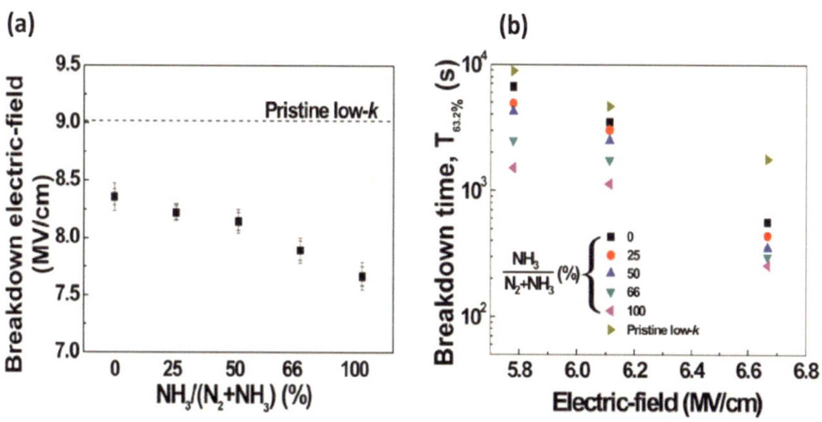

Figure 16. (a) Breakdown field. (b) Dielectric breakdown time as a function of electric field of porous low-*k* dielectrics after various NH_3/N_2 plasma treatments.

plasma-induced damage. Furthermore, the reductions in the dielectric breakdown time were significant in stronger stressing electric fields. Additionally, the $T_{63.2\%}$ values of plasma-treated low-*k* dielectrics decreased as the $NH_3/(N_2 + NH_3)$ ratio increased, which is correlated well with the moisture contents in the plasma-treated dielectrics. This indicates that the moisture content in a low-*k* dielectric plays an important role in reducing the dielectric breakdown time. The low-*k* dielectric that was plasma treated with pure N_2 gas had the longest low-*k* dielectric because the formed amide-like or nitride-like layers on the surface retard low-*k* dielectric breakdown.

5. Conclusions

Low-*k* dielectric materials are essential for *RC* delay reduction to improve the performance of ICs. However, plasma-induced damage on the low-*k* dielectric materials during Cu/low-*k* interconnects fabrication is a critical issue to influence the low-*k* integrity. Plasma damage on the low-*k* dielectrics is a complicated phenomenon involving physical and chemical reactions. The resulting main negative impact is an increased dielectric constant due to water adsorption after plasma irradiation on the low-*k* dielectrics. The plasma damage depends on the used low-*k* dielectrics, plasma gases and conditions, and reactors. Therefore, the optimization in the plasma process is required to minimize the plasma damage. Additionally, the damaged low-*k* dielectrics can be recovered by the removal of adsorbed water. This turns the hydrophilic materials to be hydrophobic. Currently, complete recovery cannot be obtained; consequently, this area requires a lot of effort to make in the future.

Author details

Yi-Lung Cheng*, Chih-Yen Lee and Chiao-Wei Haung

*Address all correspondence to: yjcheng@ncnu.edu.tw

Department of Electrical Engineering, National Chi-Nan University, Nan-Tou, Taiwan, R.O.C

References

[1] Thompson SE, Parthasarathy S. Moore's law: The future of Si microelectronics. Materials Today. 2006;**9**:20-25

[2] Bohr M. MOS transistors: Scaling and performance trends. Semiconductor International. 1995;**18**(6):75-80

[3] Isaac RD. The future of CMOS technology. IBM Journal of Research and Development. 2000;**44**:369-378

[4] Dixit GA, Havemann RH. Overview of Interconnect—Copper and Low-*k* Integration, Handbook of Semiconductor Manufacturing Technology. 2nd ed. NY: CRC Press; 2008 (Chapter 2)

[5] Havemann RH, Hutchby JA. High-performance interconnects: An integration overview. Proceedings of the IEEE. 2001;**89**:586-601

[6] Plummer JD, Deal MD, Griffin PB. Silicon VLSI Technology. Upper Saddle River, NJ: Prentice Hall; 2000

[7] Jeffery G. Process Technology for Copper Interconnects, Handbook of Thin Film Deposition. 3rd Edition, William Andrew; 2012. pp. 221-269

[8] Grill A, Gates SM, Ryan TE, Nguyen SV, Priyadarshini D. Progress in the development and understanding of advanced low *k* and ultralow *k* dielectrics for very large-scale integrated interconnects – State of the art. Applied Physics Reviews. 2014;**1**:011306-011312

[9] Grill A. Porous pSiCOH ultralow-*k* dielectrics for chip interconnects prepared by PECVD. Annual Review of Materials Research. 2009;**39**:49-69

[10] Grill A. Plasma enhanced chemical vapor deposited SiCOH dielectrics: From low-*k* to extreme low-*k* interconnect materials. Journal of Applied Physics. 2003;**93**:1785-1790

[11] Kriz J, Angelkort C, Czekalla M, Huth S, Meinhold D, Pohl A, et al. Overview of dual damascene integration schemes in Cu BEOL integration. Microelectronic Engineering. 2008;**85**:2128-2132

[12] Zhang L, Ljazouli R, Lefaucheux P, Tillocher T, Dussart R, Mankelevich YA, de Marneffe J-F, de Gendt S, Baklanov MR. Low damage cryogenic etching of porous organosilicate low-*k* materials using $SF_6/O_2/SiF_4$. ECS Journal of Solid State Science and Technology. 2013;**2**(6):N131-N139

[13] Baklanov MR, de Marneffe J-F, Shamiryan D, Urbanowicz AM, Shi H, Rakhimova TV, Huang H, Ho PS. Plasma processing of low-*k* dielectrics. Journal of Applied Physics. 2013;**113**:041101-1-041101-35

[14] Kazi H, Kelber JA. Plasma damage mechanisms in low *k* organosilicate glass and their inhibition by Ar ion bombardment. Journal of Vacuum Science and Technology B. 2010;**32**(2):021302-1-021302-7

[15] Kunnen E, Barkema GT, Maes C, Shamiryan D, Urbanowicz A, Struyf H, Baklanov MR. Integrated diffusion–recombination model for describing the logarithmic time dependence of plasma damage in porous low-*k* materials. Microelectronic Engineering. 2011; **88**:631-634

[16] Maex K, Baklanov MR, Shamiryan D, Iacopi F, Brongersma SH, Yanovitskaya ZS. Low dielectric constant materials for microelectronics. Journal of Applied Physics. 2003; **93**:8793-8841

[17] Kim CY, Navamathavan R, Lee HS, Woo JK, Hyun MT, Lee KM, Jeung WY, Choi CK. Ultraviolet irradiation effect on the properties of leakage current and dielectric break-down

of low-dielectric-constant SiOC(H) films using comb capacitor structure. Thin Solid Films. 2011;**519**:6732-6736

[18] Broussous L, Berthout G, Rebiscoul D, Rouessac V, Ayral A. Mechanical properties of a plasma-modified porous low-*k* material. Microelectronic Engineering. 2010;**87**:466-469

[19] Cheng YL, Wang YL, Lan JK, Chen HC, Lin JH, Wu YL, Liu PT, Wu YC, Feng MS. Effect of carrier gas on the structure and electrical properties of low dielectric constant SiCOH film using trimethylsilane prepared by plasma enhanced chemical vapor deposition. Thin Solid Films. 2004;**469**:178-183

[20] Chang YM, Chang WY, Leu J, Cheng YL. Effect of thermal treatment on physical, electrical properties and reliability of porogen-containing and porogen-free ultralow-*k* dielectrics. Thin Solid Films. 2013;**528**:67-71

[21] Cheng YL, Wang YL, Liu CW, Wu YL, Lo KY, Liu CP, Lan JK. Characterization and reliability of low dielectric constant fluorosilicate glass and silicon rich oxide process for deep sub-micro device application. Thin Solid Films. 2001;**398-399**:544-548

[22] Shapiro MJ, Nguyen SV, Matsuda T, Dobuzinsky D. CVD of fluorosilicate glass for ULSI applications. Thin Solid Films. 1995;**270**:503-507

[23] Lubguban Jr J, Rajagopalan T, Mehta N, Lahlouh B, Simon SL, Gangopadhyaya S. Low-*k* organosilicate films prepared by tetravinyltetramethylcyclotetrasiloxane. Journal of Applied Physics. 2002;**92**:1033-1038

[24] Gates SM, Neumayer DA, Sherwood MH, Grill A, Wang X, Sankarapandian M. Preparation and structure of porous dielectrics by plasma enhanced chemical vapor deposition. Journal of Applied Physics. 2007;**101**:094103-1-094103-8

[25] Chapelon LL, Arnal V, Broekaart M, Gosset LG, Vitiello J, Torres J. Porous ultra low *k* deposited by PECVD: From deposition to material properties. Microelectronic Engineering. 2004;**76**:1-4

[26] Ming Z, Beichao Z. Preparation of porous ultra low *k* films using different sacrificial porogen precursors for 28 nM technological node. Materials Science in Semiconductor Processing. 2015;**36**:170-178

[27] Dultsev FN, Nekrasov DV. Transformation of porous structure under vacuum ultraviolet irradiation of the films based on silicon dioxide. Thin Solid Films. 2016;**603**:249-254

[28] Vanstreels K, Ciofi I, Barbarin Y, Baklanov M. Influence of porosity on dielectric breakdown of ultralow-*k* dielectrics. Journal of Vacuum Science and Technology B. 2013;**31**: 050604-1-050604-5

[29] Amanatides E, Mataras D. Frequency variation under constant power conditions in hydrogen radio frequency discharges. Journal of Applied Physics. 2001;**89**:1556-1566

[30] Perret A, Chabert P, Jolly J, Booth J-P. Ion energy uniformity in high-frequency capacitive discharges. Applied Physics Letters. 2005;**86**:021501-1-021501-3

[31] Rakhimova TV, Braginsky OV, Ivanov VV, Kim TK, Kong JT, Kovalev AS, Lopaev DV, Mankelevich YA, Proshina OV, Vasilieva AN. Experimental and theoretical study of rf plasma at low and high frequency. IEEE Transactions on Plasma Science. 2006;**34**:867-877

[32] Smirnov E, Ferchichi AK, Huffman C, Baklanov MR. Impact of plasma exposure on organic low-*k* materials. Proceedings of SPIE. 2010;**75217**:52107-1-52107-8

[33] Pantouvaki M, Huffman C, Zhao L, Heylen N, Ono Y, Nakajima M, Nakatani K, Beyer GP, Baklanov MR. Advanced organic polymer for the aggressive scaling of low-*k* materials. Japanese Journal of Applied Physics. 2011;**50**:04DB01-1-04DB01-5

[34] Havemann RH, Antonelli GA, Arendt GK, Danek M, McKerrow AJ, Weinberg RS. Copper BEOL Solutions for Advanced Memory, Solid-State Technol. Vol. 312009. pp. 10-13

[35] Takei S. Resist poisoning studies of gap fill materials for patterning metal trenches in via-first dual damascene process. Japanese Journal of Applied Physics. 2008;**47**:8766-8770

[36] Noguchi J, Ohashi N, Jimbo T, Yamaguchi H, Takeda K, Hinode K. Effect of NH_3-plasma treatment and CMP modification on TDDB improvement in Cu metallization. IEEE Transactions on Electron Devices. 2001;**48**:1340-1345

[37] Lin MH, Lin YL, Chen JM, Yeh M-S, Chang KP, Su KC, Wang T. Electromigration lifetime improvement of copper interconnect by cap/dielectric interface treatment and geometrical design. IEEE Transactions on Electron Devices. 2005;**52**(12):2602-2608

[38] Baklanov MR, Vanhaelemeersch S, Bender H, Maex K. Effects of oxygen and fluorine on the dry etch characteristics of organic low-*k* dielectrics. Journal of Vacuum Science and Technology B. 1999;**17**:372-379

[39] Cheng YL, Lin BH, Huang SW. Effect of O_2 plasma treatment on physical, electrical, and reliability characteristics of low dielectric constant materials. Thin Solid Films. 2014; **572**:44-50

[40] Worsley MA, Bent SF, Gates SM, Fuller NCM, Volksen W, Steen M, Dalton T. Effect of plasma interactions with low-*k* films as a function of porosity, plasma, chemistry, and temperature. Journal of Vacuum Science and Technology B. 2005;**23**(2):395-405

[41] Han QY, White B, Berry IL, Waldfried C, Escorcia O. Activated he:H_2 strip of photoresist over porous low-*k* materials. Solid State Phenomena. 2005;**103-104**:341-345

[42] Lionti K, Volksen W, Magbitang T, Darnon M, Dubois G. Toward successful integration of porous low-*k* materials: Strategies addressing plasma damage. ECS Journal of Solid State Science and Technology. 2015;**4**(1):N3071-N3083

[43] Urbanowicz AM, Baklanov MR, Heijlen J, Travaly Y, Cockburn A. Damage reduction and sealing of low-*k* films by combined he and NH_3 plasma treatment. ECS Journal of Solid State Science and Technology. 2007;**10**(10):G76-G79

[44] Worsley MA, Bent SF, Gates SM, Fuller NCM, Volksen W, Steen M, Dalton T. Journal of Vacuum Science & Technology. 2005;**B23**:395-405

[45] Grill A, Neumayer DA. Structure of low dielectric constant to extreme low dielectric constant SiCOH films: Fourier transform infrared spectroscopy characterization. Journal of Applied Physics. 2003;**94**(10):6697-6707

[46] Jousseaume V, Zenasni A, Gourhant O, Favennec L, Baklanov MR. In: Baklanov MR, Ho P, Zschech E, editors. Ultra-low-*k* by CVD: Deposition and Curing. Advanced Interconnects for ULSI Technology. Wiley; 2012. pp. 35-77. (Chapter 2)

[47] Urbanowicz AM, Shamiryan D, Zaka A, Verdonck P, De Gendt S, Baklanov MR. Effects of He plasma pretreatment on low-*k* damage during Cu surface cleaning with NH_3 plasma. Journal of the Electrochemical Society. 2010;**157**(5):H565-H573

[48] Cheng YL, Huang JF, Chang YM, Leu J. Impact of plasma treatment on structure and electrical properties of porous low dielectric constant SiCOH material. Thin Solid Films. 2013;**544**:537-540

[49] Cheng YL, Tsai YS, Hung WJ, Sun CR, Lee WH. Effect of H_2/He plasma on porous low dielectric constant materials. Surface and Coatings Technology. 2016;**38**:182-188

[50] Huang JF, Bo TC, Chang WY, Chang YM, Leu J, Cheng YL. Effect of NH_3/N_2 ratio in plasma treatment on porous low dielectric constant SiCOH materials. Journal of Vacuum Science and Technology A. 2014;**32**(3):031505-031510

[51] Michelon J, Hoofman RJOM. Moisture influence on porous low-*k* reliability. IEEE Transactions on Device and Materials Reliability. 2006;**6**:169-174

[52] McPherson JW. Time dependent dielectric breakdown physics–models revisited. Microelectronics and Reliability. 2012;**52**:1753-1760

[53] Kemeling N, Matsushita K, Tsuji N, Kagami KI, Kato M, Kaneko S, Sprey H, Roest DD, Kobayashi N. A robust *k* ~2.3 SiCOH low-*k* film formed by porogen removal with UV-cure. Microelectronic Engineering. 2007;**84**:2575-2581

[54] Shi H, Huang H, Bao J, Liu J, Ho PS, Zhou Y, Pender JT, Armacost MD, Kyser D. Role of ions, photons, and radicals in inducing plasma damage to ultra low-*k* dielectrics. Journal of Vacuum Science and Technology B. 2012;**30**:011206-1-011206-1, 011206-9

[55] Urbanowicz AM, Vanstreels K, Shamiryan D, De Gendt S, Baklanov MR. Effect of porogen residue on chemical, optical, and mechanical properties of CVD SiCOH low-*k* materials. Electrochemical and Solid-State Letters. 2009;**12**:H292-H295

[56] Nichols MT, Sinha H, Wiltbank CA, Antonelli GA, Nishi Y, Shohet JL. Time-dependent dielectric breakdown of plasma-exposed porous organosilicate glass. Applied Physics Letters. 2012;**100**:112905-1-112905-4

[57] Dubois C, Sylvestre A, Chaabouni H, Farcy A. Impact of the CMP process on the electrical properties of ultra low *k* porous SiOCH. Microelectronic Engineering. 2010;**87**:333-336

[58] Cheng YL, Leon KW, Huang JF, Chang WY, Chang YM, Leu J. Effect of moisture on electrical and reliability properties for low dielectric constant materials. Microelectronic Engineering. 2014;**114**:12-16

[59] Worsley MA, Bent SF, Gates SM, Fuller NCM, Volksen W, Steen M, Dalton T. Effect of plasma interactions with low-κ films as a function of porosity, plasma chemistry, and temperature. Journal of Vacuum Science and Technology B. 2005;**23**:395-405

[60] Liu X, Gill S, Tang F, King SW, Nemanich RJ. Remote H_2/N_2 plasma processes for simultaneous preparation of low-*k* interlayer dielectric and interconnect copper surfaces. Journal of Vacuum Science and Technology B. 2012;**30**:031212-1-031212-9

[61] Posseme N, Chevolleau T, David T, Darnon M, Louveau O, Joubert O. Mechanisms of porous dielectric film modification induced by reducing and oxidizing ash plasmas. Journal of Vacuum Science and Technology B. 2007;**25**:1928-1940

[62] Yamamoto H, Takeda K, Ishikawa K, Ito M, Sekine M, Hori M, Kaminatsui T, Hayashi H, Sakai I, Ohiwa T. H_2/N_2 plasma damage on porous dielectric SiOCH film evaluated by in situ film characterization and plasma diagnostics. Journal of Applied Physics. 2011;**11**:123301

[63] Chen W, Han Q, Most R, Waldfried C, Escorcia O, Berry I. Plasma impacts to an O-SiC low-*k* barrier film. Journal of the Electrochemical Society. 2004;**151**:F182-F188

PLASMA SCIENCE AND TECHNOLOGY - BASIC FUNDAMENTALS AND MODERN APPLICATIONS

Edited by **Haikel Jelassi**
and **Djamel Benredjem**

Plasma Science and Technology - Basic Fundamentals and Modern Applications
http://dx.doi.org/10.5772/intechopen.72359
Edited by Haikel Jelassi and Djamel Benredjem

Contributors

Mikhail Tokar, Jianjun Wu, Eduard Rostomyan, Yi-Lung Cheng, . Shu-Xia, Jean Paul Allain, David Ruzic, Anton Nikiforov, Kristel Crombe, R. D' Inca, E. Faudot, H. Faugel, A. Kostic, M. Usoltceva, J.-M. Noterdaeme, Helmut Fuenfgelder, Stephane Heuraux, Jonathan Jacquot, Fabrice Louche, Roman Ochoukov, Ilya Shesterikov, Dirk Van Eester, Yasunori Ohtsu, Claude Deutsch, Irina Litovko, Alexey Goncharov, Aamir Shahzad, Muhammad Asif Shakoori, Maogang He, Siyin Zhou, Xueke Che, Wansheng Nie, Haiqing Wang, Liqiu Wei, Jingfeng Tang, Daren Yu, Daniel Engelhart, Elena Plis, W. Robert Johnston, Dale Ferguson, Russell Cooper, Ryan Hoffmann, Cheng Pan, Ju Tang

Notice

First published in London, United Kingdom, 2019 by IntechOpen
IntechOpen is the global imprint of INTECHOPEN LIMITED, registered in England and Wales, registration number: 11086078, The Shard, 25th floor, 32 London Bridge Street
London, SE19SG – United Kingdom
Printed in Croatia

British Library Cataloguing-in-Publication Data
A catalogue record for this book is available from the British Library

Additional hard copies can be obtained from orders@intechopen.com

Plasma Science and Technology - Basic Fundamentals and Modern Applications, Edited by Haikel Jelassi and Djamel Benredjem
p. cm.
Print ISBN 978-1-78985-239-4
Online ISBN 978-1-78985-240-0